# Springer Theses

Recognizing Outstanding Ph.D. Research

## Aims and Scope

The series "Springer Theses" brings together a selection of the very best Ph.D. theses from around the world and across the physical sciences. Nominated and endorsed by two recognized specialists, each published volume has been selected for its scientific excellence and the high impact of its contents for the pertinent field of research. For greater accessibility to non-specialists, the published versions include an extended introduction, as well as a foreword by the student's supervisor explaining the special relevance of the work for the field. As a whole, the series will provide a valuable resource both for newcomers to the research fields described, and for other scientists seeking detailed background information on special questions. Finally, it provides an accredited documentation of the valuable contributions made by today's younger generation of scientists.

## Theses are accepted into the series by invited nomination only and must fulfill all of the following criteria

- They must be written in good English.
- The topic should fall within the confines of Chemistry, Physics, Earth Sciences, Engineering and related interdisciplinary fields such as Materials, Nanoscience, Chemical Engineering, Complex Systems and Biophysics.
- The work reported in the thesis must represent a significant scientific advance.
- If the thesis includes previously published material, permission to reproduce this must be gained from the respective copyright holder.
- They must have been examined and passed during the 12 months prior to nomination.
- Each thesis should include a foreword by the supervisor outlining the significance of its content.
- The theses should have a clearly defined structure including an introduction accessible to scientists not expert in that particular field.

More information about this series at http://www.springer.com/series/8790

Hyung Yoon

# Palladium and Nickel Catalyzed Transformations Forming Functionalized Heterocycles

Doctoral Thesis accepted by
the University of Toronto, Toronto, Canada

Springer

*Author*
Dr. Hyung Yoon
Department of Chemistry
Yale University
New Haven, CT, USA

*Supervisor*
Mark Lautens
Davenport Chemical Laboratories
Department of Chemistry
University of Toronto
Toronto, ON, Canada

ISSN 2190-5053 ISSN 2190-5061 (electronic)
Springer Theses
ISBN 978-3-030-54076-0 ISBN 978-3-030-54077-7 (eBook)
https://doi.org/10.1007/978-3-030-54077-7



This Springer imprint is published by the registered company Springer Nature Switzerland AG
The registered company address is: Gewerbestrasse 11, 6330 Cham, Switzerland

# Supervisor's Foreword

The use of transition metal catalysis has become a fundamental aspect of synthetic organic chemistry, owing to the ability to forge novel bonds in high atom and step economy. In particular, transition metal-catalyzed domino or cascade reactions are capable of further enhancing the efficiency of synthetic sequences as multiple bonds can be formed in a single operation. In an effort to advance the field, we sought methods to generate functionalized heterocycles in a single transformation.

In 2010 and 2011, Newman and Lautens utilized catalytic reversible oxidative addition in generating heterocycles bearing $Csp^2$–Br and $Csp^3$–I bonds via palladium catalysis. Having realized a cycloisomerization reaction bearing perfect economy, our group has been devoted to expanding this carbohalogenation process. Along with advancing carbohalogenation, Hyung Yoon became interested in further functionalizing the key neopentyl Pd intermediate.

Chapter 1 of the thesis is split between the Pd- and Ni-catalyzed carboiodination reactions. Yoon initially worked alongside David Petrone in developing a diastereoselective Pd-catalyzed carboiodination reaction, which was found to be the key step in the formal synthesis of (+)-corynoline. In the final year of his doctoral degree, Yoon was the pioneer in exploring the Ni-catalyzed variant. Using a cheap and readily available Ni catalyst, nitrogen-containing heterocycles were synthesized. Perhaps the most intriguing discovery was the finding of a dual ligand system combining a bisphosphine and bisphosphine monoxide to generate enantioenriched iodinated oxindoles, which were previously inaccessible in the Pd-catalyzed variant.

Chapter 2 expands the field beyond Pd-catalyzed carboiodination of olefins to carbocyanation and carboborylation. Using similar substrates to the diastereoselective carboiodination reaction, carbocyanation furnished dihydroisoquinolinones bearing a neopentyl nitrile group in excellent yield and selectivity. Noticing that boron sources are rarely used as a trap, Yoon explored the intramolecular carboborylation. He reported the synthesis of borylated chromans as a single diastereomer.

Chapter 3 further investigates Pd-catalyzed cascade reactions utilizing carbopalladation, C–H functionalization and π-system insertion to generate spirocycles. Incorporating C–H functionalization as part of a domino process has proven to be very powerful, as it enables the metal center to interact with remote bonds that may otherwise be unreactive. Applying this method, Yoon was able to generate various functionalized spirodihydrobenzofurans and spirooxindoles. Additionally, mechanistic and computational studies suggest that the reactions proceed via an isolable spirocyclic palladacycle, wherein the subsequent benzyne or alkyne insertion occurs.

Overall, Yoon's thesis expanded on two fundamental research topics, the metal-catalyzed carboiodination reaction and domino reactions. Through his efforts and creativity, the carboiodination reaction was reinvigorated. Meanwhile, his persistence and curiosity led to the synthesis of various biologically relevant molecules and in-depth mechanistic studies.

Toronto, Canada
May 2020

Mark Lautens

# Abstract

Transition metal catalysis has been at the forefront of synthetic organic chemistry as it represents new and powerful methods to forge carbon–carbon in high atom and step economy. The Lautens group has been particularly focused on Pd and Rh catalysis in stereo- and regioselectively generating functionalized heterocycles.

In Chap. 1, a Pd- and Ni-catalyzed cycloisomerization reaction of aryl iodides to alkyl iodides, known as carboiodination, is described. In the Pd-catalyzed variant, a diastereoselective Pd-catalyzed carboiodination reaction of enantioenriched carboxamides generating dihydrosioquinolinones is explored. The transformation was the key step in the formal synthesis of (+)-corynoline. In the Ni-catalyzed variant, 3,3-disubstituted oxindoles are generated in excellent yield and an enantioselective variant employing a dual ligand system is disclosed.

In Chap. 2, a Pd-catalyzed diastereoselective anion capture cascade reaction is discussed. Building on the Pd-catalyzed carboiodination reaction en route to (+)-corynoline, a Pd-catalyzed aryl cyanation is discussed. The alkyl nitrile was synthesized in high yield and d.r. The second method generates borylated chromans as a single diastereomer in good yield. The alkyl boronate was further functionalized to showcase its utility.

In Chap. 3, a Pd-catalyzed domino process harnessing carbopalladation, C–H activation and $\pi$-system insertion to generate spirocycles is reported. In the first method, benzynes are inserted to give spirodihydrobenzofurans and spirooxindoles. A significant limitation in the first method pertains to the low regioselectivity observed. To circumvent these problems, in the second method, polarized alkynes are inserted in >20:1 rr.

**Parts of this thesis have been published in the following journal articles:**

1. Petrone, D. A.; **Yoon, H.**; Weinstabl, H.; Lautens, M. Additive effects in the Pd-Catalyzed Carboiodination of chiral N-allyl carboxamides. *Angew. Chem. Int. Ed.* **2014**, *53*, 7908.
2. **Yoon, H.**; Petrone, D. A.; Lautens, M. Diastereoselective Palladium-Catalyzed Arylcyanation/Heteroarylcyanation of Enantioenriched *N*-allylcarboxamides. *Org. Lett.* **2014**, *16*, 6420.
3. Jang, Y. J.; **Yoon, H.**; Lautens, M. Rh-Catalyzed Domino Addition-Enolate Arylation: Generation of 3-Substituted Oxindoles via Rh(III) Intermediate, *Org. Lett.* **2015**, *17*, 3895.
4. **Yoon, H.**; Jang, Y. J.; Lautens, M. Diastereoselective Pd-Catalyzed Domino-Heck Arylborylation Sequence Forming Borylated Chromans, *Synthesis*, **2016**, *48*, 1483.
5. **Yoon, H.**; Lossouarn, A.; Landau, F.; Lautens, M. Pd-Catalyzed Spirocyclization via C–H Activation and Benzyne Insertion, *Org. Lett.* **2016**, *18*, 6324.
6. **Yoon, H.**; Rölz, M.; Landau, F.; Lautens, M. Pd-Catalyzed Spirocyclization through C–H Activation and Regioselective Alkyne Insertion, *Angew. Chem. Int. Ed.* **2017**, *56*, 10920.
7. Franzoni, I.; **Yoon, H.**; García-López, J.-A.; Poblador-Bahamonde, A. I.; Lautens, M. Exploring the Mechanism of the Pd-Catalyzed Spirocyclization Reaction: A Combined DFT and Experimental Study. *Chem. Sci*, **2017**, *9*, 1496.
8. **Yoon, H.**; Marchese. A. D.; Lautens, M. Carboiodination Catalyzed by Nickel. *J. Am. Chem. Soc.* **2018**, *140*, 10950.

# Acknowledgements

First and foremost, I want to thank my supervisor Prof. Mark Lautens for giving me the opportunity to complete my doctoral studies in his group. Although I did not possess a strong background in organic chemistry, I am indebted to Mark for having faith in me and accepting me as the unexpected fourth graduate student for 2013. During the five years, I am grateful for being given the freedom to generate and explore my ideas while also being given a direction when needed. As you said before, I was a diamond in the rough and I have to thank you for your continued guidance, support and teachings to make me realize my potential. Thank you to Prof. Mark Taylor and Prof. Andrei Yudin for being my committee members for five years and also helping me throughout my graduate studies. Thank you to Prof. John Wolfe for agreeing to being my external examiner and proofreading my thesis.

I want to thank Dr. David Petrone who was my teacher, mentor, motivator, collaborator and fumehood neighbor. Thank you for taking me under your wing and showing me what it takes to become a successful chemist. The late night working on our projects while listening to 50 cent, eating at the basement (Seor Ak San) and just overall great times will not be forgotten.

I have to extend my thanks to all the Lautens group members that I have had the chance to collaborate with. To Alvin Jang, my coffee and lunch buddy, thank you for the four awesome years we had. The two collaborative projects went so smoothly and that is attributed to you. I wish you the best and see you in the USA. To Alexis Lossouarn, Felicitas Landau and Martin Rölz, thank you for all choosing to work with me during your stay in Canada. Your hard work and effort never went unnoticed. I wish you all success in your future and graduate studies. To Dr. Ivan Franzoni, thank you for teaching me to become a critical chemist and bringing chemical and personal insight during my times of need. To Austin Marchese, thank you for all your hard work and dedication. I may have demanded a lot from you during the project but thank you for persevering and seeing the project to the end.

To the postdocs that I had a chance to meet, Dr. Liher Prieto, Dr. Marcus Wegmann, Dr. Kosuke Yamamoto, Dr. Charles Loh, Dr. Juntao Ye, Dr. Cédric Bürki, Dr. Steffen Kress and Dr. Marcel Sickert, thank you all for being fantastic people and chemists. Each and every one of you was instrumental in who I am as a chemist.

Special thanks to all laboratory 3 members, past and present, Dr. David Petrone, Dr. Christine Le, Dr. Ivan Franzoni, Dr. Charles Loh, Jordan Evans, Perry Menzies, Sonia Zaichuk, Alvin Jang, Rachel Ross, Austin Marchese, Egor Larin and Bijan Mirabi for being such an awesome group of people. My days in the laboratory would not have been as enjoyable without all of you. Thank you Dr. Zafar Qureshi, Dr. Thomas Johnson, Andy Yen, Heather Lam, Andrew Whyte, José Rodríguez and Nicolas Zeidan for all the wonderful discussions whether it was regarding chemistry or daily life.

Thank you to Prof. Frank Glorius who gave me the opportunity to broaden my knowledge in photochemistry. The three months I spent in Münster were unforgettable. Thank you Felix Klauck for being a great friend and project leader. Thank you Dr. Mirco Fleige, Dr. Christian Léon, Steffen Gressies, Tobias Knecht, Maximilian Koy, Andreas Lerchen, Lena Pitzer, Lena Roling, Michael Teders, Dr. Mario Wiesenfeldt and Marco Wollenburg. I wish you all continued success.

I thank all my friends who have given me the moral support required to continue and pursue my ambitions. Lastly, I would want to thank my dad, mom and sister. I would not have made it to where I am without all the sacrifices you have made for me. Thank you from the bottom of my heart.

# Contents

# Abbreviations

| | |
|---|---|
| $[\alpha]_D^{20}$ | Specific rotation measured at 589 nm and 20 °C |
| [Si] | Generic silyl group |
| Å | Angstrom |
| Ac | Acetyl |
| aq | Aqueous |
| Ar | Aryl |
| $B_2pin_2$ | Bis(pinacolato)diboron |
| BHT | Butylated hydroxytoluene |
| BINAP | 2,2′-Bis(diphenylphosphino)-1,1′-binaphthyl |
| BINAPINE | 4,4′-Di-tert-butyl-4,4′,5,5′-tetrahydro-3,3′-bi-3*H*-dinaphtho [2,1-*c*:1′,2′-*e*]phosphepin |
| BINAPO | (2′-(diphenylphosphanyl)-[1,1′-binaphthalen]-2-yl)diphenylphosphine oxide |
| BINOL | 1,1′-Bi-2-naphthol |
| Bn | Benzyl |
| BQ | Benzoquinone |
| Bz | Benzoyl |
| Cat. | Catalytic |
| CMD | Concerted metalation–deprotonation |
| cod | 1,5-Cyclooctadiene |
| Crotyl | But-2-en-1-yl |
| Cy | Cyclohexyl |
| DART | Direct analysis in real time |
| DavePhos | 2-Dicyclohexylphosphino-2-(N,N-dimethylamino)biphenyl |
| dba | Dibenzylideneacetone |
| DCM | Dichloromethane |
| DG | General directing group |
| DIAD | Diisopropyl azodicarboxylate |
| DIBAL-H | Diisobutylaluminium hydride |

| | |
|---|---|
| Difluorophos | 5,5′-Bis(diphenylphosphino)-2,2,2′,2′-tetrafluoro-4,4′-bi-1,3-benzodioxole |
| DIOP | 4,5-Bis(diphenylphosphino-methyl)-2,2-dimethyl-1,3-dioxolane |
| DMA | Dimethylacetamide |
| DMAP | 4-Dimethylaminopyridine |
| DME | Dimethoxyethane |
| DMF | *N,N*-Dimethylformamide |
| DMPU | *N,N*-Dimethylpropyleneurea |
| DMSO | Dimethyl sulfoxide |
| dppf | 1,1′-Ferrocenediyl-bis(diphenylphosphine) |
| dppm | Bis(diphenylphosphino)methane |
| dr | Diastereomeric ratio |
| EDG | Generic electron donating group |
| ee | Enantiomeric excess |
| ESI | Electrospray ionization |
| Et | Ethyl |
| EtOAc | Ethyl acetate |
| EWG | Generic electron withdrawing group |
| FG | Generic functional group |
| g | Gram |
| Galvinoxyl | 2,6-Di-tert-butyl-α-(3,5-di-tert-butyl-4-oxo-2,5-cyclohexadien-1-ylidene)-p-tolyloxy |
| HBpin | Pinacolborane |
| Het | Generic heteroarene |
| HMDS | Hexamethyldisilazane |
| HMPA | Hexamethylphosphoramide |
| HRMS | High-resolution mass spectrometry |
| *i*Pr | Isopropyl |
| KIE | Kinetic isotope effect |
| LG | Leaving group |
| M | Generic metal |
| M.P. | Melting point |
| Me | Methyl |
| Mes | Mesityl |
| MOM | Methoxymethyl |
| Ms | Mesylate |
| MW | Microwave |
| NBS | *N*-bromosuccinamide |
| *n*Bu | Butyl |
| NMDPP | Neomenthyldiphenylphosphine |
| NMR | Nuclear magnetic resonance spectroscopy |
| Nu | Generic nucleophile |
| PEG3400 | Poly(ethylene glycol) dithiol |
| Ph | Phenyl |
| PhH | Benzene |

| | |
|---|---|
| PhMe | Toluene |
| PivOH | Pivalic acid |
| PMB | Paramethoxybenzyl |
| PMP | 1,2,2,6,6-Pentamethylpiperidine |
| QPhos | 1,2,3,4,5-Pentaphenyl-1′-(di-tert-butylphosphino)ferrocene |
| R | Generic chemical group |
| rr | Regioisomeric ratio |
| RT | Room temperature |
| SPhos | 2-Dicyclohexylphosphino-2′,6′-dimethoxybiphenyl |
| TBAB | Tetrabutylammonium bromide |
| TBAC | Tetrabutylammonium chloride |
| TBDMS | Tert-butyldimethylsilyl |
| *t*Bu | Tert-butyl |
| Tempo | (2,2,6,6-Tetramethylpiperidin-1-yl)oxyl |
| Tf | Triflyl |
| TFP | Trifurylphosphine |
| TMEDA | Tetramethylethylenediamine |
| TMS | Trimethylsilyl |
| Ts | Tosyl |
| v | Volume |
| X | Generic halogen |
| XPhos | 2-Dicyclohexylphosphino-2′,4′,6′-triisopropylbiphenyl |

# List of Figures

# List of Tables

# List of Schemes

# Chapter 1
# Carbohalogenation Catalyzed by Palladium and Nickel

## 1.1 Pd-Catalyzed Cross-Coupling Reactions

Palladium-catalyzed cross-coupling reactions have revolutionized the field of synthetic organic chemistry as they provide a mean to construct otherwise challenging novel carbon–carbon bonds. Owing to the impact cross-coupling reactions had in academic and industrial labs, the Nobel Prize in Chemistry was awarded to Professors Richard Heck, Ei-ichi Negishi and Akira Suzuki in their pioneering work in developing Pd-catalyzed C–C bond forming reactions [2].

Traditionally, a cross-coupling reaction occurs between an organohalide and an organometallic species (Scheme 1.1). Upon completing the transformation, a new carbon–carbon or carbon–heteroatom is formed alongside metal byproducts.

The general mechanism proceeds with an oxidative addition into the carbon halogen bond generating the $Pd^{II}$ intermediate **1.5**. Transmetallation with an organometallic reagent or pseudo organometallic reagent gives **1.6** and reductive elimination furnishes the new carbon–carbon bond (Scheme 1.2).

### *1.1.1 Pd-Catalyzed Mizoroki-Heck Reaction*

Similar to the Pd-catalyzed cross coupling reactions, the Pd-catalyzed Mizoroki-Heck reaction employs organohalides; however, instead of an organometallic coupling partner, alkenes and alkynes are used (Scheme 1.3). The transformation releases the new olefin **1.9** and with the aid of the added base in the reaction, the HX is neutralized, allowing the catalyst to turn over.

---

Parts of this chapter have been reproduced with permission from [Petrone, D. A.; Yoon, H.; Weinstabl, H.; Lautens, M. *Angew. Chem. Int. Ed.* **2014**, *53*, 7908.] and [Yoon, H.; Marchese, A. D.; Lautens, M. *J. Am. Chem. Soc.*, **2018**, *140*, 10950.



H. Yoon, *Palladium and Nickel Catalyzed Transformations Forming Functionalized Heterocycles*, Springer Theses, https://doi.org/10.1007/978-3-030-54077-7_1

**Scheme 1.1** General Pd-catalyzed cross-coupling reaction

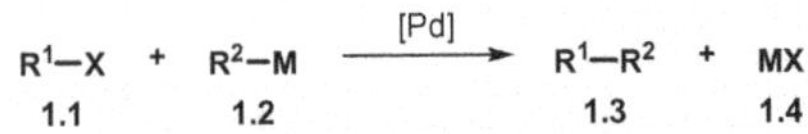


**Scheme 1.2** Pd-catalyzed cross-coupling mechanism

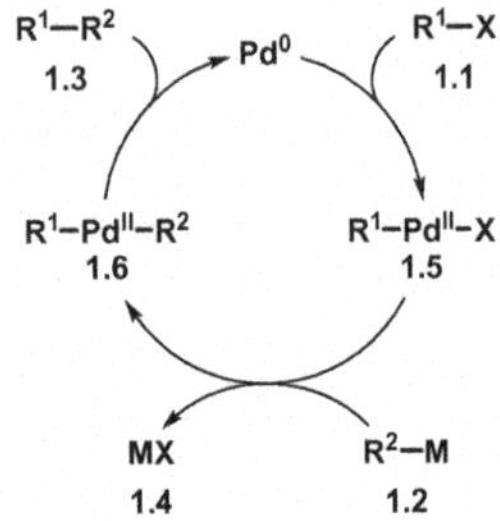


**Scheme 1.3** General Pd-catalyzed Mizoroki-Heck reaction

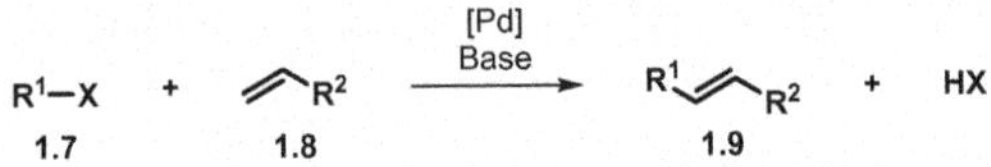


**Scheme 1.4** Pd-catalyzed Mizoroki-Heck reaction mechanism

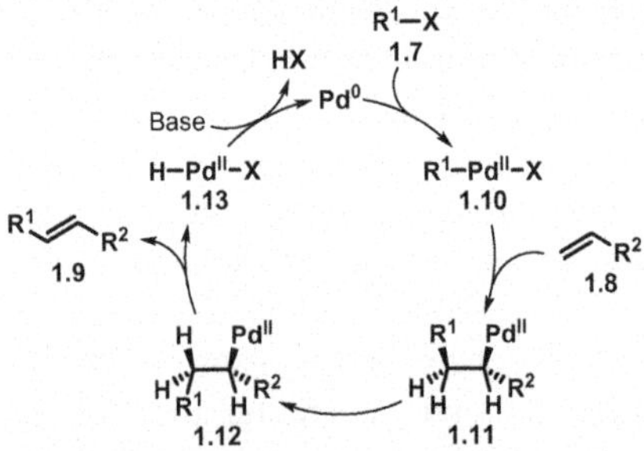


The reaction proceeds via an oxidative addition to organohalide **1.7** followed by a *syn*-carbopalladation of olefin **1.8**. Internal C–C bond rotation brings a β-hydrogen *syn* to metal which then undergoes β-hydride elimination to generate **1.9**. The base induces reductive elimination of HX from $Pd^{II}$ intermediate **1.13** to regenerate the catalyst [3] (Scheme 1.4).

### *1.1.2 Oxidative Addition*

In most $Pd^0$-catalyzed methodologies, oxidative addition is considered to the first step in transformation. Typically, aryl halides are used; however, diazonium salts and triflates have also been shown to undergo oxidative addition. The mechanism proceeds by a three center transition state which yields *cis*-isomer **1.14** (Scheme 1.5).

**Scheme 1.5** Oxidative addition mechanism

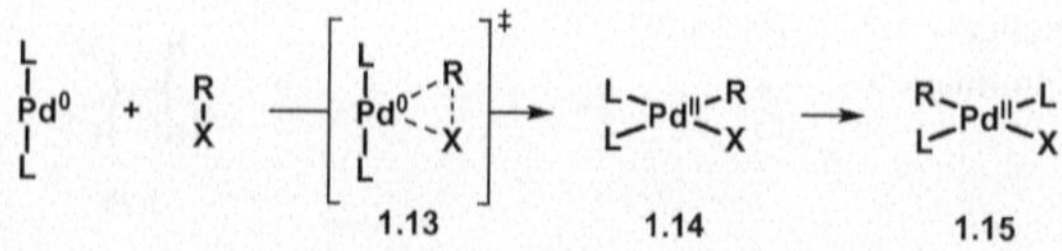


The *cis* isomer has destabilizing interactions and therefore spontaneously isomerizes to give the *trans* metal species **1.15**.

### *1.1.3 Carbopalladation (Migratory Insertion)*

Carbopalladation, commonly known as migratory insertion, is the addition of an organopalladium species across an unsaturated C–C bond in *syn* orientation. The regioselectivity of the generated alkene is dictated by this step. Several factors have been shown to influence the carbopalladation step including steric and electronic factors. The new C–C bond tends to form on the more electron-deficient carbon of the olefin. In the presence of an electron withdrawing functional group, the C–C bond forms on the β-position; however, in the presence of an electron donating group, the C–C typically forms on the α-position (Scheme 1.6a). Additionally, in the case of an intramolecular cyclization, exo-cyclizations are preferred (Scheme 1.6b) [4].

#### 1.1.3.1 β-Hydride Elimination

The β-hydride elimination is considered to be the microscopic reverse of a migratory insertion and is the termination step of the Mizoroki-Heck reaction [5]. In the presence of a β-hydrogen *syn* to the metal center, the elimination proceeds through a four-membered transition state (Scheme 1.7). As the oxidation state of the metal is not changed in this reaction, base is typically required to promote the reductive elimination of HX salts and regenerate the catalyst.

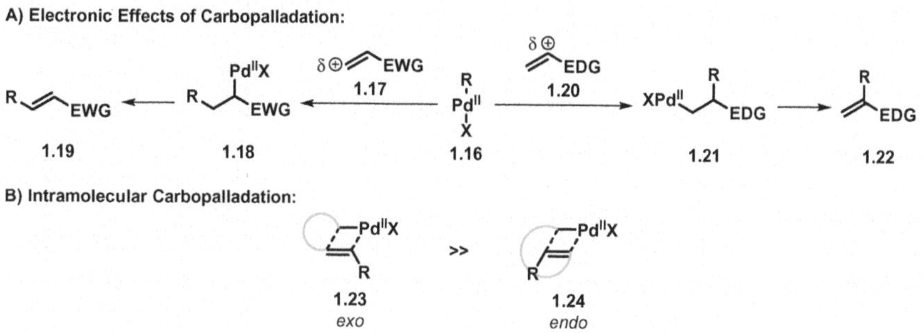


**Scheme 1.6** Regioselectivity of the inter- and intramolecular carbopalladation

Scheme 1.7 β-Hydride elimination

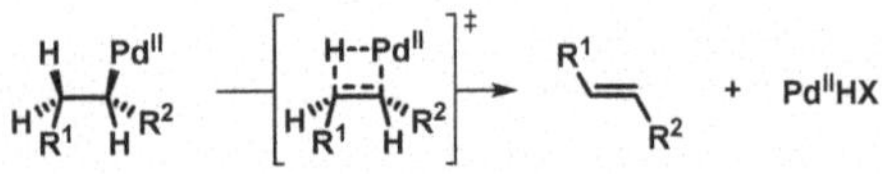

1.25 1.26

### 1.1.4 *Transmetallation*

Transmetallation is the process in which a ligand from one metal is exchanged with a ligand on another metal. The organometallic and pseudo organometallic reagents used in cross coupling reactions include organoboronates (Suzuki), stannanes (Stille), zinc (Negishi), magnesium (Kumada), and alkynyl cuprates (Sonogashira) [6]. The transmetallation step is not fully understood; however, it has been postulated that the transitions states are either via a cyclic four-membered transition state or assisted by additives as seen in the Suzuki reaction [7].

### 1.1.5 *Reductive Elimination*

Reductive elimination is the other termination sequence for a catalytic reaction besides β-hydride elimination. It is also the microscopic reverse reaction to oxidative addition. As it is the reverse of oxidative addition, the reaction proceeds through a three membered transition state forging a new C–C or C–heteroatom bond while lowering the oxidation state of the metal center [8].

## 1.2 Intramolecular Pd-Catalyzed Heck Reaction

The intramolecular Heck reaction is a powerful transformation as it circumvents the major drawbacks associated with the regioselectivity of the carbopalladation and β-hydride elimination of the intermolecular variant. Additionally, heterocycles of ranging sizes and complexity can be synthesized diastereo- and enantioselectively.

The first report of the intramolecular Heck reaction was made by Mori and Ban in 1977 where they synthesized indole **1.28** from the respective aryl iodide **1.27** (Scheme 1.8a). The generation of the indole was a result of the reinsertion of the Pd–H species generated from the initial β-hydride elimination. Subsequently after, the Heck group reported the intramolecular Heck reaction forming oxindole **1.30**. Notably, the authors found that using 1,1-disubstituted bromacrylamide **1.31** gave the six-*endo* product exclusively as the five-*exo* product is a catalytic dead end (Scheme 1.8b).

Since these findings, a significant influx of reports on the intramolecular Heck reaction have been made to demonstrate approaches to generate unique heterocycles, as key steps in total syntheses, and identifying enantioselective variants [9].

**Scheme 1.8** Pd-catalyzed intramolecular Heck reaction generating indoles, oxindoles and isoquinolinones

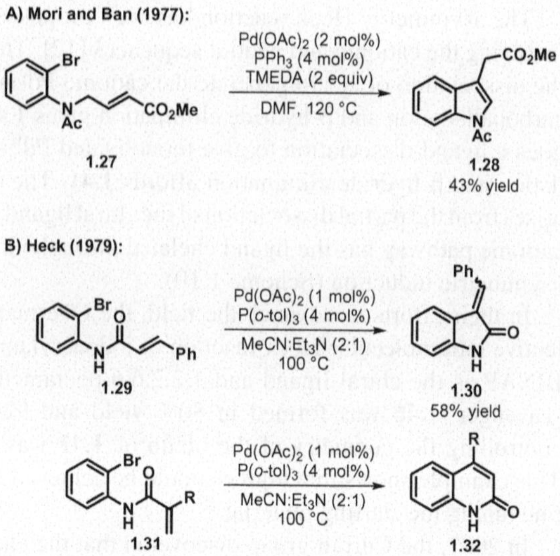


## 1.2.1 *Asymmetric Intramolecular Pd-Catalyzed Heck Reaction*

The asymmetric intramolecular Heck reaction was pioneered by the Overman and Shibasaki group. In 1989, both groups reported the first enantioselective intramolecular Heck reactions (Scheme 1.9). The Shibasaki group used bidentate ligand (*R*)-BINAP and $Ag_2CO_3$ to generate the cationic Pd-species, which gave **1.34** in 74% yield and 46% ee (Scheme 1.9a) [10]. Alternatively, the Overman group accessed the spirocycle **1.36** in 90% yield and 45% ee via the formation of a cationic Pd-species generated from the vinyl triflate (Scheme 1.9b) [11].

**Scheme 1.9** Seminal reports on the enantioselective intramolecular Pd-catalyzed Heck reaction

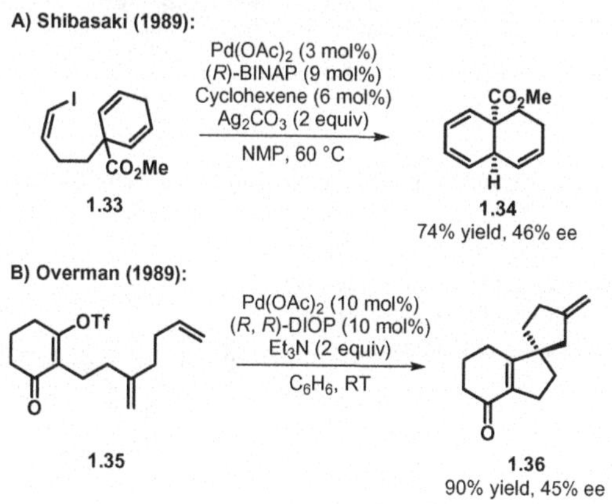

The asymmetric Heck reaction has been proposed to proceed via two pathways, involving the cationic and neutral sequences [12]. The cationic pathway begins with the dissociation of $X^-$ to generate the cationic Pd-intermediate **1.38**. A subsequent carbopalladation and β-hydride elimination gives **1.41**. The neutral pathway undergoes a ligand dissociation to give monoligated $Pd^{II}$-intermediate **1.42**. Carbopalladation and β-hydride elimination affords **1.41**. The difference in enantioselectivity arises from the partial dissociation of the chiral ligand in the neutral pathway while the cationic pathway has the ligand chelated throughout the entire process maximizing asymmetric induction (Scheme 1.10).

In their efforts to advance the field, the Overman group reported an enantioselective intramolecular Heck reaction of iodoacrylamide **1.45** (Scheme 1.11). Using BINAP as the chiral ligand and 1,2,2,6,6-pentamethylpiperidine (PMP) as the HI scavenger, **1.46** was formed in 80% yield and 92% ee. The authors found that controlling the geometry of the olefin of **1.45** was essential in obtaining high ee. This example shows that high ee could be achieved through the neutral pathway by fine tuning the starting material.

In 2007, the Curran group discovered that the enantioselectivity observed in the intramolecular Heck reaction may arise from an enantioselective oxidative addition (Scheme 1.12) [13]. They showed that using enantiopure atropisomers of iodoacrylamide **1.47**, oxindole **1.48** was generated in 85.5:14.5 er. The authors proposed that in the report by Overman, the aryl iodide rotates freely around the nitrogen and the chiral $Pd^0$ species, selectively oxidatively adds and proceeds to give product **1.46**.

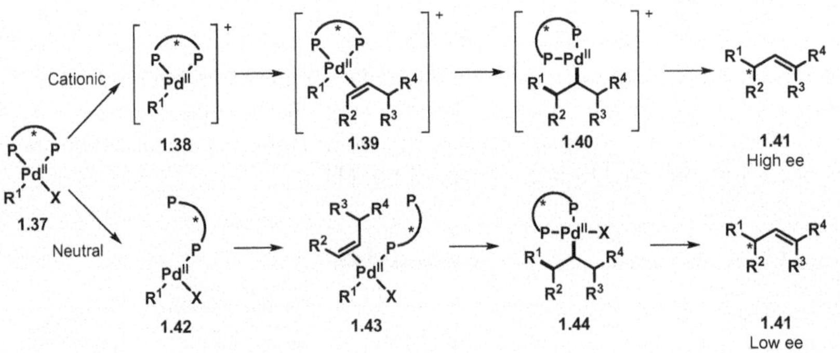


**Scheme 1.10** Proposed mechanism of the enantioselective Heck reaction

**Scheme 1.11** Enantioselective intramolecular Heck reaction generating oxindoles

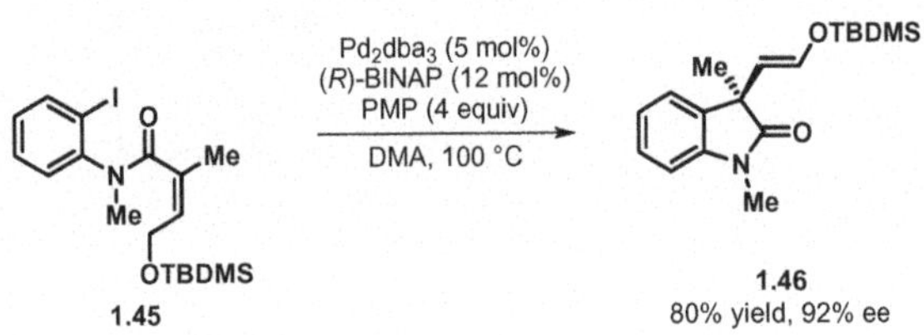

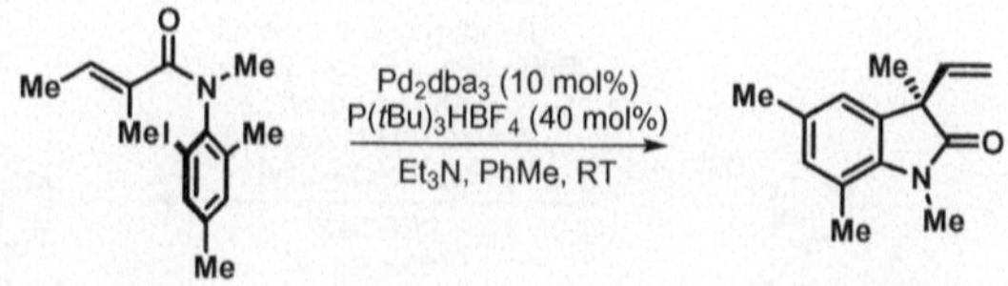


**1.47**
99.5:0.5 er

**1.48**
85.5:14.5 er

**Scheme 1.12** Enantiospecific intramolecular Heck reaction

The key difference between the two mechanisms is that in Curran's proposed route, the enantiodetermining step is the oxidative addition.

## 1.3 Reductive Elimination of Carbon–Halogen Bonds

### *1.3.1 Pd-Catalyzed Reductive Elimination of Carbon–Halogen Bonds*

The oxidative addition into carbon–halogen bonds has been believed to be irreversible as the reductive elimination of C–X bonds is energetically unfavourable. Factors which influence reductive elimination include electronic and steric factors. The reductive elimination of electron poor metal centers are more facile than the electron rich counterpart as the metal gains electron density throughout the reaction. In contrast to electronic factors, Hartwig and Roy exploited steric effects to promote the reductive elimination of aryl halides from the respective aryl $Pd^{II}$ intermediate **1.49** using $P(tBu)_3$ (Scheme 1.13) [14]. Subjecting the oxidative addition complex **1.49** to excess $P(tBu)_3$ resulted in the formation of the aryl halide. Despite $P(tBu)_3$ having strong σ-bond donating character, the implication is that the steric effects of $P(tBu)_3$ can overpower the electronic effects.

After these finding, the authors were able to observe the monoligated Pd species **1.50** (Scheme 1.14) [15]. Employing the monoligated species, a more facile reductive

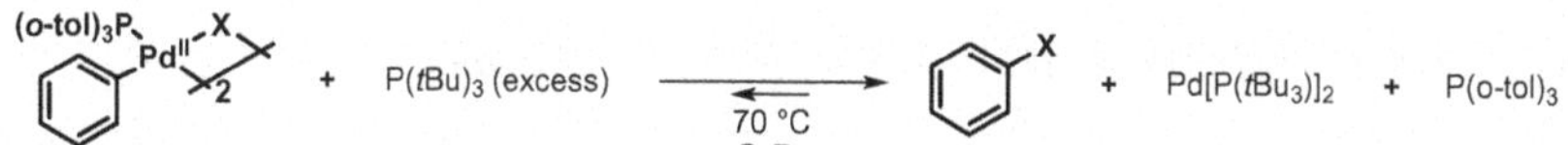


**1.49**

**Scheme 1.13** Stoichiometric studies on the reductive elimination of aryl halides

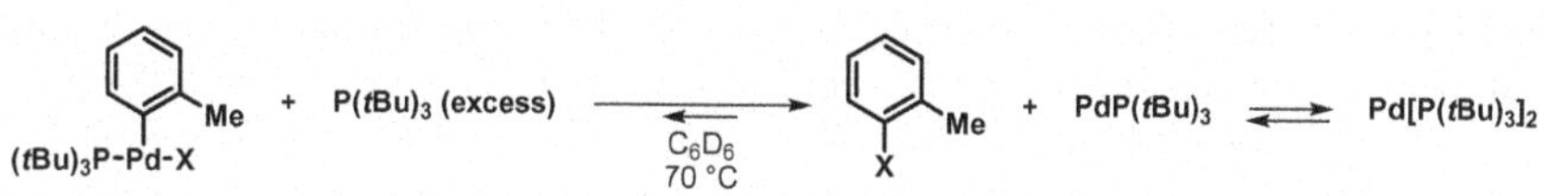


**1.50**

**Scheme 1.14** Stoichiometric studies on the reductive elimination of monoligated $Pd^{II}$ species

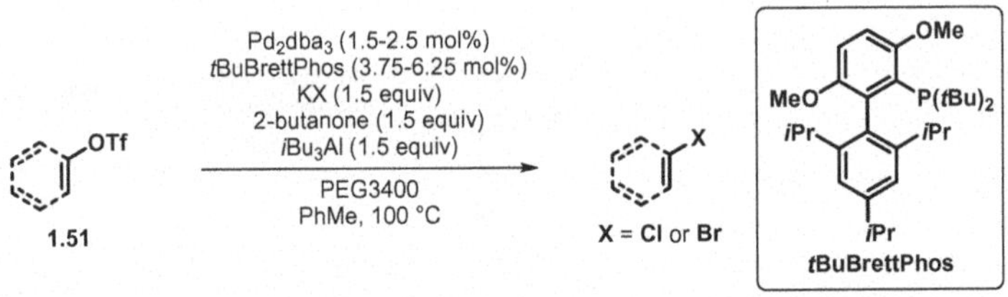


**Scheme 1.15** Pd-catalyzed synthesis of aryl and vinyl chlorides and bromides from aryl and vinyl triflates

**Scheme 1.16** Pd-catalyzed synthesis of 2-bromoindole

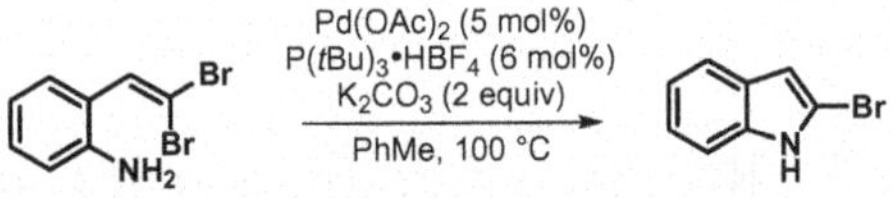


elimination was observed compared to the dimer counterpart. After extensive kinetic experiments, the authors proposed that the reaction undergoes reductive elimination from the three-coordinate complex and later regenerates the bisligated complex.

In 2010, the Buchwald group disclosed the Pd-catalyzed Finkelstein reaction of aryl or vinyl triflates to their respective chlorides or bromides [16]. Employing *t*BuBrettPhos as the ligand, PEG as the phase transfer catalyst and exogenous KBr or KCl, furnished aryl bromides and chlorides in good yield (Scheme 1.15). The authors proposed the transformation proceeded through a monoligated aryl palladium halide species prior to reductive elimination.

Applying the discoveries made by Hartwig on the reductive elimination of C–X bonds, our group in 2010 reported the Pd-catalyzed synthesis of 2-bromoindole (**1.53**) from the respective *gem*-dibromoolefin **1.52** (Scheme 1.16) [17]. Employing $P(tBu)_3$ as the ligand, selective C–N bond formation was possible. We found that $P(tBu)_3$ was essential for the transformation as it allows the catalyst to correct itself if inserted into the wrong C–Br bond via reductive elimination.

## 1.3.2 *Pd-Catalyzed Carboiodination Reactions*

Based on our report generating 2-bromoindoles via reversible oxidative addition of C($sp^2$)–Br bonds, we reported the first Pd-catalyzed cycloisomerization reaction forming heterocycles wherein the carbon–iodine bond was regenerated at the end of the catalytic cycle (Scheme 1.17a) [18]. This transformation was termed "carboiodination" as a C–I bond was added to an unsaturated C–C bond. To avoid the more favourable β-hydride elimination, aryl iodide **1.54** bearing a pendant 1,1-disubstituted olefin was used. During the ligand screen, we found bulky, electron rich phosphine ligands were optimal for the transformation. Among the ligands

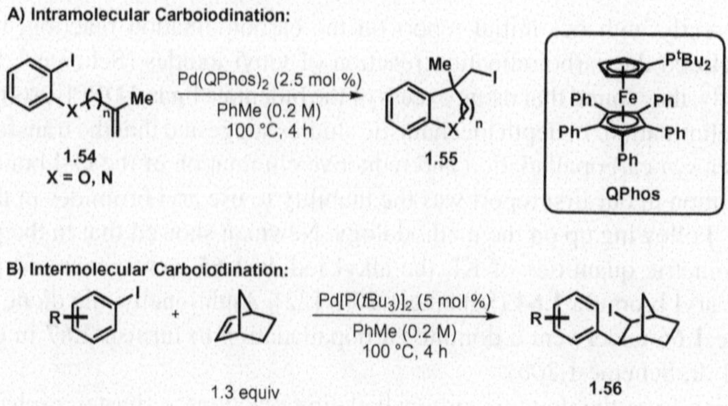


**Scheme 1.17** Pd-catalyzed carboiodination reaction

screened, QPhos [19] was optimal for the transformation while the only other ligands which exhibited reactivity were $P(tBu)_3$ and $PPh(tBu)_2$. Catalyst loadings as low as 2.5 mol% gave alkyl iodide **1.55** in excellent yield. Within this report, we also demonstrated the intermolecular variant using norbornene and aryl iodides (Scheme 1.17b).

The mechanism was computationally elucidated in collaboration with the Houk group (Scheme 1.18) [20]. Similar to traditional Heck reactions, the reaction proceeds via an oxidative addition into the aryl iodide forming **1.58**. Isomerization and coordination of the alkene with and *syn*-carbopalladation yields the alkyl palladium intermediate **1.60**. Lastly, reductive elimination via a three membered transition state generates the Pd-bound alkyl iodide which ultimately releases **1.61** and regenerates the catalyst. The rate determining step was found to be the reductive elimination (24.9 kcal $mol^{-1}$) while the C–Br and C–Cl reductive elimination was found to be higher.

**Scheme 1.18** Mechanism of the Pd-catalyzed carboiodination reaction

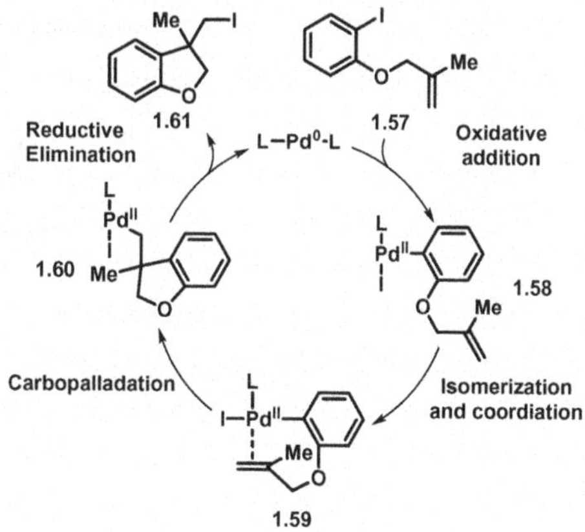

Concurrently with our initial report on the carboiodination reaction, the Tong group disclosed the carboiodination reaction of vinyl iodides (Scheme 1.19) [21]. Interestingly, they found that using excess of the bidentate ligand DPPF promoted the reductive elimination. In depth mechanistic studies suggested that the transformation undergoes a *syn*-carbopalladation and reductive elimination of the C–I bond.

A limitation in our first report was the inability to use aryl bromides in the transformation. Following up on the methodology, Newman showed that in the presence of stoichiometric quantities of KI, the alkyl iodide **1.65** were generated from the respective aryl bromide **1.64** (Scheme 1.20b) [22]. Additionally, the diene tethered aryl iodide **1.66** underwent a domino carbopalladation to furnish **1.67** in excellent d.r. (>20:1 dr, Scheme 1.20b).

Stemming from the domino carbopalladation reactions, a diastereoselective Pd-catalyzed carboiodination reaction forming chromans and isochromans was reported (Scheme 1.21a). The isochromans were generated in excellent yield and d.r.; the *cis*-diastereomer was obtained as the major product. Alternatively, the chromans generated the *trans*-diastereomers. In the presence of a preexisting chiral center, the carboiodination reaction did not affect the overall e.r. of the product (Scheme 1.21b). It was noted that in the presence of $Et_3N$, the transformations were higher yielding.

**Scheme 1.19** Pd-catalyzed carboiodination reaction of vinyl iodides

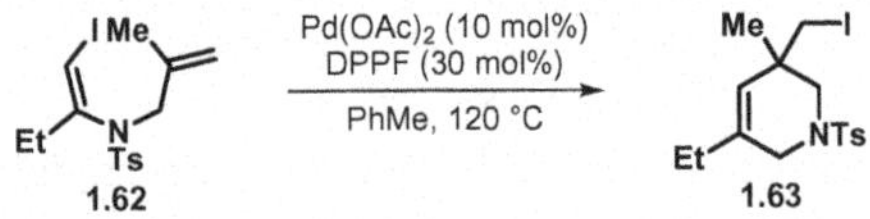


**Scheme 1.20** Pd-catalyzed halogen exchange and diastereoselective carboiodination reactions

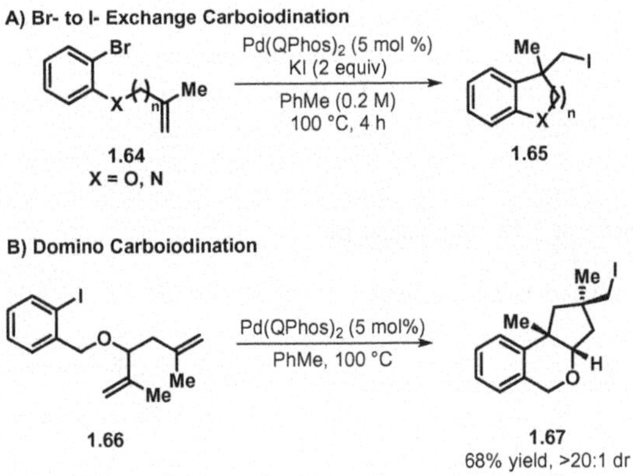

**Scheme 1.21** Diastereoselective carboiodination reaction generating chromans and isochromans

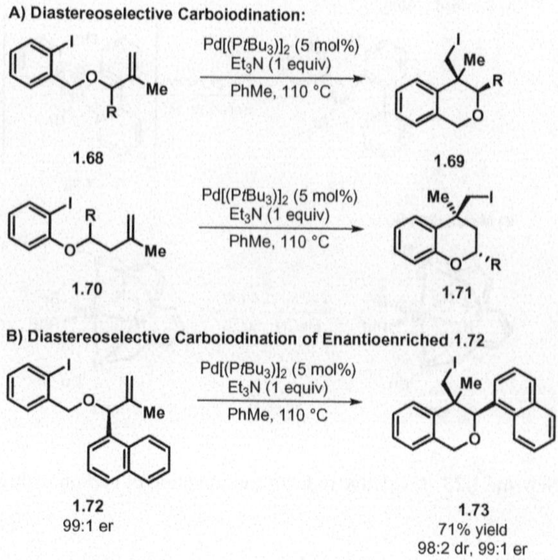


## 1.3.3 *Ni-Catalyzed Reductive Elimination of Carbon–Halogen Bonds*

Nickel-catalyzed carbon–halogen bond forming reactions have been known since the 1970s. In 1975, Cramer reported the Ni-catalyzed halogen exchange of bromobenzene to chlorobenzene using $NiCl_2$ and exogenous LiCl (Scheme 1.22a) [23]. Alternatively, in 1980, Takagi and Inokawa reported the Ni-catalyzed halogen exchange of chloro- or bromobenzene to iodobenzenes (Scheme 1.22b) [24]. Using Zn dust as the reducing agent, the authors were able to generate $Ni^0$ in situ. It should be noted that they observed the biaryl coupling in 15% yield. In this report, they proposed that $Ni^0$ oxidatively adds into the aryl bromide, halogen exchange with KI and a reductive elimination to generate **1.77**.

**Scheme 1.22** Ni-catalyzed halogen exchange reactions

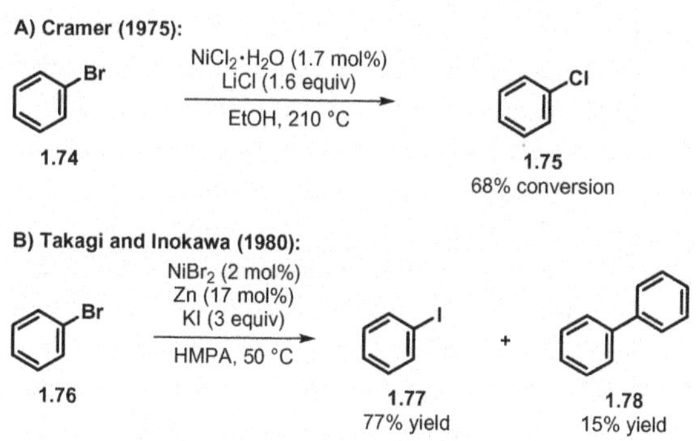

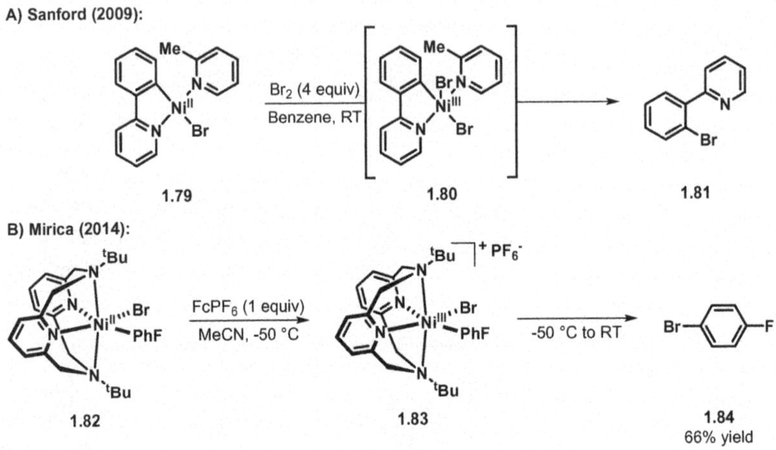


Scheme 1.23 Oxidative reductive elimination of carbon–halogen bonds

The next major breakthrough was in 2009 when Sanford reported the oxidatively induced carbon–halogen forming reaction (Scheme 1.23a) [25]. The authors proposed the addition of exogenous bromine oxidizes the $Ni^{II}$ species to $Ni^{III}$ and promotes the reductive elimination of the C–Br bond. To reinforce the oxidation to $Ni^{III}$, Miricia reported the stepwise formation of **1.84** via the oxidation of **1.82** to $Ni^{III}$ complex **1.83** (Scheme 1.23b). Upon oxidation, the complex produced **1.84** at room temperature [26].

### *1.3.4 Nickel Catalyzed Transformations with Perfect Atom Economy*

Although the Ni-catalyzed carboiodination has yet to be realized, similar reactions wherein perfect atom economy is maintained have been reported. In particular, Hiyama and coworkers in 2004 reported the first Ni-catalyzed aryl cyanation of alkynes using aryl nitriles (Scheme 1.24) [27]. The reaction is proposed to proceed through an oxidative addition into the C–CN bond to give $Ni^{II}$ intermediate **1.89**. A carbonickelation affords the vinyl $Ni^{II}$ species **1.90** and reductive elimination of the cyanide yields **1.87** and releases the catalyst. In their subsequent reports, they found that the addition of Lewis acids enabled the expansion of the scope including alkyl nitriles [28]. The Lewis acid was proposed to coordinate to the nitrile, weakening the C–CN and promoting a more facile oxidative addition.

The same group reported the intramolecular Ni-catalyzed arylcyanation of aryl nitrile **1.91** (Scheme 1.25a) [29]. A variety of heterocycles were accessed using $Ni(COD)_2$, $PMe_3$ and $AlMe_2Cl$. As the methodology generates a stereocenter, the enantioselective variant was also explored. They found that using (*R*, *R*)-*i*Pr-Foxap

**Scheme 1.24** Ni-catalyzed aryl cyanation of alkynes originating from aryl nitriles

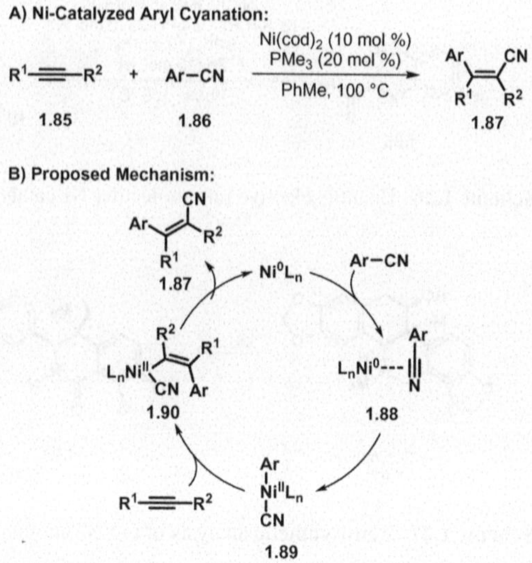


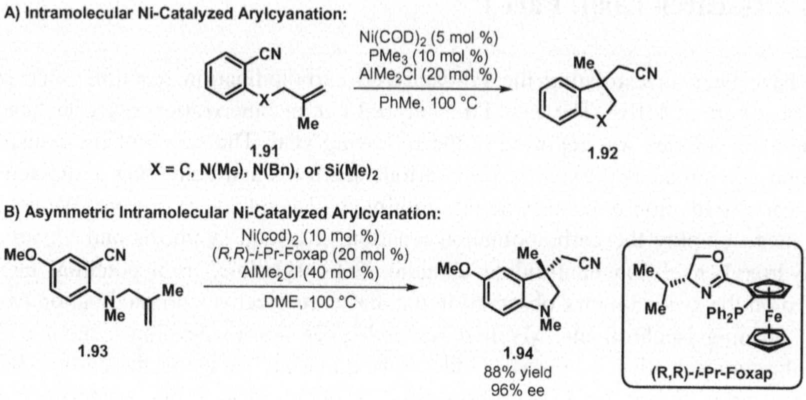


**Scheme 1.25** Intramolecular Ni-catalyzed arylcyanation

as the ligand yielded the alkyl nitrile **1.94** in excellent yield and ee (88% yield, 96% ee, Scheme 1.25b).

Simultaneously, Watson and Jacobsen reported the enantioselective intramolecular arylcyanation of **1.95** (Scheme 1.26) [30]. Using the bench stable $Ni^{II}$ precursor, $NiCl_2{\cdot}DME$, Zn was added as a reducing agent to generate $Ni^0$ in situ. After extensive screening, TangPhos was found to be optimal giving products in up to 97% ee.

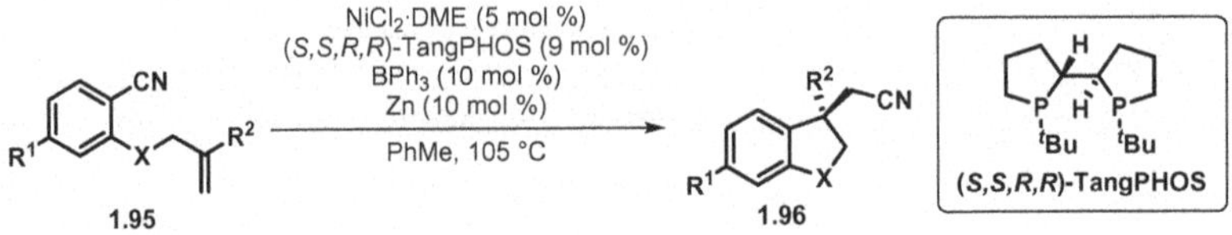


**Scheme 1.26** Enantioselective intramolecular Ni-catalyzed arylcyanation

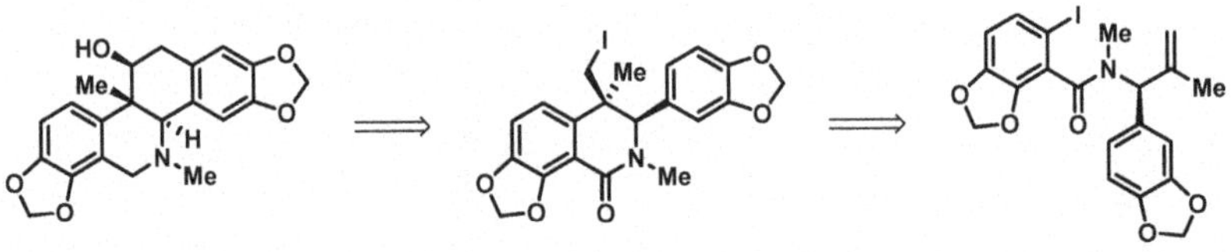


**Scheme 1.27** Retrosynthetic analysis of (+)-corynoline via the carbohalogenated product **1.99**

## 1.4 Research Goal: Part 1

We have been investigating the Pd-catalyzed carboiodination reaction since our initial report in 2010. The first Pd-catalyzed cycloisomerization reaction generating alkyl iodides was reported in the following year. The scope of the transformation was broadened by accessing various scaffolds and achieving a site selective carboiodination reaction wherein multiple C–I bonds were present. Weinstabl sought to employ the carboiodination reaction in a total synthesis and identified a hydroxybenzo[*c*]phenanthridine alkaloid, (+)-corynoline, as a potential target based on the *syn*-geometry observed in the diastereoselective carboiodination reaction forming isochromans. My first research goal was to develop a method for the diastereoselective synthesis of dihydroisoquinolinones using the carboiodination reaction (Scheme 1.27). Additionally, the effect of an amine additive on the diastereoselectivity was investigated [1].

## 1.5 Results and Discussion: Diastereoselective Carboiodination

### 1.5.1 *Starting Material Preparation*

To generate the desired carboxamides in an enantioenriched manner, alkyl or aryl aldehydes were condensed with Ellman's sulfonamide to generate **1.100**. A highly diastereoselective addition of the in situ generated alkyl zinc reagent to **1.100** gave 2° allylic amine sulfinamide **1.101**. Alkylation using LiHMDS followed by removal

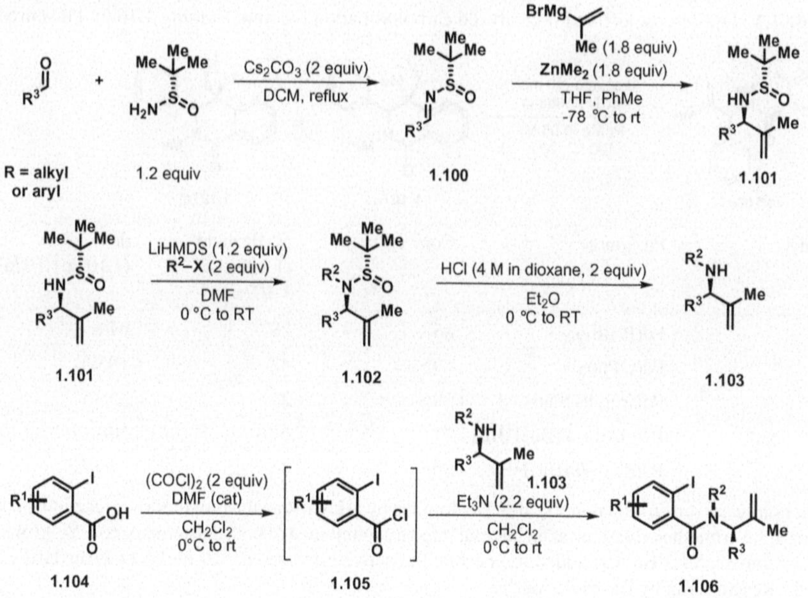


**Scheme 1.28** General route to carboxamide **1.106**

of SO$^t$Bu using HCl in dioxane afforded the highly enantioenriched amine **1.103**. Lastly, a Schotten-Baumann amide coupling reaction furnished carboxamide **1.106** (Scheme 1.28).

### *1.5.2 Optimization*

Carboxamide **1.106a** was used for the optimization of the diastereoselective carboiodination reaction (Table 1.1). Using the reaction conditions from the previously reported diastereoselective variant, **1.106a** was subjected to $Pd(PtBu_3)_2$ and $Et_3N$ in PhMe at 100 °C to give **1.107a** in 28% yield and 94:6 dr (Table 1.1, entry 1). Using $Pd(QPhos)_2$ improved the yield and dr to 42% yield and > 95:5 dr (entry 2). The in situ generation of either $Pd^0$ active catalysts led to unsatisfactory results (entries 3–5).

Various bases were screened and their effect on the reaction was measured (Table 1.2). In the absence of base, **1.107a** and **1.107a'** was formed in good yield, albeit in low d.r. (entry 1). Intrigued by the effect of an amine base on the reaction, we explored other tertiary amines. Using $Cy_2NMe$ or $BuNMe_2$ gave lower yield but significantly higher d.r. (entries 2–4). The use of bulky amine bases such as Hünig's base or PMP gave the highest observed yield and d.r. (entries, 5 and 6).

**Table 1.1** Diastereoselective Pd-catalyzed carboiodination reaction forming **1.107a**: Pd sources

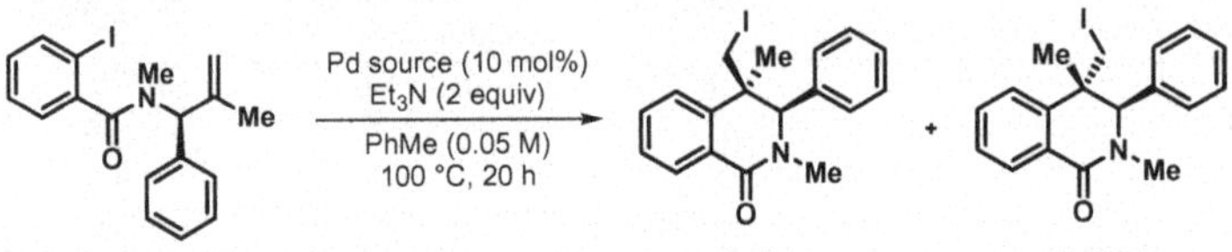


| Entry | Pd Source | Conv (%)[a] | NMR yield (1.107a + 1.107a')[a,b] | dr (1.107a:1.107a')[a] |
|---|---|---|---|---|
| 1 | $Pd(PtBu_3)_2$ | 85 | 28 | 94:6 |
| 2 | $Pd(QPhos)_2$ | >95 | 42 | >95:5 |
| 3[c,e] | PdCrotyl(QPhos)Cl | >95 | 29 | 85:15 |
| 4[d,e] | $Pd(OAc)_2/PtBu_3HBF_4$ | 77 | N/A | N/A |
| 5[d,e] | $Pd(dba)_2/PtBu_3HBF_4$ | >95 | 46 | 83:17 |

Reactions were run on 0.05 mmol scale. [a]Determined by $^1H$ NMR analysis of the crude reaction mixture using 1,3,5-trimethoxybenzene as an internal standard (combined of both diastereomers). [b]Yield of both diastereomers. [c]KO*t*Bu was added to generate the active $Pd^0$ source. [d]20 mol% of $PtBu_3HBF_4$ salt used. [e]Reaction run by David Petrone

**Table 1.2** Diastereoselective Pd-catalyzed carboiodination reaction forming **1.107a**: Bases

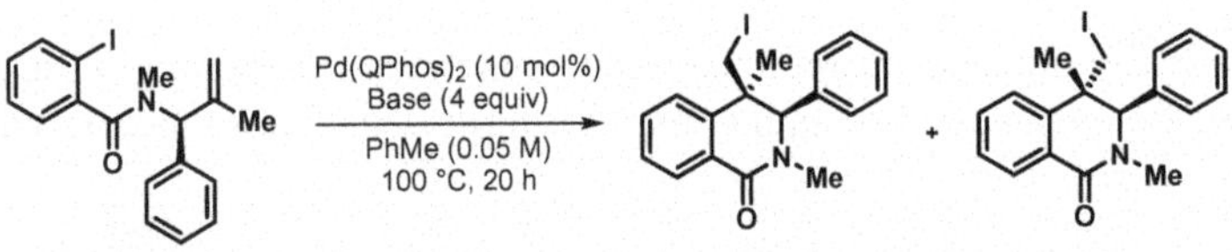


| Entry | Bases | Conv (%)[a] | NMR yield (1.107a + 1.107a')[a,b] | dr (1.107a:1.107a')[a] |
|---|---|---|---|---|
| 1 | None | >95 | 79 | 81:19 |
| 2 | $Et_3N$ | >95 | 46 | 96:4 |
| 3 | $Cy_2NMe$ | >95 | 40 | 95:5 |
| 4 | $BuNMe_2$ | >95 | 43 | 97:3 |
| 5 | $iPr_2NEt$ | >95 | 67 | >95:5 |
| 6 | PMP | >95 | 71 | >95:5 |
| 7[d] | $Cs_2CO_3$ | N/A | 10 | N/A |

Reactions were run on 0.05 mmol scale. [a]Determined by $^1H$ NMR analysis of the crude reaction mixture using 1,3,5-trimethoxybenzene as an internal standard (combined of both diastereomers). [b]Yield of both diastereomers

**Table 1.3** Diastereoselective Pd-catalyzed carboiodination reaction forming **1.107a**: PMP stoichiometry

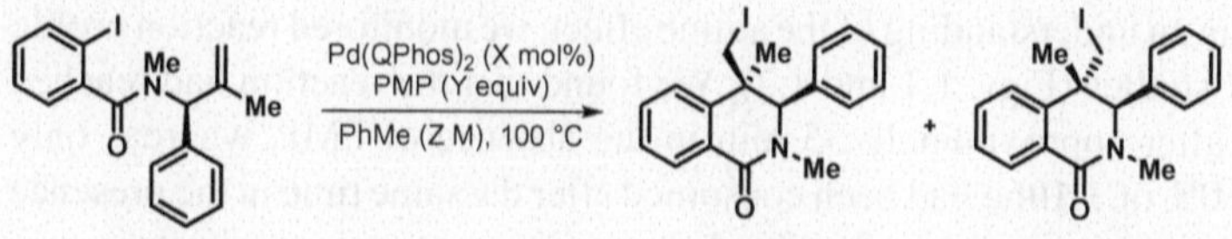


1.106a 1.107a 1.107a'

| Entry | Pd(QPhos)$_2$ Loading (mol%) | Base (equiv) Concentration (M) | NMR yield (1.107a + 1.107a')[a,b,c] | dr (1.107a:1.107a')[a] |
|---|---|---|---|---|
| 1 | 10 (10 h) | 4 (0.05 M) | >95 (90) | 92:8 |
| 2[d] | 7.5 (12 h) | 4 (0.05 M) | 76 | 92:8 |
| 3 | 7.5 (10 h) | 3.5 (0.067 M) | >95 (88) | 93:7 |
| 4 | 7.5 (10 h) | 3 (0.067 M) | (83) | 94:6 |
| 5 | 7.5 (10 h) | 2.5 (0.067 M) | (71) | 93:7 |
| 6 | 5.0 (30 h) | 4 (0.05 M) | 69 | 92:8 |
| 7 | 5.0 (22 h) | 2 (0.1 M) | >95 (88) | >95:5 |
| 8 | 5.0 (22 h) | 2.3 (0.1 M) | (83) | 94:6 |
| 9[d] | 5.0 (22 h) | 0 (0.1 M) | 95 (97) | 83:17 |

Reactions were run on 0.2 mmol scale. [a]Determined by $^1$H NMR analysis of the crude reaction mixture using 1,3,5-trimethoxybenzene as an internal standard (combined of both diastereomers). [b]Yield of both diastereomers. [c]Yield in parentheses are isolated yields. [d]Reaction run by David Petrone

Upon scaling the reaction to 0.2 mmol, **1.107a** was isolated in 90% yield and 92:8 d.r. (Table 1.3, entry 1). Lowering the catalyst loading to 7.5 mol% resulted in lower yield; however, when PMP loading was concurrently lowered and the overall concentration of the reaction was increased, product formation was restored to 88% yield. Following this trend, the optimal reaction conditions were found to be 5 mol% Pd(QPhos)$_2$ and 2 equiv of PMP at 0.1 M to give **1.107a** in 88% yield and > 95:5 d.r (entry 8).

### *1.5.3 Reaction Monitoring via $^1$H NMR Studies*[a]

In an effort to gain an understanding of the amine effect, we monitored reaction profile via in situ NMR studies (Figs. 1.1 and 1.2). We found that the reaction had reached full conversion after approximately 35 min in the absence of PMP, whereas only approximately 60% of **1.106a** had been consumed after the same time in the presence of PMP. Based on these studies, selective decomposition of a single diastereomer was not observed and supports potential amine coordination to the Pd-catalyst in the diastereodetermining step of either the oxidative addition or the carbopalladation.

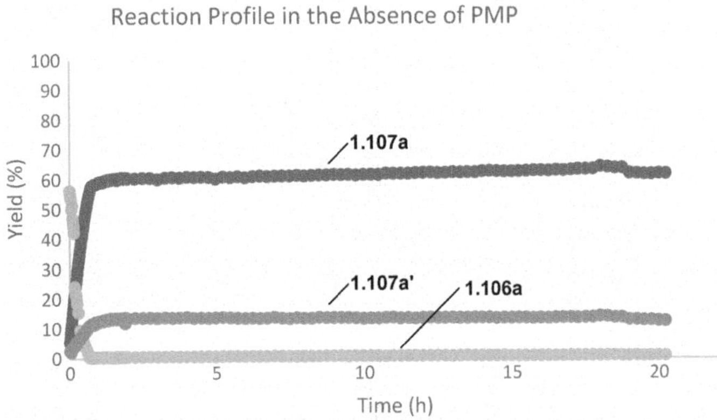


**Fig. 1.1** Reaction profile of the diastereoselective carboiodination reaction forming **1.107a** in the absence of PMP. [a]Reactions were performed by David Petrone

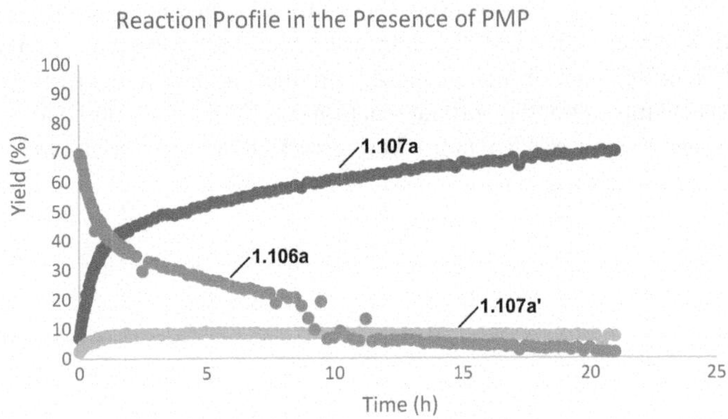


**Fig. 1.2** Reaction profile of the diastereoselective carboiodination reaction forming **1.107a** in the presence of PMP. [a]Reactions were peformed by David Petrone

### 1.5.4 Substrate Scope

The scope of the reaction furnishing enantioenriched dihydroisoquinolinones was explored (Table 1.4). In all cases, enantioenriched carboxamides (**1.106a–1.106p**) were cyclized to the corresponding products (**1.107a–1.107p**) with no erosion of e.r. Electron-rich aryl iodides (**1.106b** and **1.106c**) participated under the standard reaction conditions to provide the desired dihydroisoquinolinones in good yield and selectivity (**1.107b** and **1.107c**, 88 and 82%, 94:6 and 93:7 d.r.). Halogenated aryl iodides **1.106d** and **1.106e** underwent the transformation to the corresponding products **1.107d** and **1.107e** with identical levels of selectivity (97:3 d.r.) in 76% and 54% yields, respectively. Electron-rich substrates **1.106 g** and **1.106 h** cyclized to products **1.106 g** and **1.106 h** in 84% and 91% yield with 92:8 and 87:13 d.r. when 7.5 mol% of catalyst was used. When the *N*-protecting group was converted from −Me to −Bn (**1.106i**), the highest dr and yield was observed (95% yield and > 98:2 d.r.) Alternatively, a −Bz protected carboxamide **1.106j** was subjected to the reaction conditions using 7.5 mol% of catalyst and the corresponding product **1.107j** was obtained in lower d.r. (65:35 d.r). Substitution on the noniodinated aromatic ring appeared to have no serious deleterious effects on yield (entry 11–14). PMB ether **1.106o** was efficiently cyclized to the corresponding orthogonally protected product **1.107o** in 91% yield with 80:20 d.r. Iodothiophene **1.106p** generated the respective alkyl iodide **1.107p** in good yield and d.r.

### 1.5.5 Formal Synthesis of (+)-Corynoline

Having described the robustness of the methodology, enantiopure carboxamide **1.106 g** was further transformed to the corresponding product on gram scale according to a modified procedure (Scheme 1.29). Using increased catalyst and PMP loading (10 mol% and 3.5 equiv respectively) in the presence of 4Å molecular sieves, dihydroisquinolinone **1.107 g** was synthesized in 84% yield and 91:9 dr.

Following the formation of the core structure, nucleophilic displacement of the alkyl iodide with a cyanide source would install the last carbon atom. Due to the steric hindrance around the alkyl iodide, harsh reaction conditions were required to install the nitrile moiety (Scheme 1.30).

Originally, Petrone and Weinstabl envisioned a Houben-Hoesch reaction between the alkyl nitrile and the electron rich tethered aromatic. Unfortunately, Petrone failed to form the product employing numerous standard reaction conditions. Alternatively, we reasoned that a Friedel-Crafts type acylation would be possible using the carboxylic acid. In a two step procedure, the alkyl nitrile was reduced to the aldehyde using DIBAL-H and a Pinnick oxidation gave **1.109** in 40% yield over 2 steps (Scheme 1.31).

Acid **1.109** successfully underwent the acylation in the presence of Eaton's reagent to provide ketone **1.110** in 66% yield. The reduction of the ketone to the alcohol

**Table 1.4** Diastereoselective Pd-catalyzed carboiodination reaction. Substrate scope

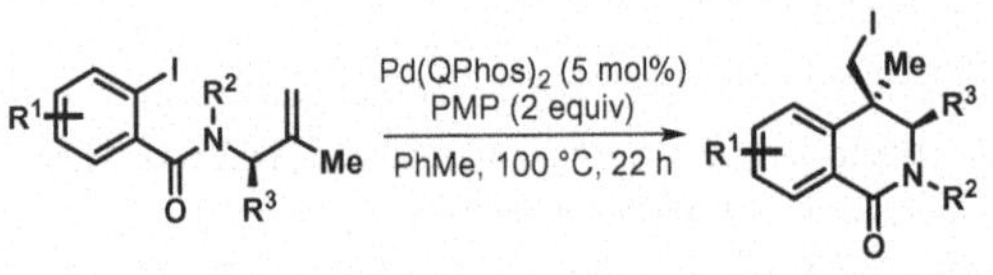


1.106 1.107

| Entry | Substrate | Product | Yield (%)[a,b] |
|---|---|---|---|
| 1 | 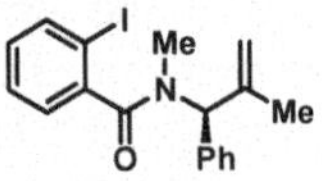 1.106a[c] | 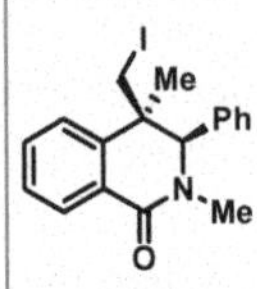 1.107a[c] | 88 (>95:5 dr) (>99:1 er) |
| 2 | 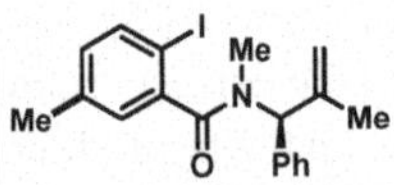 1.106b[c] | 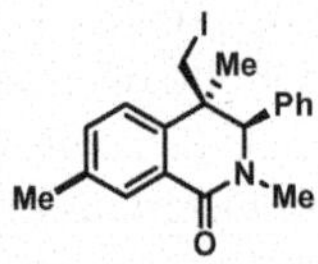 1.107b | 88 (94:6 dr) (>99:1 er) |
| 3 | 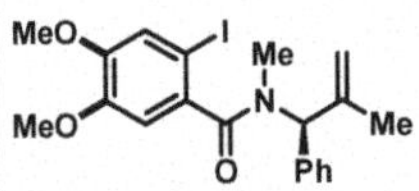 1.106c[c] | 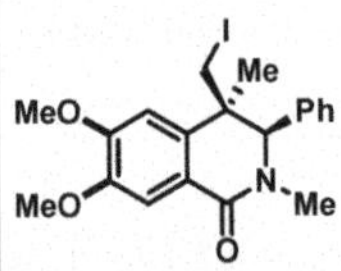 1.107c[c] | 82 (93:7 dr) (>99:1 er) |
| 4 | 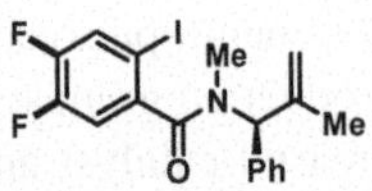 1.106d[c] | 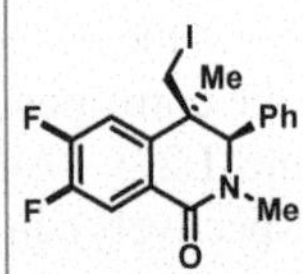 1.107d[c] | 54 (97:3 dr) (>99:1 er) |
| 5 | 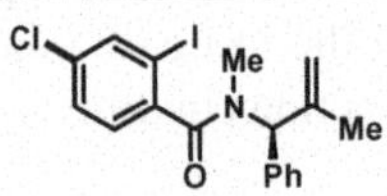 1.106e[c] | 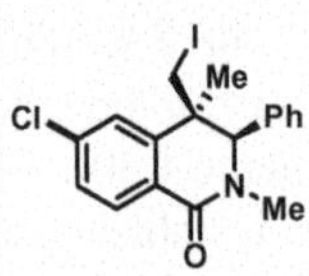 1.107e | 76 (97:3 dr) (>99:1 er) |
| 6 | 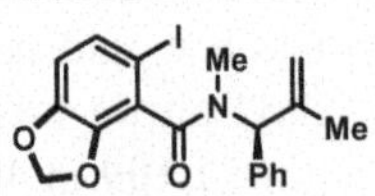 1.106f[c] | 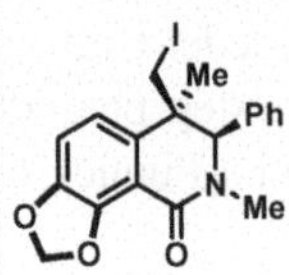 1.107f[c] | 82 (94:6 dr) (>99:1 er) |

(continued)

**Table 1.4** (continued)

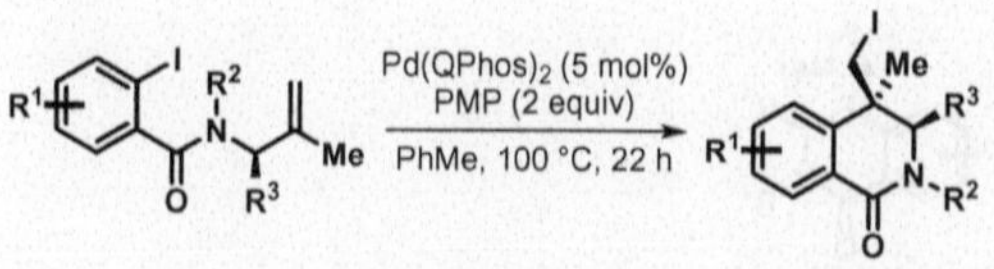


1.106 1.107

| Entry | Substrate | Product | Yield (%)[a,b] |
|---|---|---|---|
| 7 | 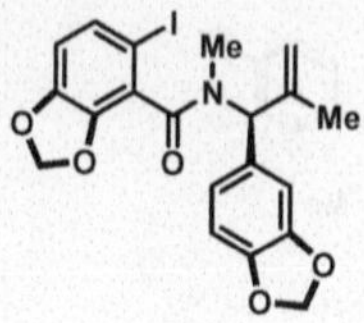 1.106g[c] | 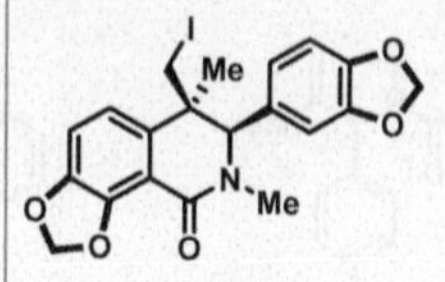 1.107g[c] | 84 (92:8 dr) (>99:1 er) |
| 8 | 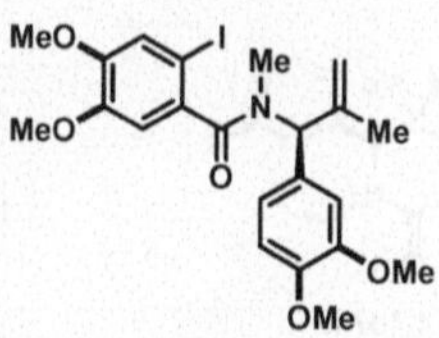 1.106h[c] | 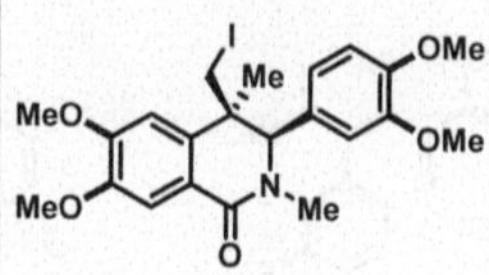 1.107h[c] | 69 (90:10 dr) (>99:1 er) |
| 9 | 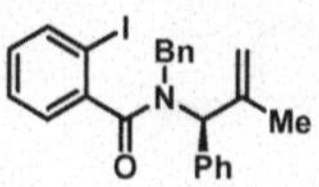 1.106i | 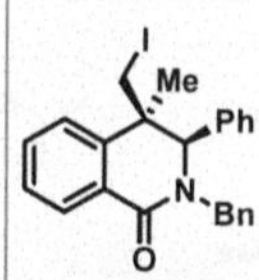 1.107i | 95 (>95:5 dr) (>99:1 er) |
| 10 | 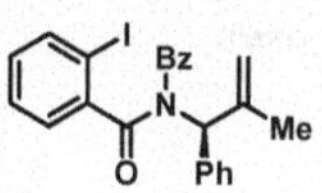 1.106j[c] | 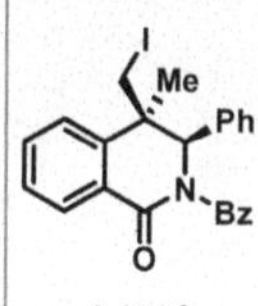 1.107j[c] | 61 (65:35 dr) (>99:1 er) |
| 11 | 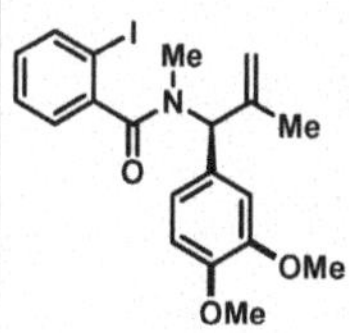 1.106k[c] | 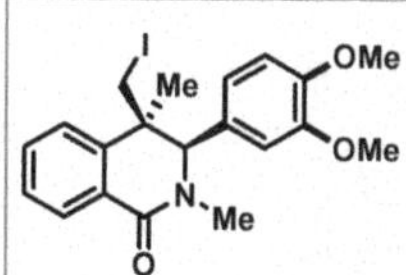 1.107k[c] | 76 (93:7 dr) (>99:1 er) |

(continued)

**Table 1.4** (continued)

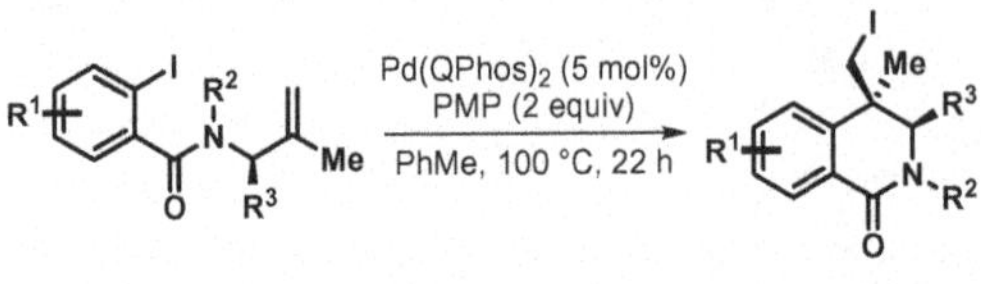


1.106 1.107

| Entry | Substrate | Product | Yield (%)[a,b] |
|---|---|---|---|
| 12 | 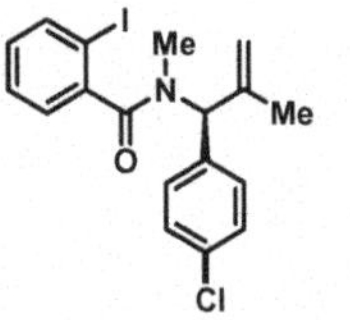 1.106l[c] | 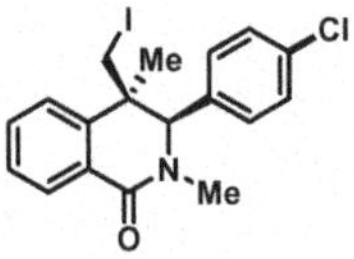 1.107l | 82 (94:6 dr) (>99:1 er) |
| 13 | 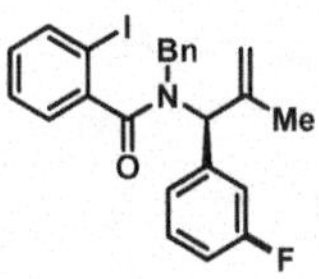 1.106m[c] | 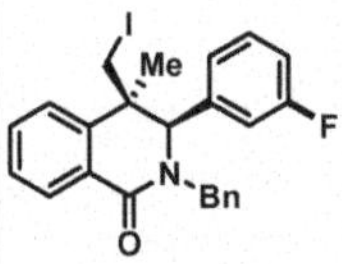 1.107m[c] | 87 (>95:5 dr) (99:1 er) |
| 14 | 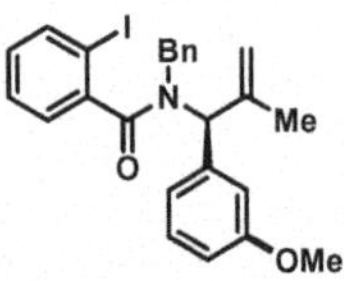 1.106n[c] | 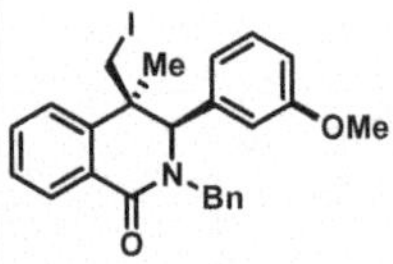 1.107n[c] | 90 (>95:5 dr) (>99:1 er) |
| 15[f] | 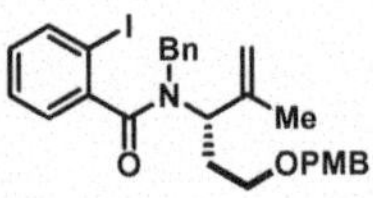 1.106o[c] | 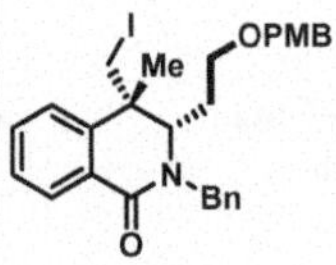 1.107o[c] | 91 (80:20 dr) (99:1 er) |
| 16 | 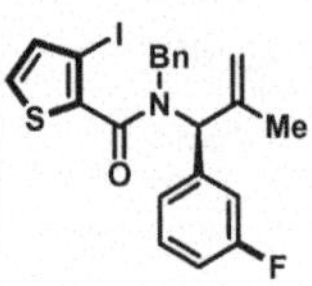 1.106p | 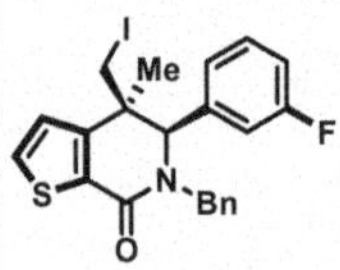 1.107p | 70 (90:10 dr) (>99:1 er) |

Reactions were run on 0.2 mmol scale. [a]Isolated yields. [b]e.r. values were determined by HPLC analysis on a chiral stationary phase. [c]This compound was prepared by David Petrone

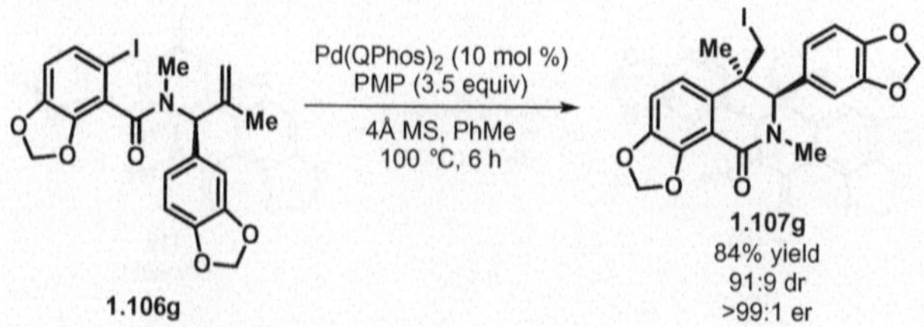


**Scheme 1.29** Diastereoselective Pd-catalyzed carboiodination reaction generating (+)-corynoline precursor **1.107 g**. Reaction was performed by David Petrone

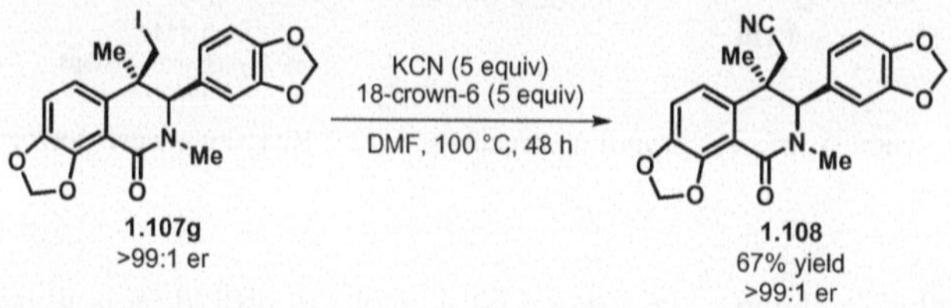


**Scheme 1.30** Nucleophilic displacement of the alkyl iodide with KCN forming **1.108**. Reaction was performed by David Petrone

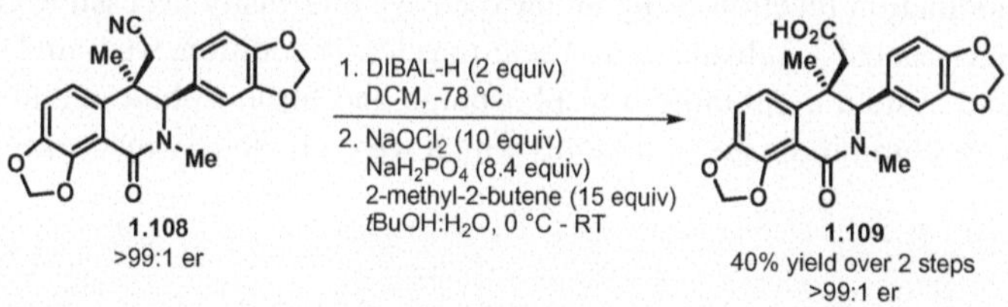


**Scheme 1.31** Reduction and oxidation of alkyl nitrile **1.108** to acid **1.109**. Reactions were run by David Petrone

proceeded in 91% yield and 95:5 dr. Subsequent mesylation and elimination gave **1.111** in 99% yield. Quinolinone **1.111** intersects a racemic total synthesis reported by Naito wherein corynoline was accessed in three steps [31] (Scheme 1.32).

## 1.6 Research Goal: Part 2

Since our seminal report in 2011, our group has made significant contributions in the Pd-catalyzed carbohalogenation reaction including the reductive elimination of C–Br and C–Cl bonds, highly diastereoselective transformations mapping onto natural products, as well as the use of HX salts as external halogen sources. My second research goal reflected our ongoing interest in advancing the reaction and we sought

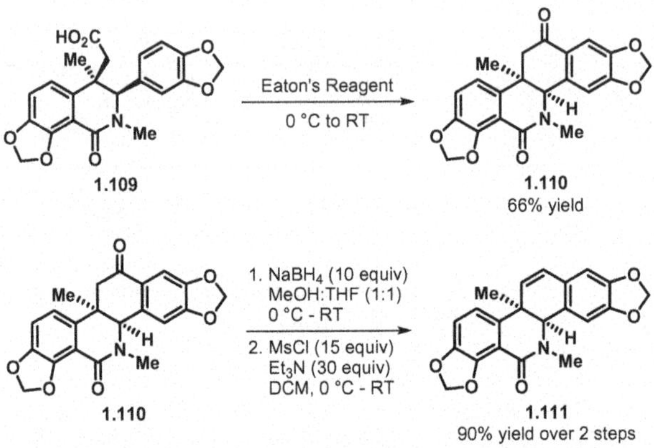


**Scheme 1.32** Synthesis of (+)-corynoline precursor **1.111**. Reactions were performed by David Petrone

to employ nickel, a cheaper base metal equivalent, to provide a new perspective in generating the respective alkyl halides. A limitation to the Pd-catalyzed methodology is the lack of an asymmetric variant and the requirement to use expensive, sterically bulky ligands. In this section, we describe the first Ni-catalyzed intramolecular carboiodination reaction using an inexpensive and easily accessible catalyst to generate halogenated 3,3-disubstituted heterocycles. In addition, we found that using a dual ligand system combining a bisphosphine and bisphosphine monoxide gave the respective chiral iodinated oxindoles which were previously inaccessible.

## 1.7 Results and Discussion: Nickel Catalyzed Carboiodination

### *1.7.1 Starting Material Preparation*

The 2-iodoacrylamides were prepared via a two-step procedure where the first step utilized a Schotten-Baumann type amide coupling of 2-phenylacryloyl chloride (**1.112**) with 2-iodoaniline to generate *N*-unsubstituted acrylamide **1.113**. Then, alkylation of these acrylamides were performed by using NaH in THF at 0 °C to generate the respective starting material (**1.114**).

The more complex substituted 2-phenylacrylic acids were synthesized according to a procedure outlined by the Singh group [32]. Diethyl oxalate undergoes a single acyl substitution with the preformed aryl Grignard reagent to generate ethyl aryl-2-glyoxalate (**1.114**). A Wittig reaction with methyl triphenylphosphonium bromide generated the acrylate and hydrolysis with LiOH gave the desired arylacrylic acid **1.112**. The alkyl substituted acrylic acids were generated from the respective malonic

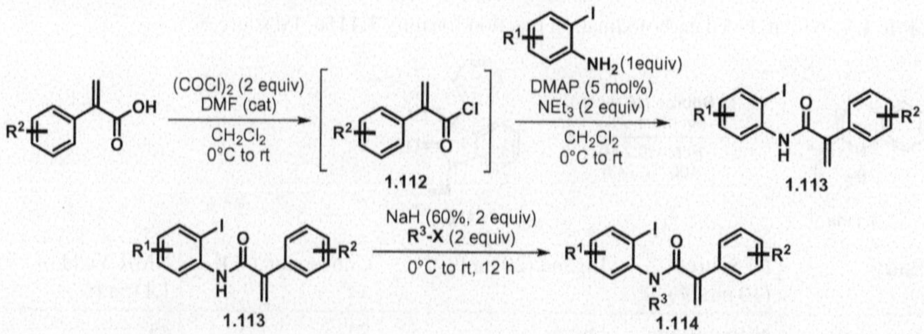

Scheme 1.33 General route to 2-iodoacrylamides

acids via a Krapcho decarboxylation and aldol condensation of formaldehyde (Scheme 1.33).

### 1.7.2 *Optimization of Racemic Variant*

At the outset of our investigation, we selected substrate **1.114a** for the subsequent optimization studies for the nickel-catalyzed carboiodination reaction (Table 1.5). Utilizing the bulky QPhos ligand previously employed for the Pd-catalyzed variant, **1.114a** was subjected to NiBr·glyme, ligand and manganese as the reducing agent to give **1.115a** in 45% yield (Table 1.5, entry 1). Employing 1,10-phenanthroline, 4,4-*t*Bu-BiPy or BiPy, which are commonly used bidentate *N,N* ligands in Ni-catalysis, failed to give the desired reaction (entries 2–4). Using $P(tBu)_3{\cdot}HBF_4$ which is the other commonly used bulky ligand in the Pd-catalyzed variant was found to be incompatible with the reaction. A possible explanation is that the K*t*BuO is incompatible with the Ni-catalyzed variant (entry 5). Fortunately, $PPh_3$ was found to be a compatible ligand and gave **1.115a** in 72% yield (entry 8). Additionally, it was found that $NiI_2$ was the superior precatalyst providing the desired product in a range of 81–92% yield (entry 10). Due to the irreproducible results, $NiI_2(PPh_3)_2$ was chosen as the optimal catalyst for this transformation (entry 11).

**Table 1.5** Ni-catalyzed carboiodination reaction forming **1.115a**: Pd sources

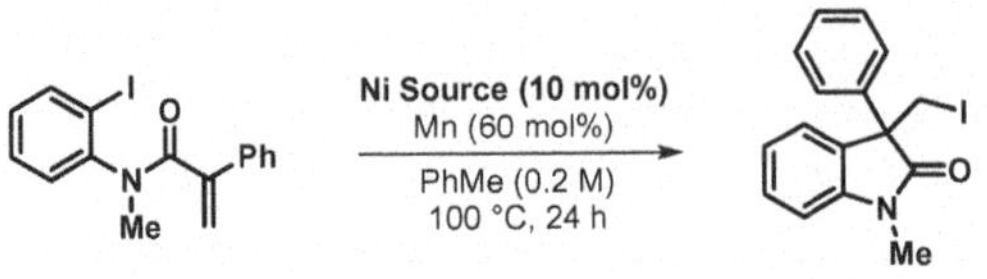


1.114a 1.115a

| Entry | Ni Source (10 mol%) | Ligand (20 mol%) | Conversion (%)[a] | NMR Yield of 1.115a (%)[a] |
|---|---|---|---|---|
| 1 | $NiBr_2$·glyme | QPhos | 82 | 45 |
| 2 | $NiBr_2$·glyme | BiPy | – | – |
| 3 | $NiBr_2$·glyme | 4,4-*t*Bu-BiPy | – | – |
| 4 | $NiBr_2$·glyme | 1,10-Phenanthroline | – | – |
| 5 | $NiBr_2$·glyme | $P(tBu)_3$·$HBF_4$ | – | – |
| 6 | $NiBr_2$·glyme | XPhos | – | – |
| 7 | $NiBr_2$·glyme | $P(o\text{-tol})_3$ | – | – |
| 8 | $NiBr_2$·glyme | $PPh_3$ | >95 | 72 |
| 9 | $Ni(acac)_2$ | $PPh_3$ | – | – |
| 10 | $NiI_2$ | $PPh_3$ | >95 | 81–92 |
| 11 | $NiI_2(PPh_3)_2$ | – | >95 | 90 |

Reactions were run on 0.2 mmol scale. [a]Determined by $^1H$ NMR analysis of the crude reaction mixtures using 1,3,5-trimethoxybenzene as internal standard

Having established the standard reaction conditions, a series of control experiments and deviations from these conditions were performed to examine the effect on the efficiency of the transformation (Table 1.6). The use of similar catalysts but with a different halide counterion (i.e. $NiBr_2(PPh_3)_2$ and $NiCl_2(PPh\cdot_3)_2)$ led to lower yields or no formation of **1.115a** (entry 2 and 3 respectively). Of note, no halogen scrambling was observed when $NiBr_2(PPh_3)_2$ was used. Removing the precatalyst or the reducing agent led to no reaction (entry 4 and 5). Interestingly, $Ni(COD)_2$ and $PPh_3$ gave the desired product in slightly lower yield in the absence of manganese (entry 6). These results suggest that the role of the manganese is exclusively the reducing agent for the in situ generation of the active catalyst. Another commonly used reducing agent such as zinc gave the product in reduced yield (entry 7). Dioxane was also found to be a suitable solvent but provided the product in lower yield (80% yield, entry 8). Finally, a lower catalyst loading or lower temperature stunted the formation of **1.115a** (entry 9 and 10) (Scheme 1.34).

**Table 1.6** Ni-catalyzed carboiodination reaction forming **1.115a**: Variation from optimized reaction conditions

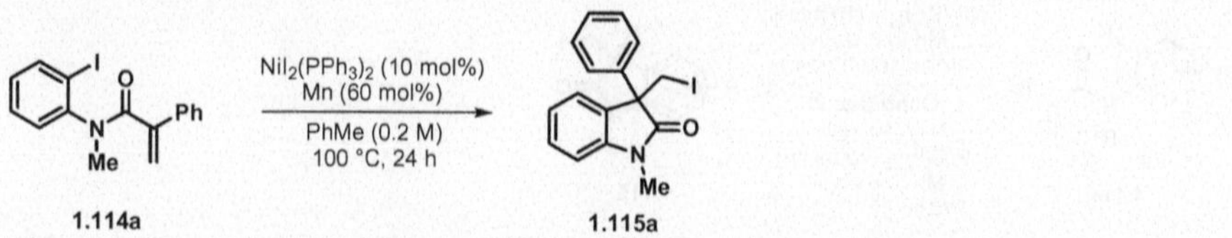


| Entry | Variation from optimized conditions | Yield 1.115a (%) |
|---|---|---|
| 1 | None | (90) |
| 2 | $NiBr_2(PPh_3)_2$ instead of $NiI_2(PPh_3)_2$ | 79 |
| 3 | $NiCl_2(PPh_3)_2$ instead of $NiI_2(PPh_3)_2$ | N.R. |
| 4 | No Ni catalyst | N.R. |
| 5 | No Mn | N.R. |
| 6[c] | $Ni(COD)_2$ and $PPh_3$ instead of $NiI_2(PPh_3)_2$ | 81 |
| 7 | Zn instead of Mn | 36 |
| 8 | Dioxane instead of PhMe | 80 |
| 9[d] | 5 mol% instead of 10 mol% | 50 |
| 10 | 80 °C instead of 100 °C | 54 |

Reactions were run on 0.2 mmol scale. [a]Determined by $^1H$ NMR analysis of the crude reaction mixtures using 1,3,5-trimethoxybenzene as internal standard. [b]Yield in parentheses are isolated yields. [c]20 mol% of $PPh_{-3}$ was added. [d]Reaction was at 0.4 M

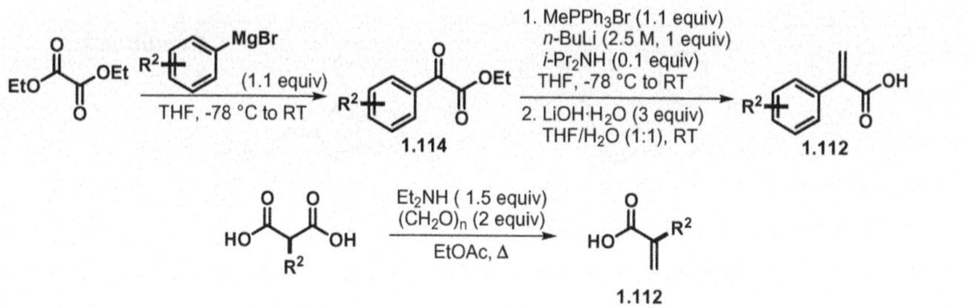


**Scheme 1.34** General route to acrylic acids

## *1.7.3 Substrate Scope of Racemic Variant*

Having the optimized conditions for the racemic reaction, we investigated the substrate scope of this reaction (Table 1.7). The weakly electron-rich acrylamides **1.114b** and **1.114c** provided **1.115b** and **1.115c** in good yield to excellent yields (85 and 92% yield respectively). Other halogens such as chloride and fluoride were

**Table 1.7** Ni-catalyzed carboiodination reaction. Substrate scope

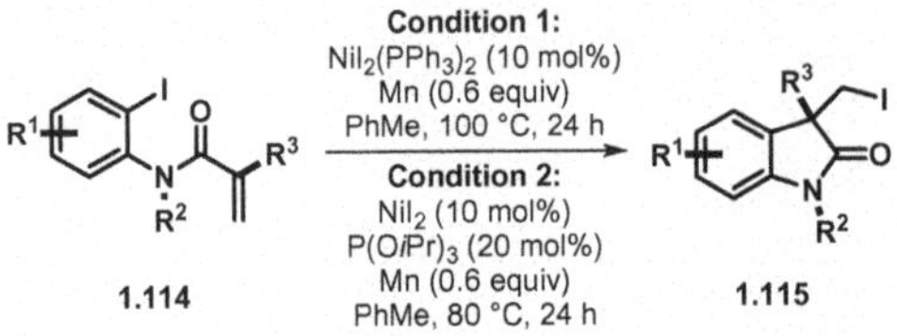


| Entry | Substrate | Product | Yield (%)[a] |
|---|---|---|---|
| 1 | 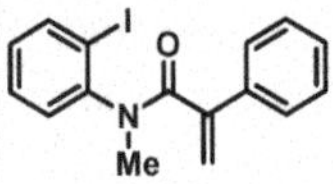 1.114a | 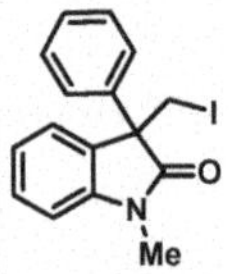 1.115a | **Condition 1**: 88 **Condition 1 (2.75 mmol)**: 86 |
| 2 | 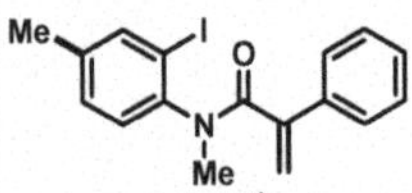 1.114b[d] | 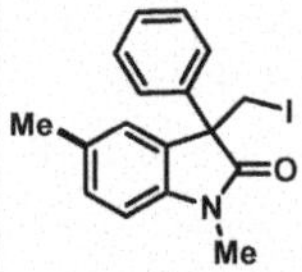 1.115b | **Condition 1**: 85 |
| 3 | 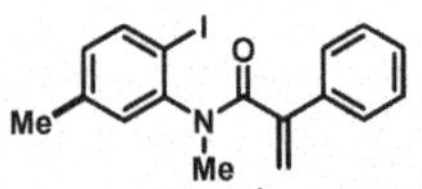 1.114c[d] | 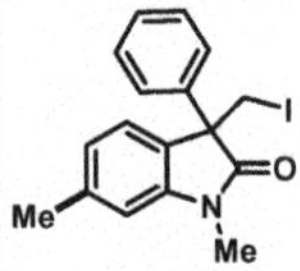 1.115c[d] | **Condition 1**: 92 |
| 4 | 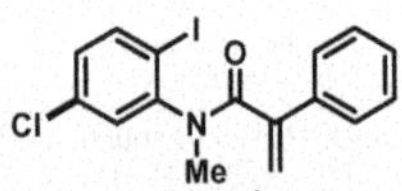 1.114d[d] | 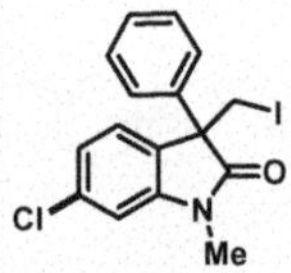 1.115d | **Condition 1**: 82 |
| 5 | 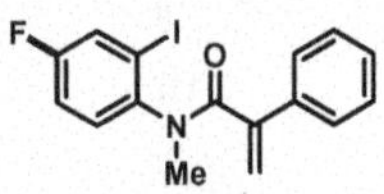 1.114e[c] | 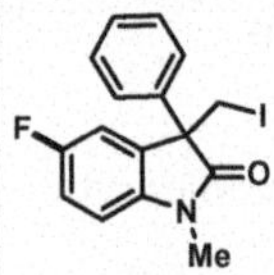 1.115e | **Condition 1**: 82 |
| 6[b,c] | 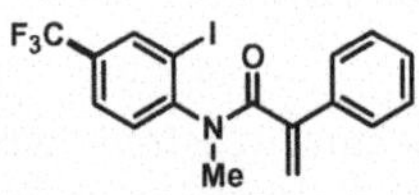 1.114f[c] | 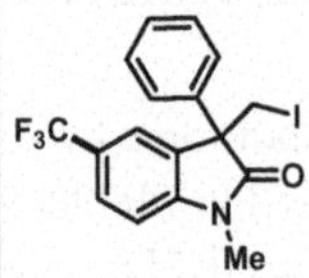 1.115f[d] | **Condition 1**: 51 |

(continued)

**Table 1.7** (continued)

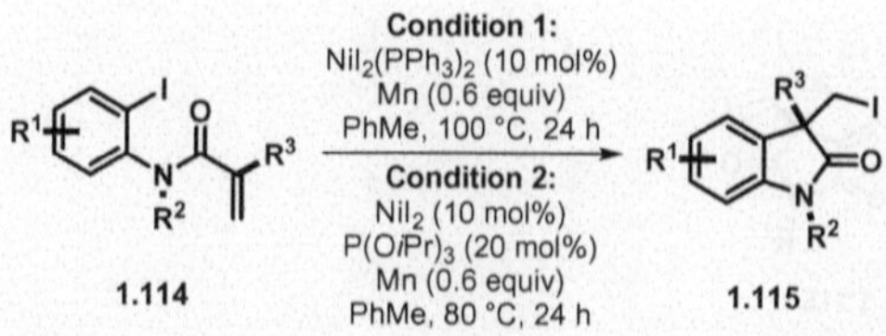


| Entry | Substrate | Product | Yield (%)[a] |
|---|---|---|---|
| 7[b] | 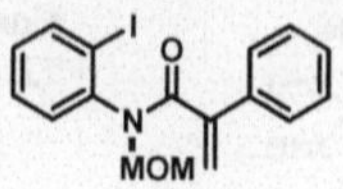 **1.114g**[c] | 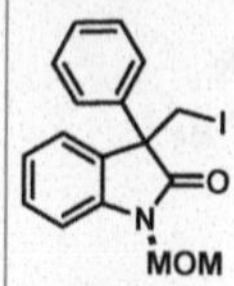 **1.115g**[d] | **Condition 1**: 68 |
| 8[b] | 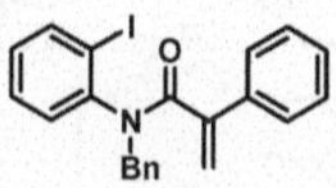 **1.114h**[c] | 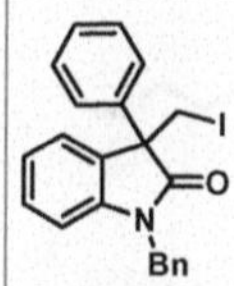 **1.115h**[d] | **Condition 1**: 91 |
| 9 | 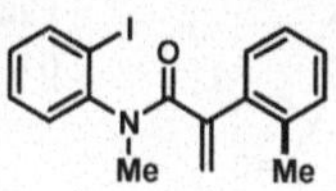 **1.114i**[c] | 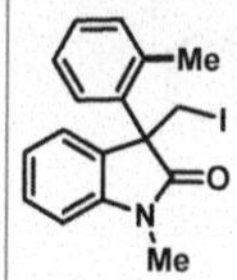 **1.115i**[d] | **Condition 1**: 38 |
| 10 | 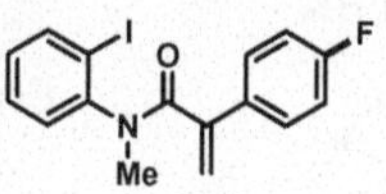 **1.114j**[c] | 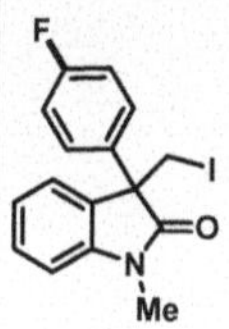 **1.115j** | **Condition 1**: 76 |
| 11 | 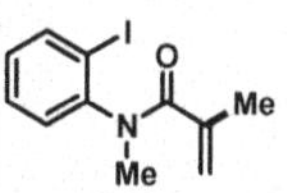 **1.114k** | 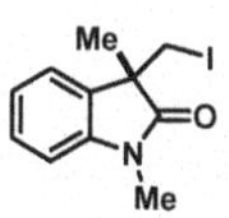 **1.115k** | **Condition 2**: 78 |
| 12 | 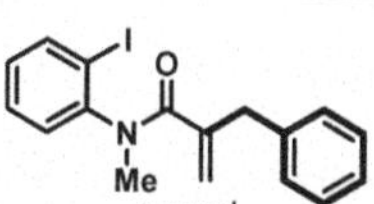 **1.114l**[d] | 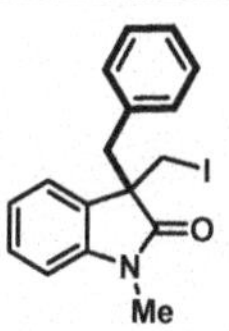 **1.115l** | **Condition 2**: 75 |

(continued)

**Table 1.7** (continued)

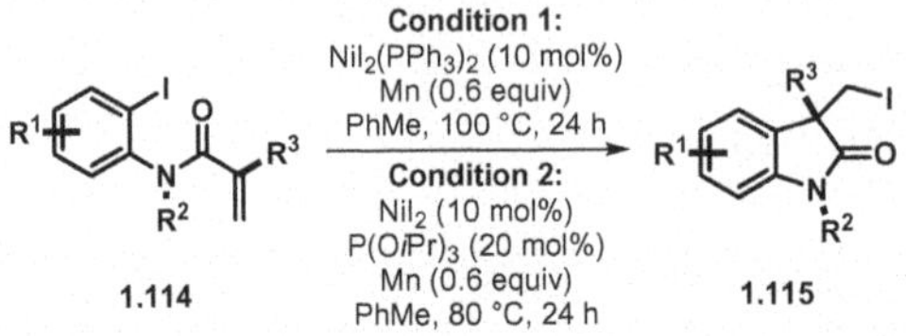


| Entry | Substrate | Product | Yield (%)[a] |
|---|---|---|---|
| 13 | 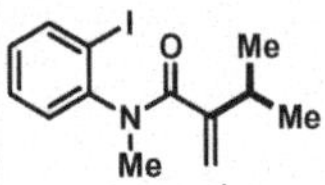 **1.114m**[d] | 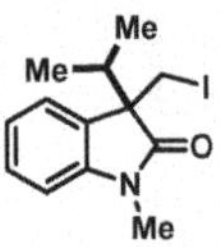 **1.115m** | **Condition 2**: 72 |
| 14[b] | 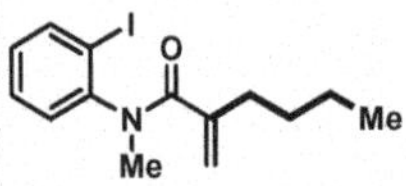 **1.114n** | 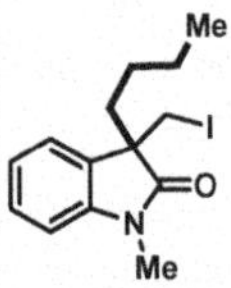 **1.115n** | **Condition 2**: 58 |
| 15 | 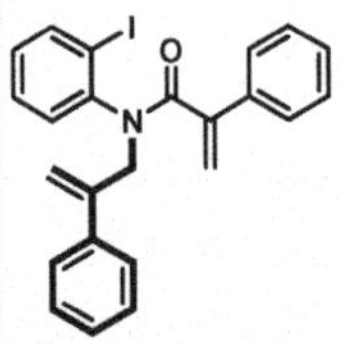 **1.114o**[d] | 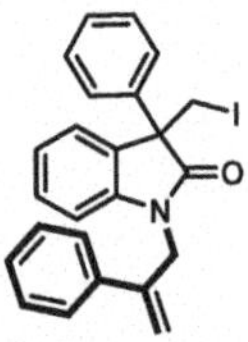 **1.115o** | **Condition 1**: 61 |
| 16 | 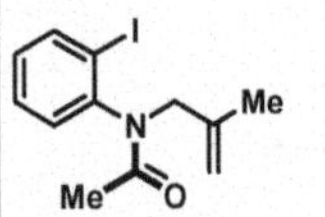 **1.114p** | 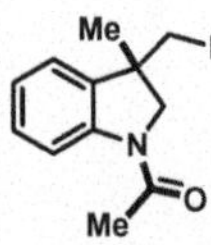 **1.115p** | **Condition 1**: 93 |
| 17 | 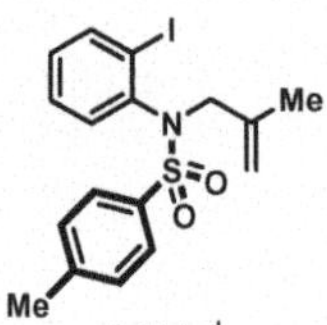 **1.114q**[d] | 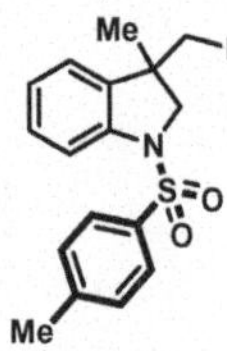 **1.115q**[d] | **Condition 1**: 89 |
| 18 | 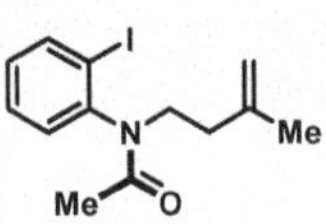 **1.114r**[d] | 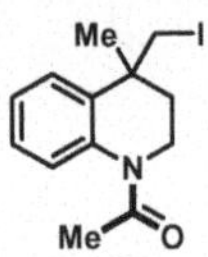 **1.115r** | **Condition 2**: 65 |

Reactions were run on 0.2 mmol scale. [a]Isolated yields. [b]Reaction run for 48 h. [c]Reaction heated to 120 °C. [d]This compound was prepared by Alexis Lossouarn [e]This compound was prepared by Austin D. Marchese

tolerated in the isomerization reaction (**1.115d** and **1.115e**, 82% yield). The trifluoromethylated acrylamide **1.114f** was found to be less reactive under the reaction conditions and required elevated temperature and longer reaction time to attain **1.115f** in moderate yield (120 °C and 48 h). Removable *N*-protecting groups such as –MOM and –Bn afforded the corresponding products in moderate to excellent yields. However, both reactions required additional time to reach full conversion (**1.115 g** and **1.115 h**, 68 and 91% yield respectively). Introduction of a 2-methylated pendant aromatic group afforded **1.115i** in diminished yield, suggesting that steric encumbrance impacts the efficiency of the cyclization (38% yield). The fluorinated pendant aromatic ring was tolerated (**1.115j**, 76% yield). Upon changing the substituent on the alkene from an aromatic group to an alkyl group, the combination of $NiI_2$ and $P(O\textit{i}Pr)_3$ was employed to give the desired cycloisomerized product in higher yields. Methyl- and benzylacrylamides **1.114 k** and **1.114 l** cyclized in good yields (78% and 75% respectively). In contrast to **1.114i**, bulkier substituent such as isopropyl afforded **1.115 m** in 72% yield. The *n*-butyl substituted substrate **1.114n** required longer reaction times to reach full conversion (48 h and 58% yield). To test the scalability of this protocol, model acrylamide **1.114a** was cyclized on gram scale to provide **1.115a** in 86% yield.

Of note, in the presence of a second acceptor where a competing 5-exo trig cyclization is possible, **1.114o** selectively cyclized on the methacryl group to yield **1.115o** exclusively. In addition, the absence of the carbonyl functionality gave no cycloisomerized product. Intrigued by these results, we sought to determine if an electron poor alkene was essential for the transformation. We prepared acetamide **1.114p** which can cyclize with the tethered methallyl group to generate the respective indoline. Subjecting **1.114p** to the standard reaction conditions gave **1.115p** in excellent yield (93% yield). In addition, replacing the acetyl group with a tosyl group also afforded the product in good yield (**1.115q**, 89% yield). These findings suggest that an oxygen atom may serve as a directing group to facilitate the oxidative addition step. Exploiting the directed oxidative addition, tetrahydroquinoline **1.115r** was obtained in 65% yield.

Aryl bromides are cheaper, stable, more readily available, and easily accessible building blocks in comparison to the corresponding aryl iodides (Table 1.8). However, under the prescribed reaction conditions, the generation of the alkyl bromide from bromoacrylamide **1.116a** proceeded in low yield. We anticipated that in the presence of an external iodide source, a halogen exchange would occur in situ to generate the respective alkyliodide (Table 1.8). Upon adding 2 equivalents of potassium iodide to the standard reaction conditions, **1.115a** was formed in 83% yield starting from **1.116a**. By using the second set of optimized conditions, the electron-rich *p*-methoxybromoacrylamide **1.116b** cyclized to give the corresponding oxindole **1.115s** in 70% yield. Additionally, the dioxol bearing oxindole **1.115t** was generated in good yield (84% yield). Analogous to the generation of **1.115 h**, the formation of the *N*-benzylated oxindole **1.115u** required longer reaction times (44% yield).

**Table 1.8** Ni-catalyzed halogen exchange-induced carboiodination reaction. Substrate scope

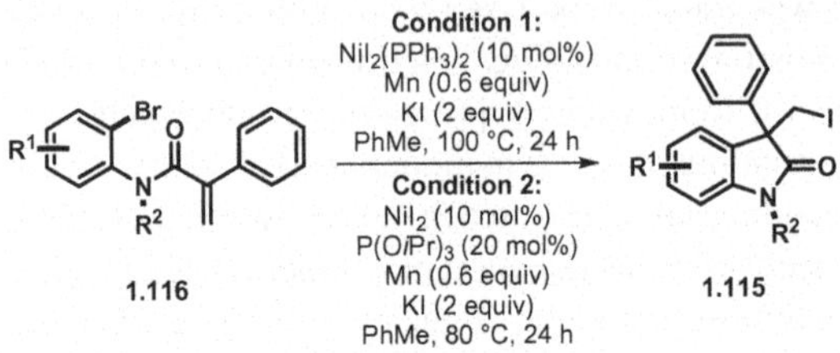


| Entry | Substrate | Product | Yield (%)[a] |
|---|---|---|---|
| 1[b] | 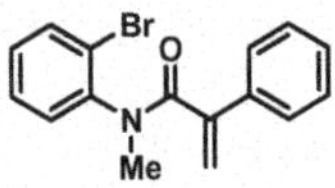 1.116a | 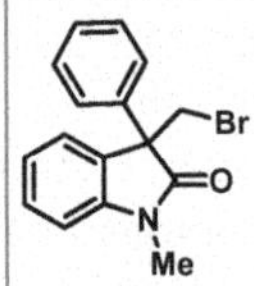 1.117 | **Condition 1**: 8 |
| 2 | 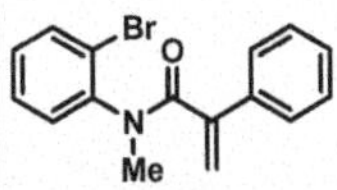 1.116a | 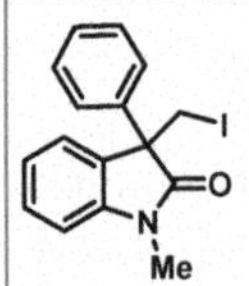 1.115a | **Condition 1**: 83 |
| 3 | 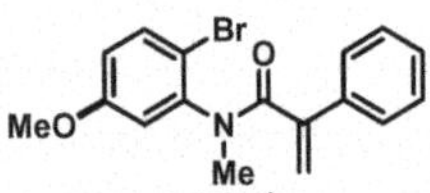 1.116b[d] | 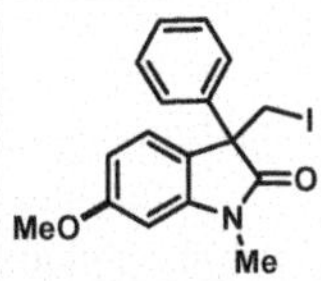 1.115s[d] | **Condition 2**: 70 |
| 4 | 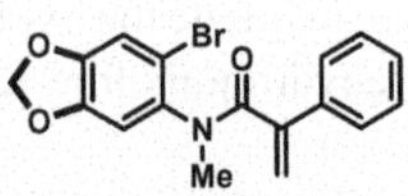 1.116c | 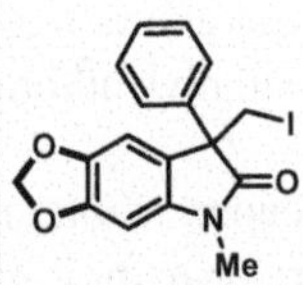 1.115t | **Condition 2**: 84 |
| 5[b] | 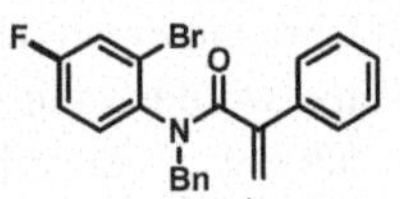 1.116d[d] | 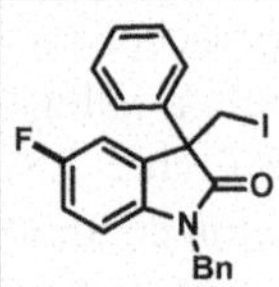 1.115u[d] | **Condition 2**: 44 |

Reactions were run on 0.2 mmol scale. [a]Isolated yields. [b]Reaction run with $NiBr_2(PPh_3)_2$ and in the absence of KI. [c]Reaction run for 48 h. [d]This compound was prepared by Austin D. Marchese

### 1.7.4 Optimization of Enantioselective Variant

To further showcase the benefits of employing nickel in carboiodination, the enantioselective variant was pursued (Table 1.9). Unfortunately, typical aryl phosphine ligands were not compatible with the reaction and failed to generate or induce chirality (entries 1–8). The proof of concept was achieved when BINOL derived phosphite **L1** gave 15% yield and 56.5:43.5 e.r. (entry 9). After screening BINOL derived phosphoramidites, Monophos was found to be the optimal ligand providing **1.115a** in 45% yield and 65:35 e.r (entries 10–14). In addition, using excess ligand in the transformation gave marginal increase in e.r. (entry 11).

Serendipitously, we found that using a dual ligand system of a bisphosphine and bisphosphine monoxide was essential in achieving higher enantioselectivity (Table 1.10). Individually, (*S*)-BINAP and (*S*)-BINAPO failed to generate product; however, a combination of the two ligands afforded **1.115a**, in appreciable yield and e.r (entry 1–7). Using a 9:1 mixture of (*S*)-BINAP to (*S*)-BINAPO gave **1.115a**

**Table 1.9** Ni-catalyzed carboiodination reaction forming **1.115a**: Chiral ligands

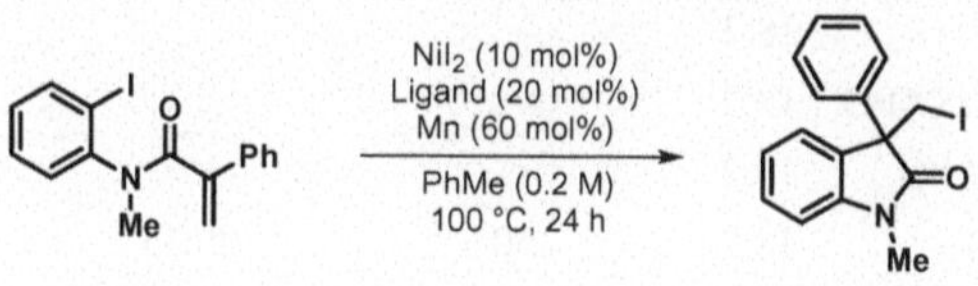


| Entry | Ligand | NMR yield of 1.115a (%)[a] | e.r.[b] |
|---|---|---|---|
| 1 | (*S*)-BINAP | – | – |
| 2 | (*S*)-TolBINAP | – | – |
| 3 | (*S*)-Tolyl-Garphos | – | – |
| 4 | (*S*)-Cl-OMe-BIPHEP | – | – |
| 5 | *t*Bu-PHOX | – | – |
| 6 | *t*Bu-PyOX | – | – |
| 7[c] | SL-J002-1 | >95 | 50:50 |
| 8[c] | (*R*)-Xyl-SDP | – | – |
| 9 | L1 | 15 | 56.5:43.5 |
| 10[c] | L2 | 45 | 65:35 |
| 11[c,d] | L2 | 40 | 70:30 |
| 12[c] | L3 | 8 | 58:42 |
| 13[c] | L4 | – | – |
| 14 | L5 | 13 | 65:35 |

Reactions were run on 0.2 mmol scale. [a]Determined by $^{1}H$ NMR analysis of the crude reaction mixtures using 1,3,5-trimethoxybenzene as internal standard. [b]e.r. values were determined by HPLC analysis on a chiral stationary phase. [c]This reaction was performed by Austin D. Marchese

**Table 1.10** Ni-catalyzed carboiodination reaction forming **1.115a**: Dual Chiral ligands

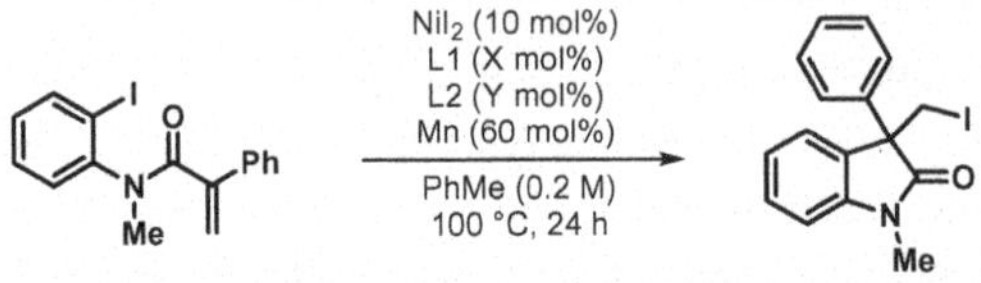


1.114a 1.115a

| Entry | Bisphosphine (L1):Bisphosphine monoxide (L2) | mol% (X:Y) | NMR Yield of 1.115a (%)[a] | e.r.[b] |
|---|---|---|---|---|
| 1 | (*S*)-BINAP | – | – | – |
| 2 | (*S*)-BINAPO | – | – | – |
| 3 | (*S*)-BINAP:(*S*)-BINAPO | 10:10 | 36 | 60:40 |
| 4 | (*S*)-BINAP:(*S*)-BINAPO | 16:4 | 60 | 81:19 |
| 5 | (*R*)-BINAP:(*S*)-BINAPO | 16:4 | 25 | 28:72 |
| 6 | (*S*)-BINAP:(*S*)-BINAPO | 18:2 | 42 | 87:13 |
| 7 | (*S*)-BINAP:(*S*)-BINAPO | 19:1 | 30 | 90:10 |
| 8 | (*S*)-BINAPINE:(*S*)-BINAPO | 18:2 | – | – |
| 9 | (*S*)-NMDPP:(*S*)-BINAPO | 18:2 | – | – |
| 10 | (*S*)-Cl-MeO-BIPHEP:(*S*)-BINAPO | 18:2 | – | – |
| 11 | (*S*)-Tolyl-Garphos:(*S*)-BINAPO | 18:2 | – | – |
| 12 | (*S*)-TolBINAP:(*S*)-BINAPO | 18:2 | 50 | 89:11 |

Reactions were run on 0.2 mmol scale. [a]Determined by $^1$H NMR analysis of the crude reaction mixtures using 1,3,5-trimethoxybenzene as internal standard. [b]e.r. values were determined by HPLC analysis on a chiral stationary phase. [c]This reaction was performed by Austin D. Marchese

in 42% yield and 87:13 e.r (entry 6). After screening chiral phosphine ligands in conjunction with (*S*)-BINAPO, (*S*)-TolBINAP generated the oxindole in the highest combination of yield and e.r (entry 12).

## 1.7.5 Substrate Scope of Enantioselective Variant

Utilizing the 9:1 mixture of (*S*)-TolBINAP:(*S*)-BINAPO, model substrate **1.115a** was isolated in 50% yield and 89:11 e.r. Employing a more electron rich acrylamide **1.114c** gave **1.115c** in increased yield and good e.r. (57% yield, 89:11 e.r.). In contrast to the racemic variant which required elevated temperature and longer reaction time, generation of the trifluoromethylated oxindole **1.115f** proceeded in good yields and moderate e.r. (84% yield, 78:22 e.r.). Switching the vinyl aromatic group to an alkyl group gave moderate yields and lower e.r. (65:35 e.r.) (Table 1.11).

**Table 1.11** Enantioselective Ni-catalyzed carboiodination reaction: Substrate scope

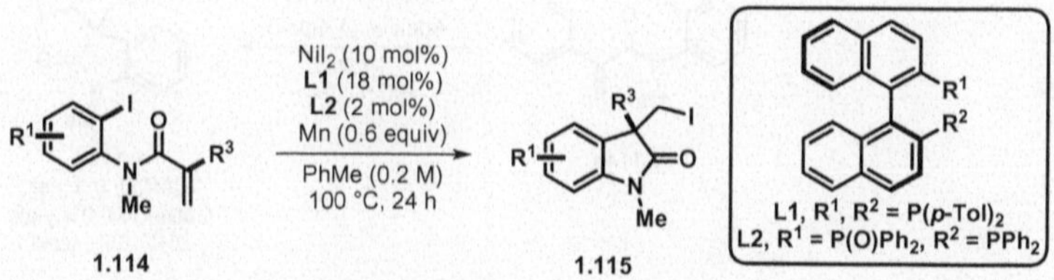


1.114 1.115

| Entry | Substrate | Product | Yield (%)[a,b,c] |
| --- | --- | --- | --- |
| 1 | 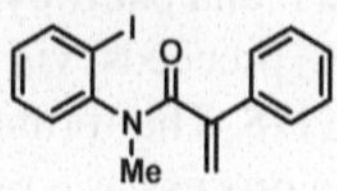 1.114a | 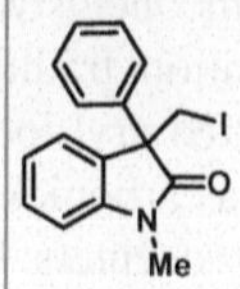 1.115a | 50 (89:11 e.r.) |
| 2 | 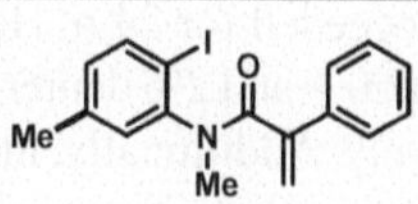 1.114c[c] | 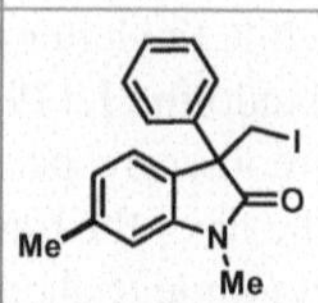 1.115c[c] | 57 (89:11 e.r.) |
| 3 | 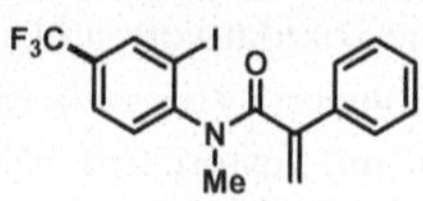 1.114f[d] | 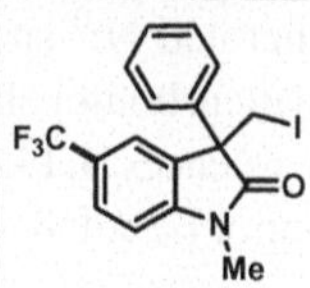 1.115f | 84 (78:22 e.r.) |
| 4[d] | 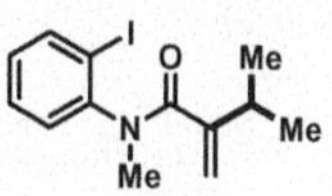 1.114m[c] | 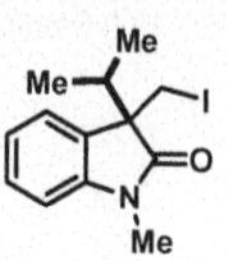 1.115m[c] | 51 (65:35 e.r.) |

Reactions were run on 0.2 mmol scale. [a]Isolated yields. [b]e.r. values were determined by HPLC analysis on a chiral stationary phase. [c]This compound was prepared by Austin D. Marchese. [d]This compound was prepared by Alexis Lossouarn

## 1.8 Mechanistic Studies

In an attempt to elucidate the mechanism of the transformation, we initially investigated the possibility of a radical pathway. The stoichiometric addition of TEMPO stopped product formation; however, no radical trap adducts were detected and inhibition may be due to the formation of an inactive catalyst [33]. In addition, galvinoxyl also inhibited the reaction but the presence of BHT in the reaction did not impede product formation (Scheme 1.35). As no adducts were observed from these

**Scheme 1.35** Radical trap additive studies

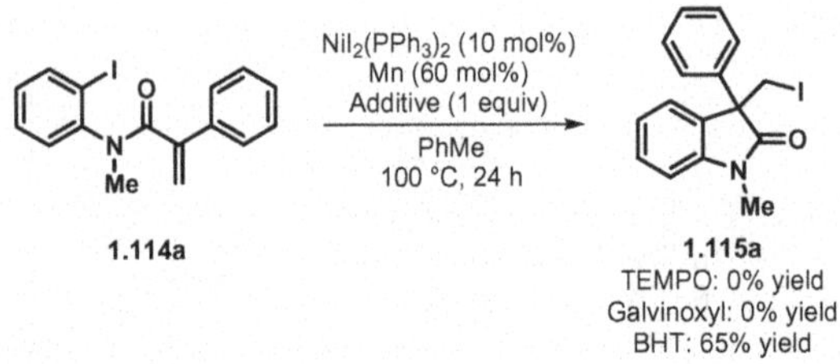


experiments, these results are inconclusive and a radial pathway cannot be excluded.

Knowing that carbometalation traditionally proceeds via a *syn*-addition, we synthesized the mono-deuterated aryl iodide **1.118**. The formation of **1.119** under high diastereocontrol implies a carbonickelation process as a radical addition to the olefin would epimerize the stereocenters. Understanding that epimerization may also occur on the diastereomers generated, two reactions were performed where one reaction was stopped at 10 h while the latter proceeded for 24 h. Halting the reaction at 10 h formed the deuterated indoline **1.119** in 60% and 2% deuterium incorporation of the major and minor diastereomers respectively. Additionally, increasing the reaction time significantly affects the dr of the transformation suggesting that the Ni-species is reinserting into the alkyl iodide (Scheme 1.36).

Based on these observations and the enantioinduction previously reported, we propose that the in situ generated $Ni^0$ species oxidatively adds into iodoacrylamide **1.114a** forming **1.120**. A *syn*-carbonickelation across the tethered olefin forms **1.121** and reductive elimination releases **1.115a** and restores the Ni-catalyst. A radical pathway is unlikely but cannot be ruled out (Scheme 1.37).

**Scheme 1.36** Ni-catalyzed carboiodination reaction of monodeuterated substrate **1.119**. Reaction was run by Austin D. Marchese

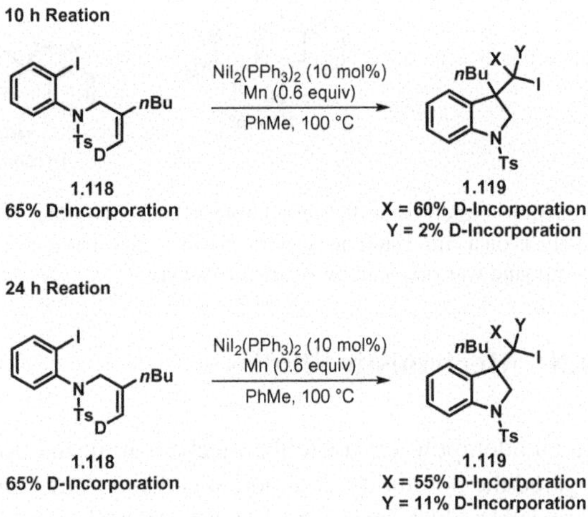

**Scheme 1.37** Proposed mechanism of the Ni-catalyzed carboiodination reaction

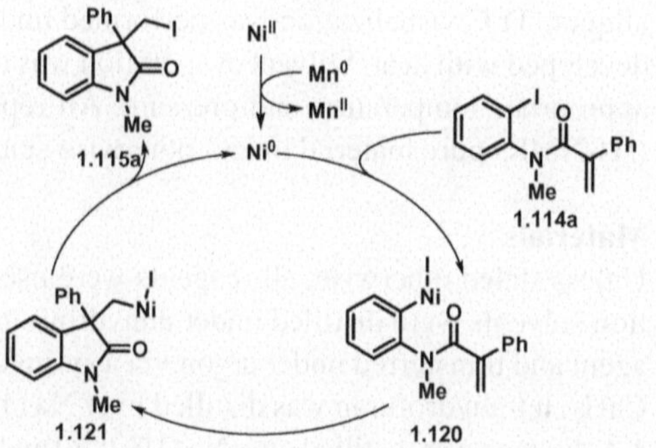


## 1.9 Conclusion

We have developed a diastereoselective $Pd^0$-catalyzed carboiodination of enantioenriched *N*-allyl carboxamides which allows convenient access to enantioenriched dihydroisoquinolinones. We have found that the addition of PMP, a bulky amine base, significantly increases the diastereoselectivity of this transformation. Having discovered this effect, the products were generated in good to excellent yields and high diastereocontrol. We have applied this methodology to the asymmetric formal synthesis of (+)-corynoline which represents the first application of the carboiodination methodology towards the synthesis of a natural product.

We have reported the first nickel-catalyzed cycloisomerization reaction involving the regeneration of a valuable carbon-halogen bond. Two sets of precatalyst combinations were found to promote this transformation ($NiI_2(PPh_3)_2$ or $NiI_2$ and $P(OiPr)_3$). Diverse functional groups were tolerated and the resulting oxindoles were obtained in good to excellent yields. Additionally, the Ni-catalyzed halogen-exchange carboiodination reaction starting from aryl bromides to the respective alkyl iodides in the presence of KI is also reported. An enantioselective variant was identified employing (*S*)-TolBINAP and (*S*)-BINAPO to attain good e.r. In spite of the significant progress, a limited scope of heterocycles was explored. Future work could involve approaches to other heterocycles, developing a diastereoselective variant and further optimizing the enantioselective reaction.

## 1.10 Experimental

### General Reaction Conditions

All non-aqueous reactions were performed in flame dried round bottom flasks sealed with a fitted rubber septum under an inert atmosphere of argon unless otherwise stated. All reactions were magnetically stirred and elevated temperatures were reported as the temperature of the surrounding oil .bath. Reactions were monitored by thin layer chromatography (TLC) or by crude $^1$H-NMR analysis of a worked up aliquot. TLC visualization was performed under a UV lamp or $KMnO_4$/CAM stain developed with heat. Solvent evaporation was conducted by rotary evaporation at the appropriate temperature and pressure. All reported yields reflect spectroscopically ($^1$H-NMR) pure material unless otherwise stated.

**Materials**

Unless stated otherwise, all reagents were used as received and the following reaction solvents were distilled under anhydrous conditions over the appropriate drying agent and transferred under argon via a syringe. Dichloromethane was distilled over $CaH_2$, tetrahydrofuran was distilled over Na (1% w:v) and benzophenone (1% w:v), 1,4-dioxane was distilled over Na (1% w:v) and benzophenone (1% w:v), and triethylamine was distilled over KOH. Dimethylformamide was distilled over 5Å molecular sieves and stored over 5Å molecular sieves (water content was kept lower than 50 ppm). 1,3,5-trimethoxybenzene was crushed into a fine powder by a mortar and pestle, dried overnight *in vacuo* and stored in a desiccator.

**Analysis**

$^1$H-NMR and $^{13}$C-NMR spectra of catalytic starting material and products were obtained on the Agilent DD2 500 equipped with a 5 mm Xsens Cold Probe. The starting material precursor $^1$H-NMR and $^{13}$C-NMR spectra were obtained from one of the following spectrometers: Varian NMR system 400, Bruker Avance III 400, Varian Mercury 400 or Varian Mercury 300. All $^{19}$F-NMR spectra were obtained on the Varian Mercury 300 and Varian Mercury 400. Measurements were carried out at 23 °C and chemical shifts (δ) are reported as parts per million (ppm). The solvent resonance was used as the internal standard for $^1$H-NMR (Chloroform at 7.26 ppm) and $^{13}$C-NMR (Chloroform at 77.0 ppm). The *J* values are reported in hertz (Hz) and are rounded off to the nearest 0.5 Hz. Multiplicities are abbreviated as follows: singlet (s), doublet (d), triplet (t), quartet (q), multiplet (m) and broad (br). All accurate mass values were obtained from the following spectrometers: Agilent 6538 Q-TOF (ESI) and JEOL AccuTOF-DART. Melting points were obtained on a Fisher-Johns Melting Point Apparatus and uncorrected. Infrared (IR) spectra were obtained as a neat film or dissolved in $CHCl_3$ on a NaCl disk using a Shimadzu FTIR-8400S FT-IR spectrometer.

## General Procedures

### General procedure 1 (GP1)

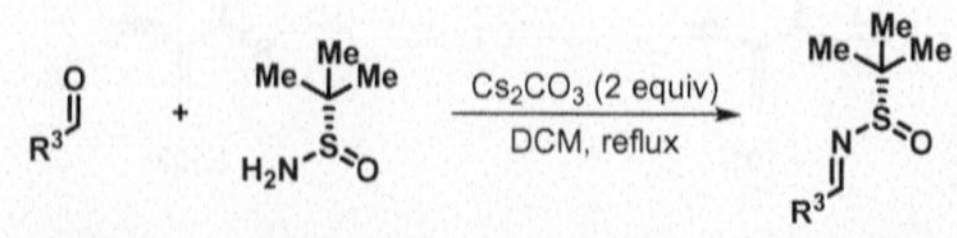


R = alkyl or aryl 1.2 equiv 1.100

(*S*)-2-methylpropane-2-sulfinamide (1.2 equiv), anhydrous $Cs_2CO_3$ (2 equiv) and the aldehyde (1 equiv) were added to a flame dried flask. Freshly distilled DCM was added the reaction was heated at reflux until TLC analysis indicated full conversion of aldehyde starting material. The mixture was cooled to room temperature, and the contents were filtered over Celite® which was subsequently washed through with DCM. The collected organic fraction was concentrated *in vacuo* and the crude sulfinamide purified by flash column chromatography.

### General Procedure 2 (GP2)

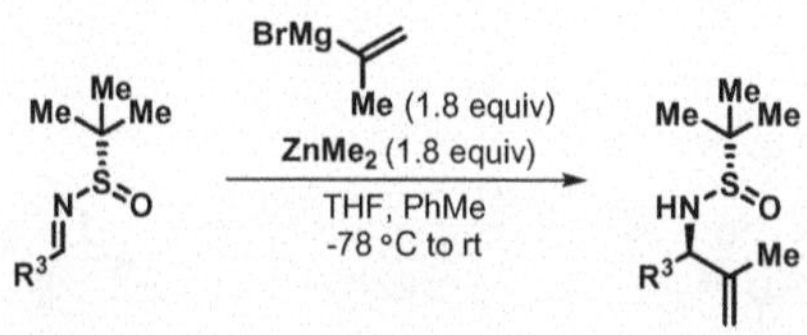


1.100 1.101

Synthesized according to a modified procedure of H. W. Lam and co-workers [34]. Two flame dried flasks were purged with argon. One flask was charged with isopropenyl magnesium bromide (0.5 M in THF, 1.8 equiv). Then a solution of dimethylzinc (2 M in toluene, 1.8 equiv) was added drop-wise over 10 min at room temperature was stirred for 30 min before cooling to −78 °C. The second flask was charged with **1.100** (1 equiv), and purged with argon for 10 min. The sulfinamide was dissolved in dry THF and cooled to −78 °C. The generated zinc reagent was slowly added dropwise via cannula to the second flask containing **1.100**, using a positive pressure of argon. The resulting solution was stirred for 3 h before slowly warming to room temperature. Once TLC analysis indicated full and clean conversion of starting material, the reaction was cooled to 0 °C and quenched by the addition of saturated ammonium chloride. The mixture was extracted with EtOAc (3×) and the combined organic layers were washed with brine, dried over $Na_2SO_4$ and concentrated *in vacuo*. The allylic sulfinamides **1.101** was purified by flash column chromatography.

## General Procedure 3 (GP3)

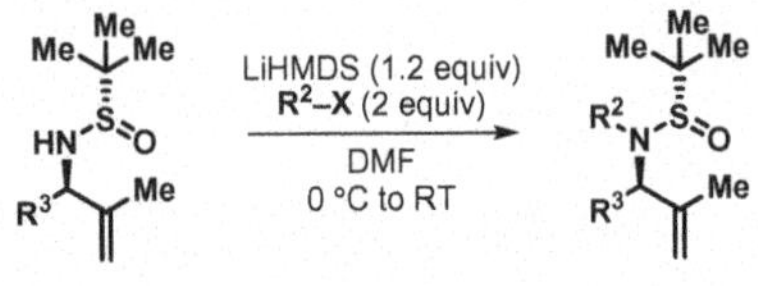


**1.101** **1.102**

A flame dried flask was charged with the allylic sulfinamide (1 equiv, **1.101**) and was dissolved in anhydrous DMF (0.3 M). The resultant solution was cooled to 0 °C and LiHMDS (1M in THF, 1.2 equiv) was added drop wise over 10 min. The solution was stirred for 30 min before warming to room temperature. The solution was re-cooled to 0 °C and the alkyl halide (2 equiv) was added drop wise over 10 min. The resulting solution was warmed to room temperature and was stirred until TLC analysis indicated full conversion of starting material. The reaction was quenched at 0 °C by the addition of saturated ammonium chloride The mixture was extracted with EtOAc (3×) and the combined organic layers were washed with brine, dried over $Na_2SO_4$ and concentrated *in vacuo*. The alkylated allylic sulfinamide **1.102** was obtained via flash column chromatography.

## General Procedure 4 (GP4)

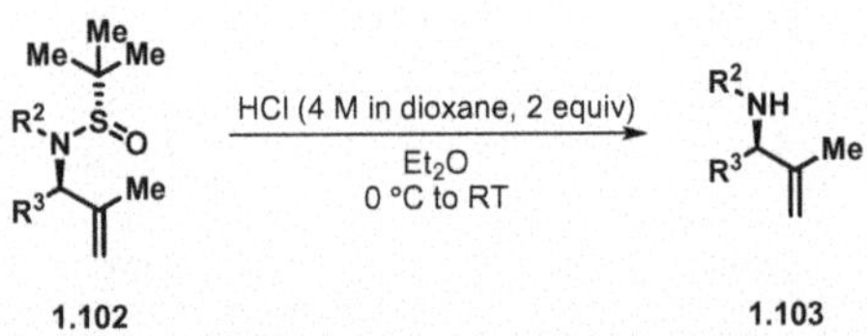


**1.102** **1.103**

A flame dried flask was charged with the alkylated allylic sulfonamide **1.102** (1 equiv) and was dissolved in anhydrous $Et_2O$. The solution was cooled to 0 °C and anhydrous HCl in 1,4-dioxane (4M, 2 equiv) was added drop wise via syringe. The resulting mixture was stirred for 30 min before warming to room temperature. The white precipitate was filtered and the filter cake was washed with $Et_2O$. The filter cake was dissolved through the frit into a new flask using distilled water. The aqueous filtrate was slowly basified to pH 10 using 1M aqueous $NH_4OH$. The mixture was extracted with $Et_2O$ (3×) and the combined organic layers were washed with brine, dried over $Na_2SO_4$ and concentrated *in vacuo*. The chiral allylic amine **1.103** was used without further purification.

## General Procedure 5 (GP5)

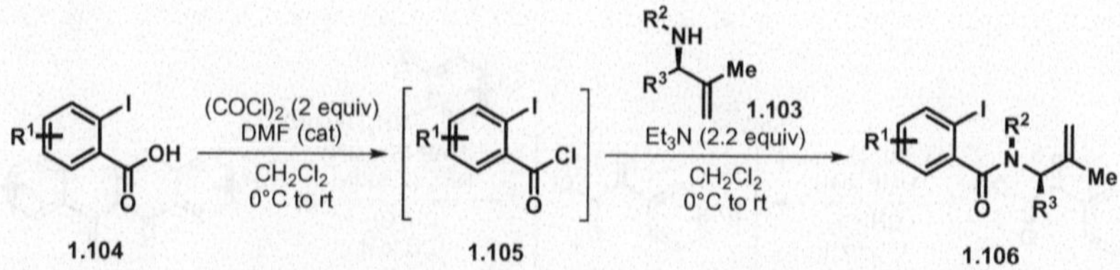

A solution of *o*-iodobenzoic acid **1.104** (1.0 equiv) and DMF (4 drops) in $CH_2Cl_2$ (0.40M) was prepared and cooled to 0 °C. A bubbler was attached to the vessel and $(COCl)_2$ (2 equiv) was added dropwise. After 5 min, the reaction was allowed to warm to room temperature and was stirred for 1 h. The acyl chloride was concentrated *in vacuo* and redissolved in $CH_2Cl_2$. A solution of the chiral allyl amine **1.103** (1.1 equiv) and $NEt_3$ (2.0 equiv) was added dropwise for 10 min at 0 °C. The reaction was stirred at this temperature for 10 min and then warmed to room temperature where it was stirred for an additional 10 min. The reaction was quenched with a saturated $NaHCO_3$ solution and extracted with EtOAc (3×). The combined organic layers were washed with brine, dried over $Na_2SO_4$ and concentrated *in vacuo*. The chiral *N*-allyl carboxamide **1.106** was purified using flash column chromatography.

## General procedure 6 (GP6)

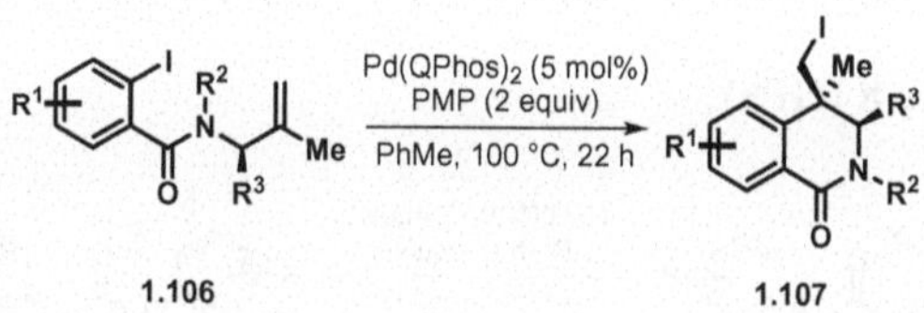

To a flame dried two dram vial under argon atmosphere, **1.106** (1 equiv) and $Pd(QPhos)_2$ (5 mol%) were added. PhMe (0.2M) was added, 1,2,2,6,6-pentamethylpiperidine (2 equiv) was added using a Hamilton syringe, and a Teflon line screw cap was fitted. The vial was sealed with Teflon tape and placed in a preheated oil bath at the indicated temperature and time. The reaction was filtered through a plug of silica gel using EtOAc and concentrated *in vacuo*. The pure product **1.107** was obtained via silica gel flash column chromatography using the indicated mobile phase.

**General procedure 7 (GP7)**

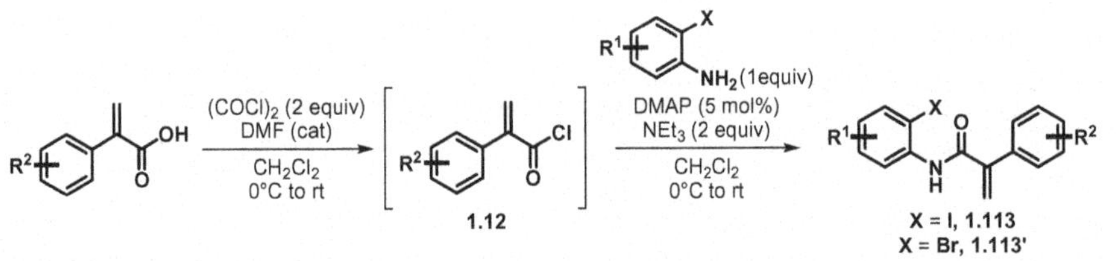


A solution of substituted 2-phenylacrylic acid (1.0 equiv) and DMF (4 drops) in $CH_2Cl_2$ (0.40M) was prepared and cooled to 0 °C. A bubbler was attached to the vessel and $(COCl)_2$ (2 equiv) was added dropwise. After 5 min, the reaction was allowed to warm to room temperature and was stirred for 1 h. The acyl chloride was concentrated *in vacuo* and redissolved in $CH_2Cl_2$. A solution of the substituted 2-iodoaniline (1.0 equiv), DMAP (0.05 equiv) and $NEt_3$ (2.0 equiv) was prepared in $CH_2Cl_2$ (0.50M) and cooled to 0 °C. The acyl chloride solution was added dropwise into the vessel containing the substituted 2-iodoaniline. After 5 min, the reaction was allowed to warm to room temperature and was stirred overnight. The reaction was quenched with a saturated $NaHCO_3$ solution and extracted with EtOAc (3×). The combined organic layers were washed with brine, dried over $Na_2SO_4$ and concentrated *in vacuo*. The crude unsubstituted acrylamide was passed through a plug of silica gel prior to proceeding to **GP8**.

**General procedure 8 (GP8)**

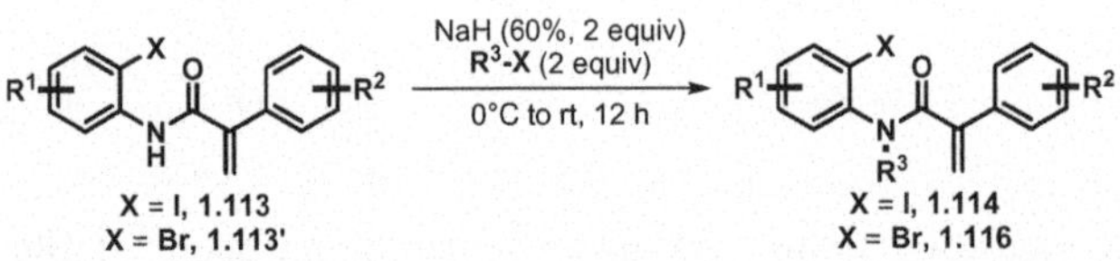


A solution of the unsubstituted acrylamide (1.0 equiv) in THF (0.20M) was prepared and cooled to 0 °C. NaH (60%w/w, 2.0 equiv) was added to the solution and the mixture was stirred for 15 min before adding $R^3$-X (2 equiv) dropwise. The reaction was allowed to warm at room temperature after 10 min and was stirred for 2 h to overnight. The reaction was quenched with cold water and extracted with EtOAc (2×). The combined organic layers were washed with brine, dried with $Na_2SO_4$ and concentrated *in vacuo*. The reaction was purified via silica gel flash chromatography using the indicated mobile phase.

**General procedure 9 (GP9)**

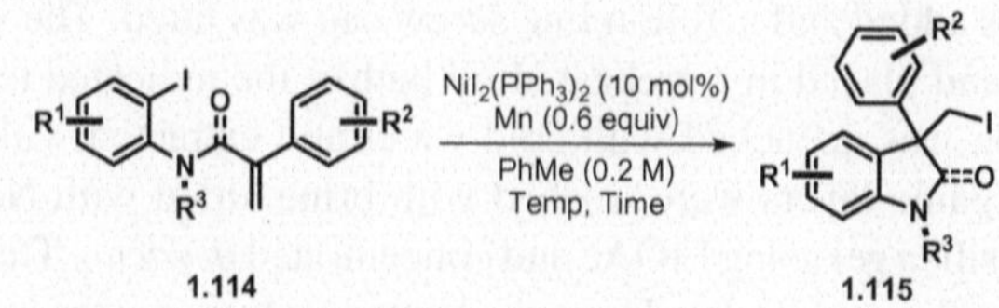


To a flame dried two dram vial under argon atmosphere, iodoacrylamide **1** (1 equiv), $NiI_2(PPh_3)_2$ (10 mol% Ni), and Mn (0.6 equiv) were added. PhMe (0.2M) was added, a Teflon line screw cap was fitted. The vial was sealed with Teflon tape and placed in a preheated oil bath at the indicated temperature and time. The reaction was quenched with cold water and extracted with EtOAc (3×). The combined organic layers were washed with brine, dried with $Na_2SO_4$, filtered through a plug of silica gel using EtOAc and concentrated *in vacuo*. The pure products were obtained via silica gel flash column chromatography using the indicated mobile phase.

**General Procedure 10 (GP10)**

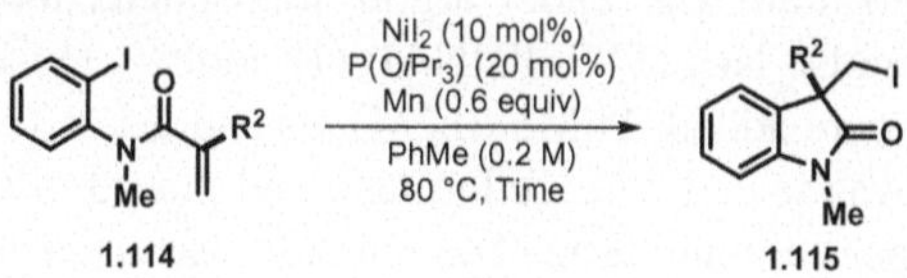


To a flame dried two dram vial under argon atmosphere, iodoacrylamide **1.114** (1 equiv), $NiI_2$ (10 mol% Ni), and Mn (0.6 equiv) were added. PhMe (0.2M) and P(O*i*Pr)$_3$ was added sequentially and a Teflon line screw cap was fitted. The vial was sealed with Teflon tape and placed in a preheated oil bath at the indicated temperature and time. The reaction was quenched with cold water and extracted with EtOAc (3×). The combined organic layers were washed with brine, dried with $Na_2SO_4$, filtered through a plug of silica gel using EtOAc and concentrated *in vacuo*. The pure oxindoles were obtained via silica gel flash column chromatography using the indicated mobile phase.

**General procedure 11 (GP11)**

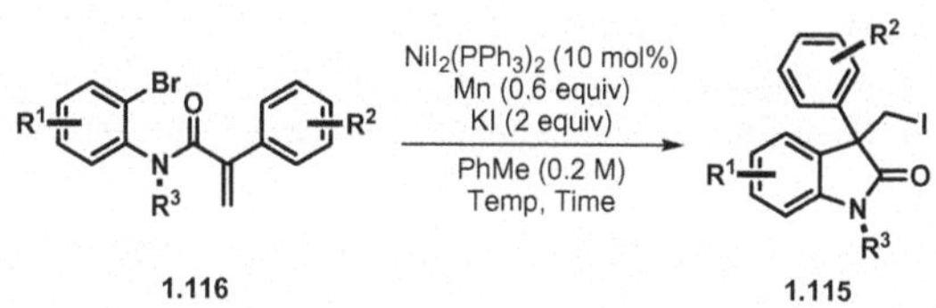

To a flame dried two dram vial under argon atmosphere, bromoacrylamide **1.116** (1 equiv), $NiI_2(PPh_3)_2$ (10 mol% Ni), Mn (0.6 equiv), and KI (2 equiv) were added. PhMe (0.2M) was added and a Teflon line screw cap was fitted. The vial was sealed with Teflon tape and placed in a preheated oil bath at the indicated temperature and time. The reaction was quenched with cold water and extracted with EtOAc (3×). The combined organic layers were washed with brine, dried with $Na_2SO_4$, filtered through a plug of silica gel using EtOAc and concentrated *in vacuo*. The pure products were obtained via silica gel flash column chromatography using the indicated mobile phase.

**General procedure 12 (GP12)**

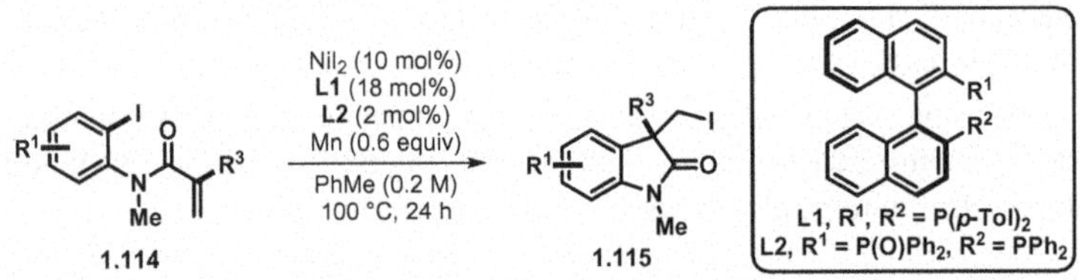


To a flame dried two dram vial under argon atmosphere, iodoacrylamide **1.114** (1 equiv), $NiI_2$ (10 mol% Ni), (*S*)-TolBINAP (18 mol%), (*S*)-BINAPO (2 mol%) and Mn (0.6 equiv) were added. PhMe (0.2M) was added, a Teflon line screw cap was fitted. The vial was sealed with Teflon tape and placed in a preheated oil bath at the indicated temperature and time. The reaction was quenched with cold water and extracted with EtOAc (3×). The combined organic layers were washed with brine, dried with $Na_2SO_4$, filtered through a plug of silica gel using EtOAc and concentrated *in vacuo*. The pure products were obtained via silica gel flash column chromatography using the indicated mobile phase.

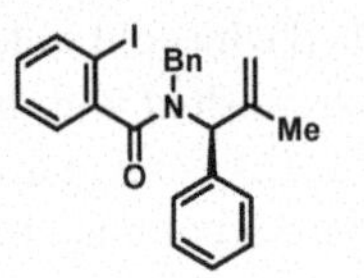

(*S*)-*N*-benzyl-2-iodo-*N*-(2-methyl-1-phenylallyl)benzamide (1.106i)

Synthesized according to **GP5** using 2-iodobenzoic acid (143 mg, 0.58 mmol) and (*S*)-*N*-benzyl-2-methyl-1-phenylprop-2-en-1-amine (150 mg, 0.64 mmol). The aryl iodide was purified by flash column chromatography using hexanes:DCM: EtOAc (12:8:0.75 v:v:v) and was obtained as a clear and light yellow oil (153 mg, 0.33 mmol, 57%).

*$^1H$ NMR analysis showed the pure desired compound to be present as a complex mixture of 4 rotamers, and only approximate integration values are displayed.*

**$^1$H NMR** (500 MHz, $CDCl_3$) δ 7.92–7.64 (m, 1H), 7.50–6.80 (m, 13H), 6.53 (d, *J* = 7.5 Hz, ~ 0.3H), 6.13–5.94 (m, ~ 0.15H), 5.46–3.92 (m, ~ 4.5H), 1.82 (s, ~ 0.5H), 1.61 (dd, *J* = 1.5, 1.0 Hz, ~ 0.3H), 1.60–1.58 (m, ~ 0.5H), 1.34 (s, ~ 1.7H). **$^{13}$C NMR**

(126 MHz, $CDCl_3$) δ 172.4, 171.7, 144.0, 143.3, 142.2, 141.9, 139.8, 139.4, 138.6, 138.3, 138.0, 134.9, 130.8, 130.4, 130.1, 130.1, 128.7, 128.6, 128.5, 128.4, 128.3, 128.3, 128.2, 128.1, 127.9, 127.9, 127.6, 127.6, 127.4, 127.3, 127.1, 127.0, 126.7, 126.2, 118.7, 113.9, 92.6, 92.3, 68.4, 47.6, 47.4, 23.1, 21.3. **IR** (neat, $cm^{-1}$) 3085, 3062, 3028, 2969, 1641, 1602, 1584, 1495, 1467, 1452, 1429, 1404, 1362, 1324, 1311, 1294, 1159, 1077. **HRMS** (ESI+) Calc'd for $C_{24}H_{23}INO$ 468.0816, found 468.0816.

The enantiomeric ratio was determined by HPLC: Chiracel AS column, *n*-hexanes/*i*-PrOH = 92.5:7.5, flow rate: 1 mL/min. $[\alpha]_D^{20} = -33.8$ (c = 0.5, $CHCl_3$).

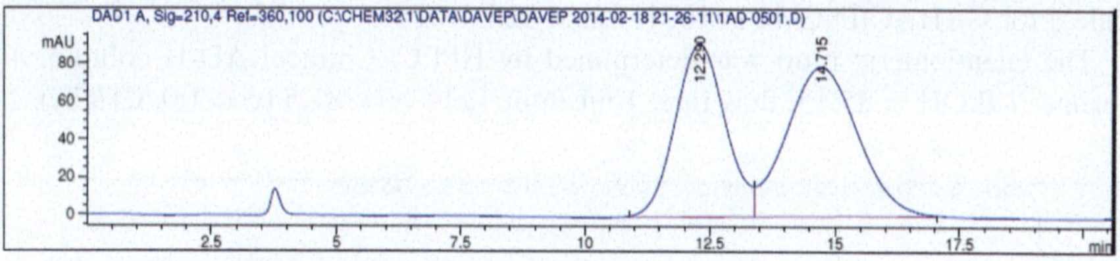


Signal 1: DAD1 A, Sig=210,4 Ref=360,100

| Peak # | RetTime [min] | Type | Width [min] | Area [mAU*s] | Height [mAU] | Area % |
|---|---|---|---|---|---|---|
| 1 | 12.290 | BV | 1.0681 | 6351.72314 | 91.47583 | 45.0565 |
| 2 | 14.715 | VB | 1.4479 | 7745.50830 | 78.85874 | 54.9435 |

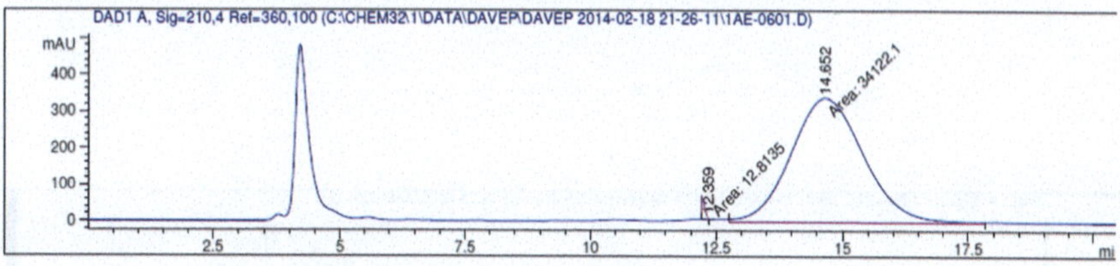


Signal 1: DAD1 A, Sig=210,4 Ref=360,100

| Peak # | RetTime [min] | Type | Width [min] | Area [mAU*s] | Height [mAU] | Area % |
|---|---|---|---|---|---|---|
| 1 | 12.359 | MM | 0.0186 | 12.81350 | 11.47438 | 0.0375 |
| 2 | 14.652 | MM | 1.6706 | 3.41221e4 | 340.40930 | 99.9625 |

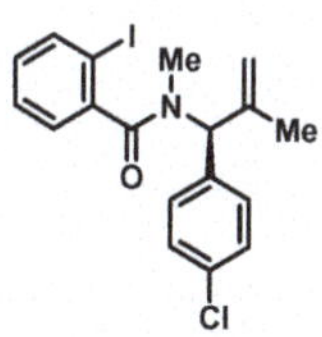


(*S*)-*N*-(1-(4-chlorophenyl)-2-methylallyl)-2-iodo-*N*-methylbenzamide (1.106 l)

Synthesized according to **GP5** using 2-iodobenzoic acid (459 mg, 1.85 mmol) and (*S*)-1-(4-chlorophenyl)-*N*,2-dimethylprop-2-en-1-amine (400 mg, 2.03 mmol). The aryl iodide was purified by flash column chromatography using hexanes:DCM:EtOAc (6:4:0.75 v:v:v) and was obtained as a clear, colourless oil (789 mg, 1.84 mmol, 99%).

*$^1$H NMR analysis showed the pure desired compound to be present as a complex mixture of 4 rotamers, and only approximate integration values are displayed.*

**$^1$H NMR** (500 MHz, $CDCl_3$) δ 7.93–7.74 (m, 1H), 7.42–7.17 (m, 5H), 7.14–6.92 (m, ~ 1.8H), 6.66 (dd, $J$ = 7.5, 1.5 Hz, ~ 0.2H), 6.36 (s, ~ 0.6H), 5.26–4.51 (m, ~ 2.4H), 3.15 (s, ~ 0.4H), 2.86 (s, ~ 0.7H), 2.69–2.50 (m, ~ 1.9H), 1.95–1.76 (m, ~ 2.3H), 1.50–1.40 (m, ~ 0.7H). **$^{13}$C NMR** (126 MHz, $CDCl_3$) δ 171.8, 171.2, 170.9, 142.8, 142.6, 142.1, 141.7, 141.5, 139.7, 139.5, 139.3, 137.6, 134.3, 134.1, 133.3, 131.3, 131.3, 130.4, 130.3, 130.1, 128.9, 128.8, 128.7, 128.5, 127.8, 127.5, 126.9, 126.9, 126.6, 118.6, 115.7, 114.7, 113.6, 92.8, 92.7, 92.0, 67.4, 67.2, 61.2, 33.4, 33.1, 30.5, 30.3, 22.1, 21.4, 21.1. **IR** (neat, $cm^{-1}$) 3087, 3051, 2993, 2972, 2940, 2913, 2852, 1640, 1490, 1436, 1391, 1337, 1259, 1180, 1091, 1073, 1034. **HRMS** (ESI+) Calc'd for $C_{18}H_{18}ClINO$ 426.01071, found 426.01074.

The enantiomeric ratio was determined by HPLC: Chiracel AD-H column, *n*-hexanes/*i*-PrOH = 85:15, flow rate: 1 mL/min. $[\alpha]_D^{20} = -81.5$ (c = 1.0, $CHCl_3$).

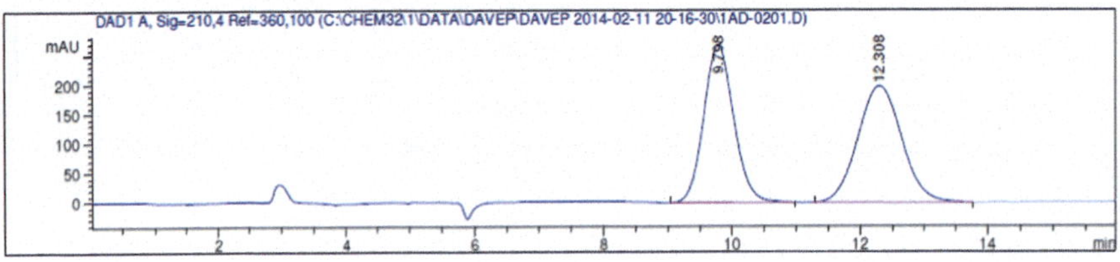


Signal 1: DAD1 A, Sig=210,4 Ref=360,100

| Peak # | RetTime [min] | Type | Width [min] | Area [mAU*s] | Height [mAU] | Area % |
|---|---|---|---|---|---|---|
| 1 | 9.798 | BB | 0.5088 | 8878.47070 | 270.60934 | 48.2806 |
| 2 | 12.308 | BB | 0.7452 | 9510.86133 | 199.13530 | 51.7194 |

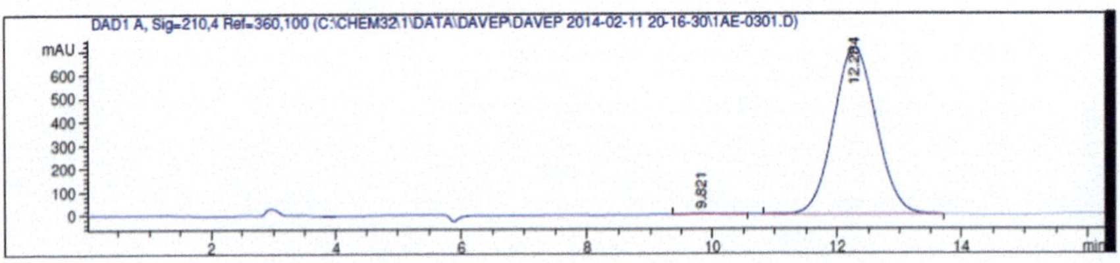


Signal 1: DAD1 A, Sig=210,4 Ref=360,100

| Peak # | RetTime [min] | Type | Width [min] | Area [mAU*s] | Height [mAU] | Area % |
|---|---|---|---|---|---|---|
| 1 | 9.821 | BB | 0.4308 | 108.88960 | 3.44462 | 0.3218 |
| 2 | 12.284 | BB | 0.7346 | 3.37270e4 | 712.11829 | 99.6782 |

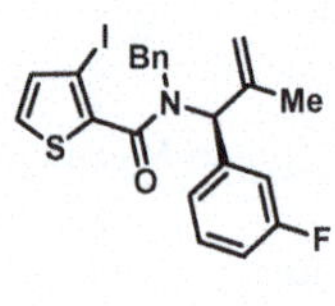


(*S*)-*N*-benzyl-*N*-(1-(3-fluorophenyl)-2-methylallyl)-3-iodothiophene-2-carboxamide (1.106p)

Synthesized according to **GP5** using 3-iodothiophene-2-carboxylic acid (128 mg, 0.5 mmol) and (*S*)-*N*-benzyl-1-(3-fluorophenyl)-2-methylprop-2-en-1-amine (140 mg, 0.6 mmol). The heteroaryl iodide was purified by flash column chromatography using hexanes:EtOAc (3:1 v:v) and was obtained as a white solid (195 mg, 0.41 mmol, 82%, M.P.: 100–101 °C).

$^1$H NMR analysis showed the pure desired compound to be present as a complex mixture of 4 rotamers, and only approximate integration values are displayed

**$^1$H NMR** (500 MHz, CDCl3) δ 7.24–6.90 (m, 10H), 6.86 (tdd, *J* = 8.5, 2.5, 1.0 Hz, 1H), 5.68–4.86 (m (br), 4H), 4.34 (d, *J* = 15.5 Hz, 1H), 1.74–1.45 (m(br), 3H). **$^{13}$C NMR** (126 MHz, $CDCl_3$) δ 166.2, 163.7, 161.7, 142.8, 138.4, 137.4, 136.1, 135.1, 129.9, 129.8, 127.8, 127.7, 127.5, 126.6, 125.7, 125.7, 117.0, 116.8, 115.1, 114.9, 79.4, 21.7. **$^{19}$F NMR** (564 MHz, CDCl3) δ -112.80-113.10 (m). **IR** (neat, $cm^{-1}$) 3107, 3080, 3063, 3034, 2940, 2928, 1628, 1616, 1589, 1489, 1441, 1427, 1395, 1316, 1283, 1262, 1246, 1234, 1152, 1138, 910, 874, 862. **HRMS** (DART) Calc'd for $C_{22}H_{20}FINOS$ 492.02,943, found 492.02,949. $[\alpha]_D^{20}$ −37.731 (c = 0.58, $CHCl_3$).

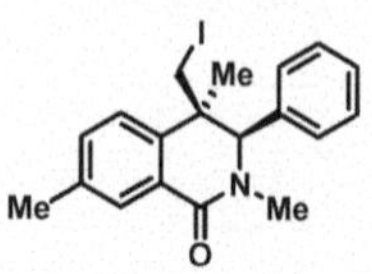


(3*R*,4*R*)-4-(iodomethyl)-2,4,7-trimethyl-3-phenyl-3,4-dihydroisoquinolin-1(2*H*)-one (1.107b)

Synthesized according to **GP6** using (*S*)-2-iodo-*N*,5-dimethyl-*N*-(2-methyl-1-phenylallyl)benzamide (81.0 mg, 0.2 mmol). The alkyl iodide was purified by flash column chromatography using hexanes:EtOAc (3:1 v:v) and was obtained as a white solid (72 mg, 0.18 mmol, 88%, 94:6 dr (*cis:trans*)). **Major Diastereomer:** Isolated as a white solid (M.P.: 47–49 °C).

**$^1$H NMR** (500 MHz, $CDCl_3$) δ 8.04–8.03 (m, 1H), 7.30–7.27 (m, 1H), 7.25–7.21 (m, 1H), 7.16–7.11 (m, 2H), 7.13–7.09 (m, 1H), 7.09–.05 (m, 2H), 4.31 (s, 1H), 3.91 (d, *J* = 10.0 Hz, 0H), 3.05 (s, 3H), 2.94–2.90 (m, 1H), 2.43 (s, 3H), 1.60 (d, 1.0 Hz, 3H). **$^{13}$C NMR** (126 MHz, $CDCl_3$) δ 163.7, 137.4, 137.1, 136.2, 132.9, 129.3, 128.6, 128.5, 128.3, 123.1, 72.1, 40.0, 34.3, 31.4, 21.0, 16.4. **IR** ($cm^{-1}$, film) 3070, 3028, 2966, 2861, 1647, 1608, 1573, 1482, 1452, 1394, 1374, 1353, 1298, 1261, 1195, 1087. **HRMS** (DART) Calc'd for $C_{19}H_{21}INO$, 406.06678; found, 406.06645.

The enantiomeric ratio was determined by HPLC: Chiracel AD-H column, *n*-hexanes/*i*-PrOH = 95:5, flow rate: 1 mL/min. $[\alpha]_D^{20}$ = + 216.8 (c = 0.5, $CHCl_3$).

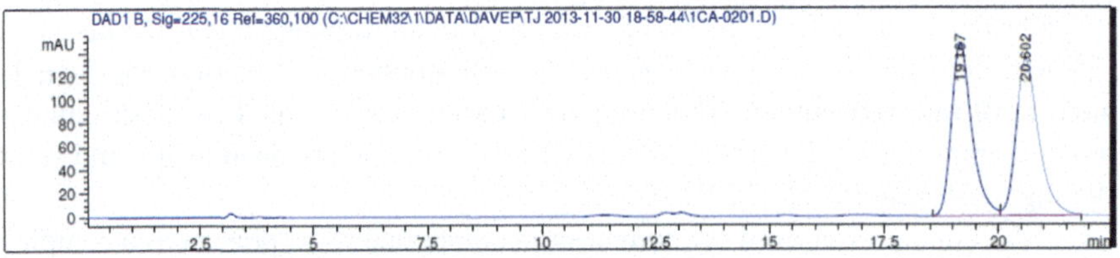


Signal 2: DAD1 B, Sig=225,16 Ref=360,100

| Peak # | RetTime [min] | Type | Width [min] | Area [mAU*s] | Height [mAU] | Area % |
|---|---|---|---|---|---|---|
| 1 | 19.167 | BV | 0.4724 | 4647.43311 | 148.77739 | 49.8708 |
| 2 | 20.602 | VB | 0.5268 | 4671.51709 | 133.97400 | 50.1292 |

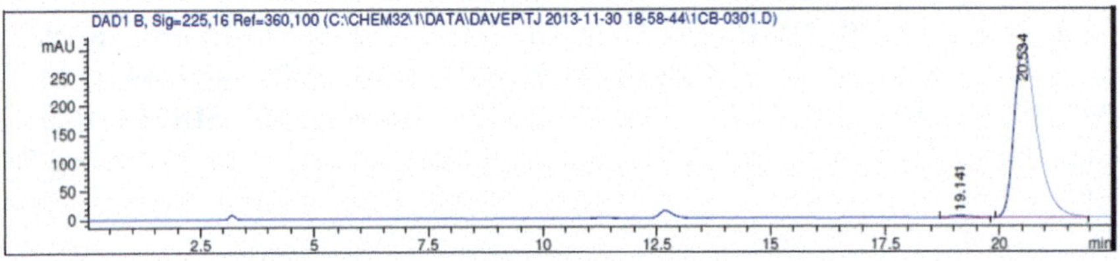


Signal 2: DAD1 B, Sig=225,16 Ref=360,100

| Peak # | RetTime [min] | Type | Width [min] | Area [mAU*s] | Height [mAU] | Area % |
|---|---|---|---|---|---|---|
| 1 | 19.141 | BB | 0.4567 | 101.12936 | 3.30807 | 0.9412 |
| 2 | 20.534 | BB | 0.5252 | 1.06434e4 | 306.46530 | 99.0588 |

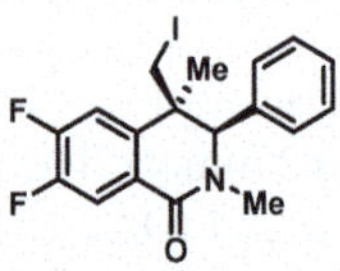


(3*R*,4*R*)-6,7-difluoro-4-(iodomethyl)-2,4-dimethyl-3-phenyl-3,4-dihydroisoquinolin-1(2*H*)-one (1.107d)

Synthesized according to **GP6** using (*S*)-4,5-difluoro-2-iodo-*N*-methyl-*N*-(2-methyl-1-phenylallyl)benzamide (81.0 mg, 0.2 mmol), $Pd(QPhos)_2$ (7.5 mol%) and 1,2,2,6,6-pentamethylpiperidine (3.5 equiv). The alkyl iodide was purified by flash column chromatography using hexanes:EtOAc (3:1 v:v) and was obtained as a white solid (43.5 mg, 0.11 mmol, 54%, 97:3 dr (*cis:trans*)). **Major Diastereomer:** Isolated as a white solid (M.P.: 111–113 °C).

**$^1$H NMR** (500 MHz, $CDCl_3$) δ 8.05 (dd, $J$ = 10.5, 8.5 Hz, 1H), 7.30–7.26 (m, 1H), 7.21–7.16 (m, 2H), 7.09–7.04 (m, 3H), 4.34 (s, 1H), 3.74 (d, $J$ = 10.0 Hz, 1H), 3.04 (s, 3H), 2.91 (dd, $J$ = 10.0, 1.0 Hz, 1H), 1.62 (d, $J$ = 1.0 Hz, 3H). **$^{13}$C NMR** (125 MHz, $CDCl_3$) δ 161.8 (s), 152.6 (dd, $J$ = 255.0, 13.0 Hz), 149.5 (dd, $J$ = 250.0, 13.0 Hz), 137.6 (dd, $J$ = 6.0, 4.0 Hz), 135.5, 128.9, 128.6, 128.4, 126.1 (dd, $J$ = 5.5, 3.5 Hz), 118.4 (dd, $J$ = 19.0, 1.5 Hz), 113.3 (d, $J$ = 19.5 Hz), 72.0, 40.1 (d, $J$ = 1.0 Hz), 34.4, 31.3 (d, $J$ = 1.0 Hz), 14.5. **$^{19}$F NMR** (564 MHz, $CDCl_3$) δ-130.57 – –130.68 (m), –137.94 – –138.03 (m). **IR** ($cm^{-1}$, film) 3059, 3030, 3004, 2968,

2926, 2861, 1652, 1616, 1608, 1505, 1487, 1452, 1394, 1355, 1314, 1291, 1245, 1208, 1190, 1173, 1080. **HRMS** (DART) Calc'd for $C_{18}H_{17}F_2INO_3$, 428.03229; found, 428.03243.

The enantiomeric ratio was determined by HPLC: Chiracel AD-H column, *n*-hexanes/*i*-PrOH = 95:5, flow rate: 1 mL/min. $[\alpha]_D^{20} = +246.8$ (c = 1.0, $CHCl_3$).

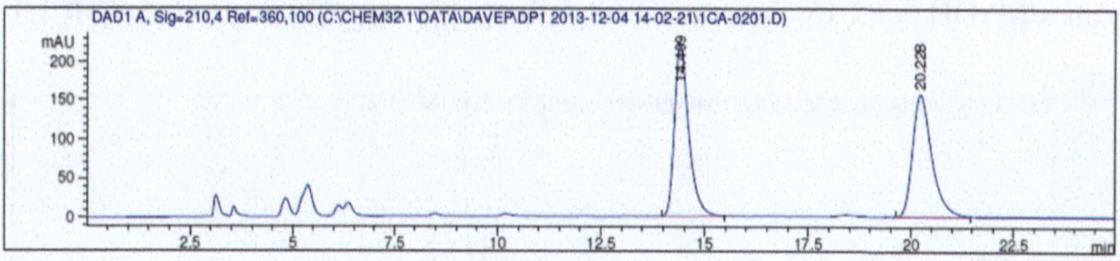


Signal 1: DAD1 A, Sig=210,4 Ref=360,100

| Peak # | RetTime [min] | Type | Width [min] | Area [mAU*s] | Height [mAU] | Area % |
|---|---|---|---|---|---|---|
| 1 | 14.409 | BB | 0.3568 | 5246.69824 | 221.15370 | 50.4530 |
| 2 | 20.228 | BB | 0.4977 | 5152.47168 | 155.87173 | 49.5470 |

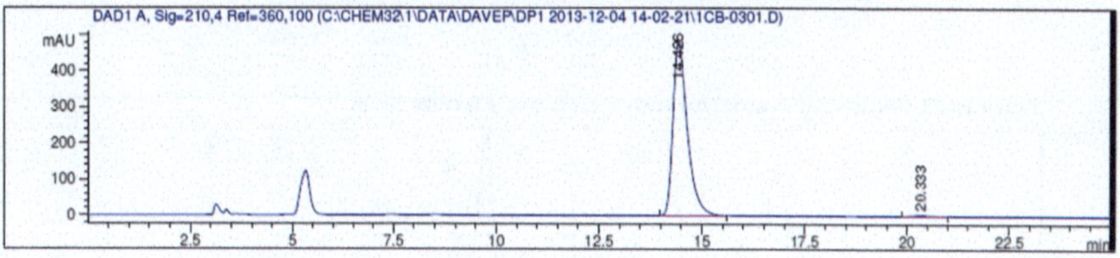


Signal 1: DAD1 A, Sig=210,4 Ref=360,100

| Peak # | RetTime [min] | Type | Width [min] | Area [mAU*s] | Height [mAU] | Area % |
|---|---|---|---|---|---|---|
| 1 | 14.426 | BB | 0.3680 | 1.16990e4 | 480.66937 | 99.1350 |
| 2 | 20.333 | BB | 0.4397 | 102.08034 | 3.36692 | 0.8650 |

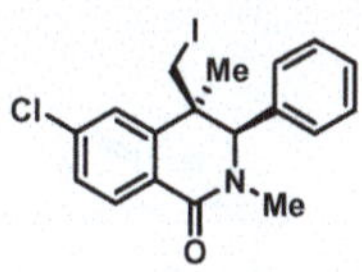


(3*R*,4*R*)-6-chloro-4-(iodomethyl)-2,4-dimethyl-3-phenyl-3,4-dihydroisoquinolin-1(2*H*)-one (1.107e)

Synthesized according to **GP6** using (*S*)-4-chloro-2-iodo-*N*-methyl-*N*-(2-methyl-1-phenylallyl)benzamide (85.0 mg, 0.2 mmol). The alkyl iodide was purified by flash column chromatography using hexanes:DCM:EtOAc (6:4:0.75 v:v:v) and was obtained as a white solid (64.3 mg, 0.15 mmol, 76%, 97:3 dr (*cis:trans*)). **Major Diastereomer:** Isolated as a white solid (M.P.: 59–61 °C).

**$^1$H NMR** (500 MHz, $CDCl_3$) δ 8.16 (dd, *J* = 8.5, 0.5 Hz, 1H), 7.41 (dd, *J* = 8.5 2.0 Hz, 1H), 7.28–7.24 (m, 1H), 7.22 (d, *J* = 2.0 Hz, 1H), 7.19–7.14 (m, 2H), 7.10–7.05 (m, 2H), 4.35 (s, 1H), 3.85 (d, *J* = 10.0 Hz, 0H), 3.05 (s, 3H), 2.94–2.90

(m, 1H), 1.63 (d, $J$ = 1.0 Hz, 3H). **$^{13}$C NMR** (125 MHz, $CDCl_3$) δ 162.7, 142.0, 138.6, 135.8, 130.4, 128.8, 128.6, 128.5, 127.9, 127.3, 123.9, 72.0, 40.5, 34.3, 31.3, 15.1. **IR** ($cm^{-1}$, film) 3063, 3028, 3001, 2966, 2926, 2855, 1647, 1593, 1450, 1402, 1392, 1308, 1257, 1197, 1160, 1101, 1079, 1052, 1031. **HRMS** (DART) Calc'd for $C_{18}H_{18}ClINO$, 426.01216; found, 426.01178.

The enantiomeric ratio was determined by HPLC: Chiracel AD-H column, *n*-hexanes/*i*-PrOH = 85:15, flow rate: 1 mL/min. $[\alpha]^{20}_{D} = +272.0$ (c = 0.5, $CHCl_3$).

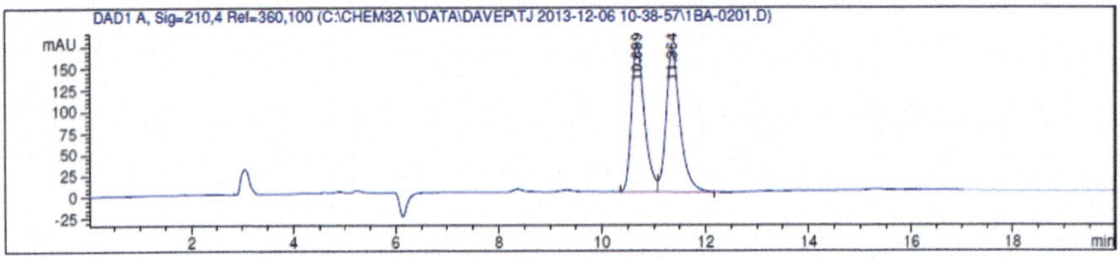


Signal 1: DAD1 A, Sig=210,4 Ref=360,100

| Peak # | RetTime [min] | Type | Width [min] | Area [mAU*s] | Height [mAU] | Area % |
|---|---|---|---|---|---|---|
| 1 | 10.689 | BV | 0.2596 | 2978.32446 | 174.09749 | 48.9248 |
| 2 | 11.364 | VB | 0.2824 | 3109.23364 | 164.60594 | 51.0752 |

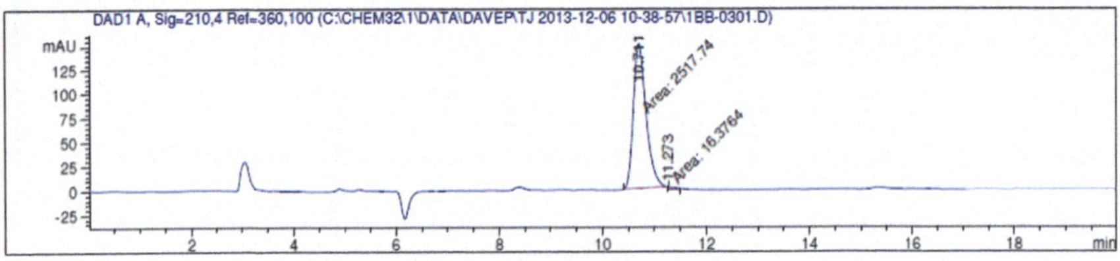


Signal 1: DAD1 A, Sig=210,4 Ref=360,100

| Peak # | RetTime [min] | Type | Width [min] | Area [mAU*s] | Height [mAU] | Area % |
|---|---|---|---|---|---|---|
| 1 | 10.711 | MM | 0.2801 | 2517.73779 | 149.81326 | 99.3538 |
| 2 | 11.273 | MM | 0.1118 | 16.37640 | 1.98540 | 0.6462 |

(3*R*,4*R*)-2-benzyl-4-(iodomethyl)-4-methyl-3-phenyl-3,4-dihydroisoquinolin-1(2*H*)-one (1.107i)

Synthesized according to **GP6** using (*S*)-*N*-benzyl-2-iodo-*N*-(2-methyl-1-phenylallyl)benzamide (93.4 mg, 0.2 mmol). The alkyl iodide was purified by flash column chromatography using hexanes:DCM:EtOAc (6:4:0.75 v:v:v) and was obtained as a white foam (88.6 mg, 0.19 mmol, 95%, 98:2 dr (*cis:trans*)).

**$^{1}$H NMR** (500 MHz, $CDCl_3$) δ 8.32–8.26 (m, 1H), 7.53–7.43 (m, 2H), 7.43–7.29 (m, 5H), 7.29 – 7.21 (m, 1H), 7.23–7.17 (m, 1H), 7.15 (ddd, $J$ = 8.0, 7.0, 1.0 Hz, 2H), 7.09–7.06 (m, 2H), 5.64 (d, $J$ = 14.0 Hz, 1H), 4.34 (s, 1H), 3.82 (d, $J$ =

10.0 Hz, 1H), 3.54 (d, $J$ = 14.5 Hz, 1H), 2.86 (dd, $J$ = 10.0, 1.0 Hz, 1H), 1.25 (d, $J$ = 1.0 Hz, 3H). **$^{13}$C NMR** (126 MHz, $CDCl_3$) δ 163.4, 140.2, 136.5, 136.3, 132.5, 129.3, 129.1, 128.9, 128.9, 128.6, 128.6, 128.4, 127.8, 127.7, 123.3, 67.4, 48.2, 39.8, 31.6, 16.0. **IR** ($cm^{-1}$, film) 3064, 3028, 2978, 2925, 1647, 1600, 1578, 1495, 1467, 1454, 1445, 1433, 1353, 1297, 1261, 1153, 1080. **HRMS** (ESI+) Calc'd for $C_{24}H_{23}INO$, 468.08243; found, 468.08175.

The enantiomeric ratio was determined by HPLC: Chiracel AD-H column, *n*-hexanes/*i*-PrOH = 90:10, flow rate: 1 mL/min.

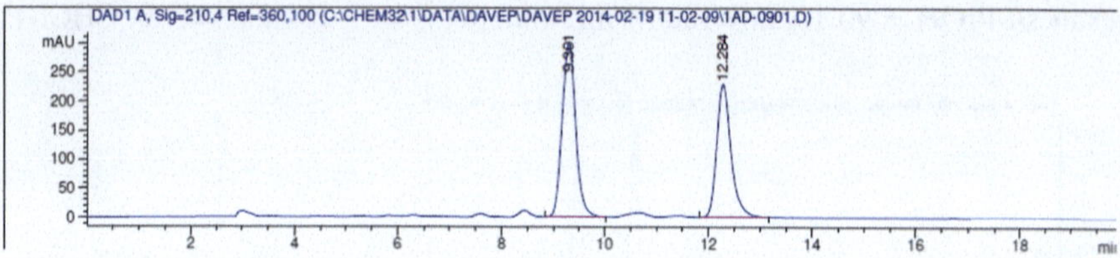


Signal 1: DAD1 A, Sig=210,4 Ref=360,100

| Peak # | RetTime [min] | Type | Width [min] | Area [mAU*s] | Height [mAU] | Area % |
|---|---|---|---|---|---|---|
| 1 | 9.301 | VB | 0.2891 | 5528.46826 | 299.87448 | 55.1911 |
| 2 | 12.284 | VB | 0.2961 | 4488.48682 | 229.54823 | 44.8089 |

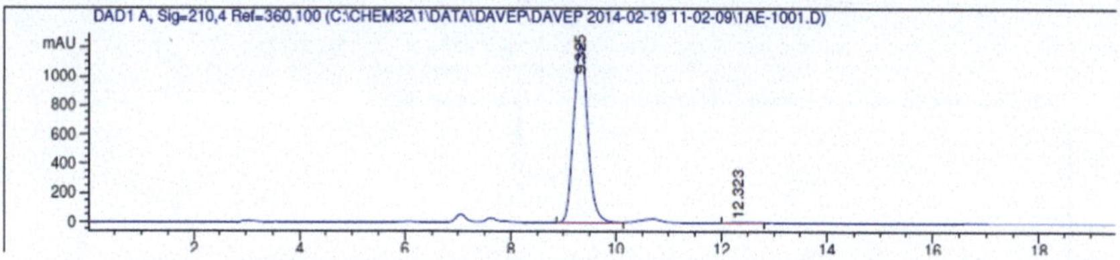


Signal 1: DAD1 A, Sig=210,4 Ref=360,100

| Peak # | RetTime [min] | Type | Width [min] | Area [mAU*s] | Height [mAU] | Area % |
|---|---|---|---|---|---|---|
| 1 | 9.325 | BB | 0.2952 | 2.30828e4 | 1228.40320 | 99.5932 |
| 2 | 12.323 | BB | 0.2804 | 94.27396 | 5.13030 | 0.4068 |

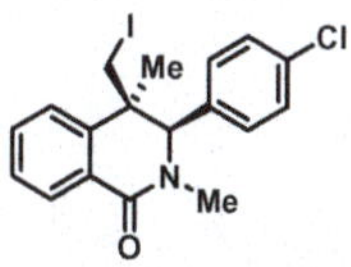


(3*R*,4*R*)-3-(4-chlorophenyl)-4-(iodomethyl)-2,4-dimethyl-3,4-dihydroisoquinolin-1(2*H*)-one (1.107 l)

Synthesized according to **GP6** using (*S*)-*N*-(1-(4-chlorophenyl)-2-methylallyl)-2-iodo-*N*-methylbenzamide (85.3 mg, 0.2 mmol). The alkyl iodide was purified by flash column chromatography using hexanes:DCM:EtOAc (12:8:1 v:v:v) and was obtained as a white solid (69.8 mg, 0.16 mmol, 82%, 98:2 dr (*cis:trans*)). **Major Diastereomer:** Isolated as a white solid (M.P.: 117–118 °C).

**$^1$H NMR** (500 MHz, $CDCl_3$) δ 8.21 (dd, $J$ = 7.5, 1.5 Hz, 1H), 7.52–7.47 (m, 1H), 7.47–7.41 (m, 1H), 7.21 (d, $J$ = 8.0 Hz, 1H), 7.12 (d, $J$ = 8.5 Hz, 2H), 7.02 (d, $J$ = 8.5 Hz, 2H), 4.33 (s, 1H), 3.94 (d, $J$ = 10.0 Hz, 1H), 3.05 (s, 3H), 2.89 (d, $J$ = 10.5 Hz, 1H), 1.62 (s, 3H). **$^{13}$C NMR** (126 MHz, $CDCl_3$) δ 163.4, 139.8, 134.8, 134.5, 132.5, 129.9, 128.9, 128.6, 128.6, 127.8, 123.2, 71.4, 40.4, 34.3, 31.3, 15.9. **IR** ($cm^{-1}$, film) 3070, 3000, 2968, 2925, 2863, 1652, 1643, 1601, 1576, 1492, 1473, 1397, 1375, 1216, 1197, 1111, 1094, 1082. **HRMS** (DART) Calc'd for $C_{18}H_{18}ClINO$, 426.01216; found, 426.01119.

The enantiomeric ratio was determined by HPLC: Chiracel AD-H column, *n*-hexanes/*i*-PrOH = 90:10, flow rate: 1 mL/min. $[\alpha]_D^{20}$ = + 287.2 (c = 1.0, $CHCl_3$).

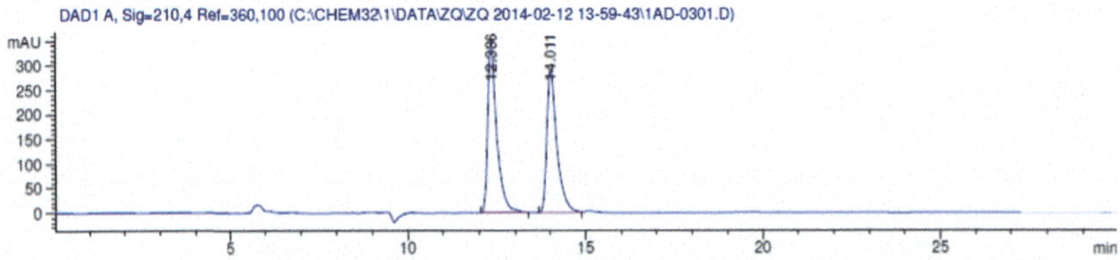


Signal 1: DAD1 A, Sig=210,4 Ref=360,100

| Peak # | RetTime [min] | Type | Width [min] | Area [mAU*s] | Height [mAU] | Area % |
|---|---|---|---|---|---|---|
| 1 | 12.336 | BB | 0.2543 | 6058.08496 | 349.34561 | 51.5402 |
| 2 | 14.011 | BB | 0.2862 | 5696.00342 | 291.14886 | 48.4598 |

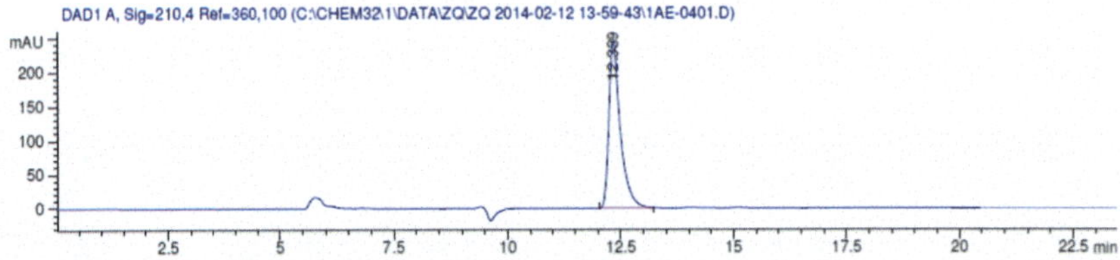


Signal 1: DAD1 A, Sig=210,4 Ref=360,100

| Peak # | RetTime [min] | Type | Width [min] | Area [mAU*s] | Height [mAU] | Area % |
|---|---|---|---|---|---|---|
| 1 | 12.339 | BB | 0.2504 | 4228.37061 | 246.12251 | 100.0000 |

(minor enantiomer not visible)

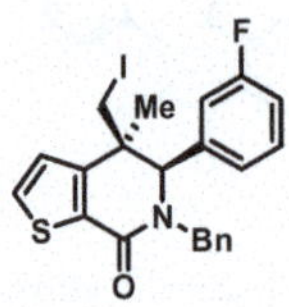


(4*R*,5*R*)-6-benzyl-5-(3-fluorophenyl)-4-(iodomethyl)-4-methyl-5,6-dihydrothieno[2,3-c]pyridin-7(4*H*)-one (1.107p)

Synthesized according to **GP6** using (*S*)-*N*-benzyl-*N*-(1-(3-fluorophenyl)-2-methylallyl)-3-iodothiophene-2-carboxamide (78.3 mg, 0.2 mmol). The alkyl iodide was purified by flash column chromatography using hexanes:EtOAc (3:1 v:v) and was obtained as a white solid (69.1 mg, 0.18 mmol, 88%, > 95:5 dr (*cis:trans*)). **Major Diastereomer:** Isolated as a white solid (M.P.: 136–137 °C).

**$^1$H NMR** (500 MHz, $CDCl_3$) δ 7.58 (d, $J$ = 5.0 Hz, 1H), 7.38–7.29 (m, 5H), 7.24–7.16 (m, 1H), 7.03–6.93 (m, 3H), 6.90–6.80 (m, 1H), 5.56 (d, $J$ = 14.5 Hz, 1H), 4.21 (s, 3H), 3.58 (d, $J$ = 10.5 Hz, 1H), 3.51 (d, $J$ = 14.5 Hz, 1H), 2.90–2.75 (m, 1H), 1.30 (d, $J$ = 1.0 Hz, 3H). **$^{13}$C NMR** (126 MHz, $CDCl_3$) δ 162.4 (d, $J$ = 247.0 Hz), 159.9, 145.1, 138.6 (d, $J$ = 6.5 Hz), 136.4, 132.2, 131.3, 130.1 (d, $J$ = 8.0 Hz), 129.1, 128.7, 127.9, 124.9, 124.2, 116.0, 115.8 (d, $J$ = 21.0 Hz), 69.3 (d, $J$ = 1.5 Hz), 47.5, 40.1, 30.6, 14.4. **$^{19}$F NMR** (564 MHz, $CDCl_3$) δ-111.5. **IR** (cm-1, film) 3064, 3031, 2966, 2929, 2875, 1645, 1600, 1575, 1473, 1457, 1395, 1297, 1258, 1197, 1158. **HRMS** (ESI+) Calc'd for C18H19INO, 392.0506; found, 392.0511. $[\alpha]_D^{20}$ = + 261.6 ($c$ = 1.0, $CHCl_3$).

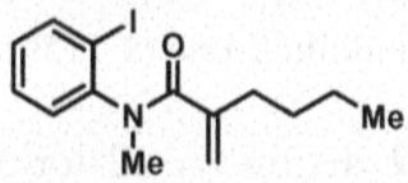


*N*-(2-iodophenyl)-*N*-methyl-2-methylenehexanamide (1.114n)

Prepared according to **GP7** and **GP8** with an overall yield of 35%. The *N*-substituted acrylamide was purified by flash column chromatography using hexanes:EtOAc (4:1 v:v) and was obtained as a pale yellow oil.

*Two rotamers were observed in a 6.7:1 ratio. The spectral data for major rotamer are reported below.*

**$^1$H NMR** (500 MHz, $CDCl_3$) δ 7.74 (dd, $J$ = 8.0, 1.5 Hz, 1H), 7.55–7.50 (m, 2H), 7.44–7.34 (m, 3H), 7.13–7.07 (m, 1H), 7.03–6.94 (m, 2H), 6.84–6.77 (m, 1H), 6.76–6.68 (m, 3H), 6.04–5.99 (m, 1H), 5.98 (dd, $J$ = 8.0, 1.5 Hz, 1H), 5.57 (s, 1H), 5.29 (s, 1H), 5.25 (s, 1H), 4.99 (s, 1H), 3.92 (d, $J$ = 15.0 Hz, 1H). **$^{13}$C NMR** (101 MHz, $CDCl_3$) δ 171.7, 146.7, 144.7, 140.1, 129.6, 129.3, 129.2, 117.3, 99.1, 36.8, 33.4, 29.8, 22.3, 13.9. **IR** (neat film, $cm^{-1}$) 2936, 2963, 2855, 1745, 1654, 1627, 1578, 1475, 1436, 1363, 1024, 918, 763, 724. **HRMS** (DART) calculated 466.06678 m/z (found 466.06599 m/z for $C_{24}H_{21}$.INO).

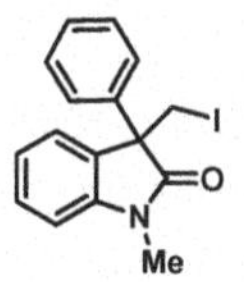


3-(iodomethyl)-1-methyl-3-phenylindolin-2-one (1.115a)

Prepared according to **GP9** using *N*-(2-iodophenyl)-*N*-methyl-2-phenylacrylamide (72.6 mg, 0.2 mmol) at 100 °C for 24 h. The oxindole was purified by flash column chromatography using pentanes:DCM:EtOAc (12:8:0.5 v:v:v) and was obtained as a white solid (65.4 mg, 0.18 mmol, 90%, M.P.: 98–99 °C).

**$^1$H NMR** (500 MHz, CDCl_3) δ 7.47–7.44 (m, 2H), 7.44–7.40 (m, 1H), 7.40–7.37 (m, 1H), 7.35–7.26 (m, 3H), 7.23–7.16 (m, 1H), 6.94 (d, *J* = 8.0 Hz, 1H), 4.04 (d, *J* = 10.0 Hz, 1H), 3.77 (d, *J* = 10.0 Hz, 1H), 3.25 (s, 3H). **$^{13}$C NMR** (126 MHz, $CDCl_3$) δ 176.2, 144.0, 137.7, 130.8, 129.1, 128.8, 128.1, 127.1, 125.0, 122.7, 108.6, 56.6, 26.6, 10.5. **IR** (neat, $cm^{-1}$) 3063, 3056, 3056, 2949, 1703, 1609, 1493, 1470, 1445, 1371, 1349, 1265, 1184. **HRMS** (ESI+) calculated 364.01983 m/z (found 364.02004 m/z for $C_{16}H_{15}$.INO).

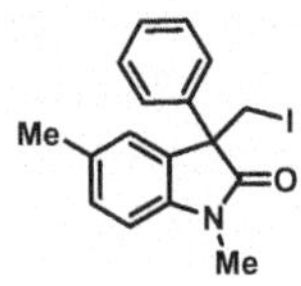


3-(iodomethyl)-1,5-dimethyl-3-phenylindolin-2-one (1.115b)

Prepared according to **GP9** using *N*-(2-iodo-4-methylphenyl)-*N*-methyl-2-phenylacrylamide (75.4 mg, 0.2 mmol) at 100 °C for 24 h. The oxindole was purified by flash column chromatography using pentanes:DCM:EtOAc (12:8:0.5 v:v:v) and was obtained as a white solid (64.0 mg, 0.17 mmol, 85%, M.P.: 125–126 °C).

**$^1$H NMR** (500 MHz, $CDCl_3$) δ 7.47–7.42 (m, 2H), 7.35–7.26 (m, 3H), 7.22–7.19 (m, 1H), 7.19–7.17 (m, 1H), 6.83 (d, *J* = 8.0 Hz, 1H), 4.03 (d, *J* = 10.0 Hz, 1H), 3.76 (d, *J* = 10.0 Hz, 1H), 3.22 (s, 3H), 2.41 (s, 3H). **$^{13}$C NMR** (126 MHz, $CDCl_3$) δ 176.1, 141.6, 137.9, 132.3, 130.9, 129.4, 128.7, 128.0, 127.1, 125.6, 108.3, 56.7, 26.6, 21.3, 10.7. **IR** (neat, $cm^{-1}$) 3060, 3023, 2969, 2920, 2855, 1707, 1618, 1601, 1496, 1447, 1348, 1239, 1186, 1073, 1031. **HRMS** (ESI+) calculated 378.03548 m/z (found 378.03550 m/z for $C_{17}H_{17}$.INO).

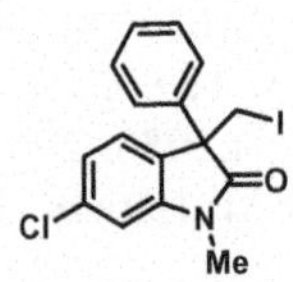


6-chloro-3-(iodomethyl)-1-methyl-3-phenylindolin-2-one (1.115d)

Prepared according to **GP3** using *N*-(5-chloro-2-iodophenyl)-*N*-methyl-2-phenylacrylamide (79.5 mg, 0.2 mmol) at 100 °C for 24 h. The oxindole was purified by flash column chromatography using pentanes:DCM:EtOAc (12:8:0.5 v:v:v) and was obtained as an off white solid (65.0 mg, 0.16 mmol, 82%, M.P.: 119–120 °C).

**$^1$H NMR** (500 MHz, $CDCl_3$) δ 7.45–7.37 (m, 2H), 7.35–7.27 (m, 4H), 7.20–7.14 (m, 1H), 6.94 (d, *J* = 2.0 Hz, 1H), 4.02 (d, *J* = 10.0 Hz, 1H), 3.73 (d, *J* = 10.0 Hz, 1H), 3.23 (s, 3H). **$^{13}$C NMR** (126 MHz, $CDCl_3$) δ 176.1, 145.2, 137.2, 135.0, 129.2, 128.9, 128.3, 127.0, 125.9, 122.6, 109.4, 56.4, 26.7, 9.9. **IR** (neat, $cm^{-1}$) 3063, 3020, 2961, 2929, 1711, 1609, 1592, 1494, 1443, 1365, 1312, 1291, 1232, 1183. **HRMS** (ESI+) calculated 397.98086 m/z (found 397.98063 m/z for $C_{16}H_{14}$·ClINO).

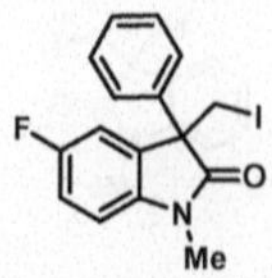

5-fluoro-3-(iodomethyl)-1-methyl-3-phenylindolin-2-one (1.115e)

Prepared according to **GP3** using *N*-(4-fluoro-2-iodophenyl)-*N*-methyl-2-phenylacrylamide (76.2 mg, 0.2 mmol) at 100 °C for 24 h. The oxindole was purified by flash column chromatography using pentanes:DCM:EtOAc (12:8:05 v:v:v) and was obtained as a white solid (65 mg, 0.16 mmol, 82%, M.P.: 118–120 °C).

**$^1$H NMR** (500 MHz, $CDCl_3$) δ 7.43–7.39 (m, 2H), 7.36–7.27 (m, 3H), 7.16–7.09 (m, 2H), 6.86 (dd, $J$ = 8.5, 4.0 Hz, 1H), 4.02 (d, $J$ = 10.0 Hz, 1H), 3.73 (d, $J$ = 10.0 Hz, 1H), 3.24 (s, 3H). **$^{13}$C NMR** (126 MHz, $CDCl_3$) δ 175.9 (d, $J$ = 1.0 Hz), 159.2 (d, $J$ = 241.5 Hz), 140.0 (d, $J$ = 2.0 Hz), 137.2, 132.4 (d, $J$ = 8.0 Hz), 128.9, 128.3, 126.9, 115.5 (d, $J$ = 23.5 Hz), 113.1 (d, $J$ = 25.0 Hz), 109.1 (d, $J$ = 8.0 Hz), 57.0 (d, $J$ = 1.5 Hz), 26.7, 9.8. **$^{19}$F NMR** (564 MHz, $CDCl_3$) δ -119.6. **IR** (neat, $cm^{-1}$) 3059, 3018, 2923, 2854, 1706, 1615, 1493, 1453, 1351, 1267, 1232, 1190, 1102, 1074. **HRMS** (ESI+) calculated 382.01041 m/z (found 382.01026 m/z for $C_{16}H_{14}\cdot FINO$).

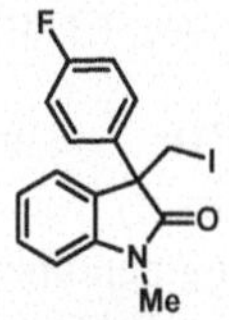

3-(4-fluorophenyl)-3-(iodomethyl)-1-methylindolin-2-one (1.115j)

Prepared according to **GP9** using 2-(4-fluorophenyl)-*N*-(2-iodophenyl)-*N*-methylacrylamide (76.2 mg, 0.2 mmol) at 100 °C for 24 h. The oxindole was purified by flash column chromatography using pentanes:DCM:EtOAc (12:8:0.5 v:v:v) and was obtained as a white solid (58.1 mg, 0.15 mmol, 76%, M.P.: 89–90 °C).

**$^1$H NMR** (500 MHz, $CDCl_3$) δ 7.47–7.41 (m, 3H), 7.41–7.38 (m, 1H), 7.21 (td, $J$ = 7.5, 1.0 Hz, 1H), 7.03–6.97 (m, 2H), 6.95 (d, $J$ = 8.0 Hz, 1H), 3.97 (d, $J$ = 10.0 Hz, 1H), 3.73 (d, $J$ = 10.0 Hz, 1H), 3.24 (s, 3H). **$^{13}$C NMR** (126 MHz, $CDCl_3$) δ 176.0 (d, $J$ = 1.0 Hz), 162.4 (d, $J$ = 248.0 Hz), 144.0, 133.4 (d, $J$ = 3.5 Hz), 130.5, 129.3, 129.0 (d, $J$ = 8.0 Hz), 125.0, 122.8, 115.6 (d, $J$ = 21.5 Hz), 108.8, 55.9, 26.6, 10.5 (d, $J$ = 1.0 Hz). **$^{19}$F NMR** (564 MHz, $CDCl_3$) δ -114.0. **IR** (neat, $cm^{-1}$) 3083, 3058, 2962, 2923, 2853, 1698, 1609, 1600, 1492, 1468, 1371, 1351, 1229, 1196, 1160, 1025. **HRMS** (ESI+) calculated 382.01041 m/z (found 382.01010 m/z for $C_{16}H_{14}\cdot FINO$).

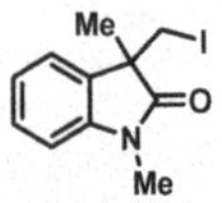

3-(iodomethyl)-1,3-dimethylindolin-2-one (1.115 k)

Prepared according to **GP10** using *N*-(2-iodophenyl)-*N*-methylmethacrylamide (60.23 mg, 0.2 mmol) for 24 h. The oxindole was purified by flash column chromatography using pentanes:DCM:EtOAc (12:8:1 v:v:v) and was obtained as a clear yellow oil (47.0 mg, 0.16 mmol, 78%). Spectroscopic data is consistent with literature.

**$^1$H NMR** (500 MHz, $CDCl_3$) δ 7.34 (td, *J* = 7.5, 1.0 Hz, 1H), 7.29–7.26 (m, 1H), 7.11 (td, *J* = 7.5, 1.0 Hz, 1H), 6.88 (d, *J* = 7.5 Hz, 1H), 3.52 (d, *J* = 10.0 Hz, 1H), 3.43 (d, *J* = 10.0 Hz, 1H), 3.24 (s, 3H), 1.52 (s, 3H). **$^{13}$C NMR** (126 MHz, $CDCl_3$) δ 177.8, 143.2, 132.6, 128.7, 122.7, 122.6, 108.3, 48.6, 26.3, 23.0, 10.9.

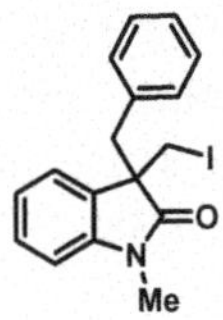

3-benzyl-3-(iodomethyl)-1-methylindolin-2-one (1.115l)

Prepared according to **GP10** using 2-benzyl-*N*-(2-iodophenyl)-*N*-methylacrylamide (76.2 mg, 0.2 mmol) for 24 h. The oxindole was purified by flash column chromatography using pentanes:DCM:EtOAc (12:8:0.5 v:v:v) and was obtained as a white solid (58.1 mg, 0.15 mmol, 76%, M.P.: 100–102 °C).

**$^1$H NMR** (500 MHz, $CDCl_3$) δ 7.30–7.23 (m, 1H), 7.20–7.15 (m, 1H), 7.13–7.00 (m, 4H), 6.88–6.78 (m, 2H), 6.64 (d, *J* = 8.0 Hz, 1H), 3.69 (d, *J* = 10.0 Hz, 1H), 3.52 (d, *J* = 10.0 Hz, 1H), 3.18 (d, *J* = 13.0 Hz, 1H), 3.13 (d, *J* = 13.0 Hz, 1H), 3.00 (s, 3H). **$^{13}$C NMR** (126 MHz, $CDCl_3$) δ 176.7, 143.7, 135.1, 130.0, 129.8, 128.7, 127.6, 126.8, 123.6, 122.3, 108.0, 54.6, 43.4, 26.0, 9.1. **IR** (neat, $cm^{-1}$) 3082, 3060, 3030, 2922, 2850, 1694, 1611, 1493, 1469, 1376, 1358, 1246, 1197, 1117, 1083, 1020. **HRMS** (ESI+) calculated 378.03548 m/z (found 378.03491 m/z for $C_{17}H_{17}$.INO).

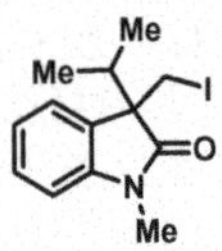

33-(iodomethyl)-3-isopropyl-1-methylindolin-2-one (1.115 m)

Prepared according to **GP10** using *N*-(2-iodophenyl)-*N*,3-dimethyl-2-methylenebutanamide (65.84 mg, 0.2 mmol) for 24 h. The oxindole was purified by flash column chromatography using pentanes:DCM:EtOAc (12:8:0.5 v:v:v) and was obtained as a white solid (48.0 mg, 0.14 mmol, 72%, M.P.: 41–42 °C).

$^1$**H NMR** (500 MHz, $CDCl_3$) δ 7.35–7.30 (m, 1H), 7.20 (d, $J$ = 7.5 Hz, 1H), 7.08 (t, $J$ = 7.5 Hz, 1H), 6.84 (d, $J$ = 7.5 Hz, 1H), 3.56 (m, 2H), 3.20 (s, 3H), 2.26 (hept, $J$ = 7.0 Hz, 1H), 0.96 (d, $J$ = 7.0 Hz, 3H), 0.81 (d, $J$ = 7.0 Hz, 3H). $^{13}$**C NMR** (126 MHz, $CDCl_3$) δ 177.6, 144.3, 130.1, 128.6, 123.4, 122.4, 108.0, 56.5, 35.4, 26.0, 18.2, 17.4, 9.6. **IR** (neat, $cm^{-1}$) 3056, 2965, 2934, 2878, 1708, 1612, 1495, 1468, 1420, 1375, 1352, 1337, 1261, 1198, 1125. **HRMS** (ESI+) calculated 330.03548 m/z (found 330.03488 m/z for $C_{13}H_{17}$.INO).

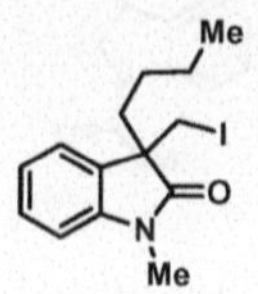


3-butyl-3-(iodomethyl)-1-methylindolin-2-one (1.115n)

Prepared according to **GP10** using *N*-(2-iodophenyl)-*N*-methyl-2-methylenehexanamide (68.6 mg, 0.2 mmol) for 48 h. The oxindole was purified by flash column chromatography using pentanes:DCM:EtOAc (12:8:1 v:v:v) and was obtained as clear yellow oil (40.0 mg, 0.12 mmol, 58%).

$^1$**H NMR** (500 MHz, $CDCl_3$) δ 7.34 (td, $J$ = 7.5, 1.0 Hz, 1H), 7.23–7.19 (m, 1H), 7.12 (td, $J$ = 7.5, 1.0 Hz, 1H), 6.87 (d, $J$ = 7.5 Hz, 1H), 3.51 (d, $J$ = 9.5 Hz, 1H), 3.41 (d, $J$ = 9.5 Hz, 1H), 3.24 (s, 3H), 2.01–1.91 (m, 1H), 1.87–1.78 (m, 1H), 1.24–1.11 (m, 2H), 1.07–0.96 (m, 1H), 0.92–0.79 (m, 1H), 0.77 (t, $J$ = 7.5 Hz, 3H). $^{13}$**C NMR** (126 MHz, $CDCl_3$) δ 177.5, 143.9, 131.1, 128.6, 122.7, 122.7, 108.1, 53.1, 37.2, 27.3, 26.2, 22.8, 13.7, 10.7. **IR** (neat, $cm^{-1}$) 3055, 3016, 2957, 2872, 1705, 1615, 1495, 1470, 1435, 1420, 1377, 1354, 1303, 1256, 1242, 1223, 1080. **HRMS** (ESI+) calculated 344.05113 m/z (found 344.05091 m/z for $C_{14}H_{19}$.INO).

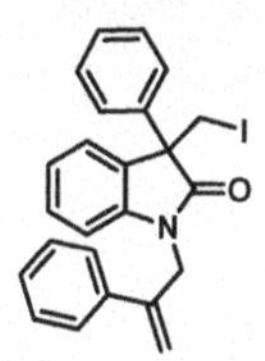


3-(iodomethyl)-3-phenyl-1-(2-phenylallyl)indolin-2-one (1.115o)

Prepared according to **GP9** using *N*-(2-iodophenyl)-2-phenyl-*N*-(2-phenylallyl)acrylamide (93.1 mg, 0.2 mmol) at 100 °C for 24 h. The oxindole was purified by flash column chromatography using pentanes:DCM:EtOAc (12:8:0.5 v:v:v) and was obtained as an off white solid (57.0 mg, 0.12 mmol, 61%, M.P.: 124–126 °C).

**$^1$H NMR** (500 MHz, $CDCl_3$) δ 7.42–7.37 (m, 2H), 7.37–7.30 (m, 2H), 7.29–7.20 (m, 8H), 7.15 (td, *J* = 7.5, 1.0 Hz, 1H), 6.98 (d, *J* = 7.5 Hz, 1H), 5.44 (s, 1H), 5.25 (s, 1H), 4.98–4.91 (m, 1H), 4.69–4.61 (m, 1H), 3.99 (d, *J* = 10.0 Hz, 1H), 3.78 (d, *J* = 10.0 Hz, 1H). **$^{13}$C NMR** (126 MHz, $CDCl_3$) δ 176.0, 143.2, 142.1, 138.2, 138.0, 131.1, 129.0, 128.8, 128.4, 128.1, 127.9, 127.0, 126.4, 124.8, 122.8, 114.5, 110.0, 56.6, 44.3, 9.8. **IR** (neat, $cm^{-1}$) 3055, 3024, 2918, 1713, 1613, 1489, 1466, 1445, 1403, 1377, 1356, 1192. **HRMS** (ESI+) calculated 466.06678 m/z (found 466.06738 m/z for $C_{24}H_{21}$.INO).

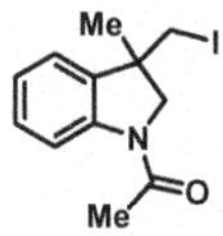


31-(3-(iodomethyl)-3-methylindolin-1-yl)ethan-1-one (1.115p)

Prepared according to **GP9** using *N*-(2-iodophenyl)-*N*-(2-methylallyl)acetamide (63.0 mg, 0.2 mmol) at 100 °C for 24 h. The indoline was purified by flash column chromatography using pentanes:EtOAc (2:1 v:v) and was obtained as a pale yellow solid (58.6 mg, 0.19 mmol, 93%, M.P.: 67–68 °C).

*Two rotamers were observed in a 5.3:1 ratio. The spectral data of the major rotamer are reported below.*

**$^1$H NMR** (500 MHz, $CDCl_3$) δ 8.18 (d, *J* = 8.0 Hz, 1H), 7.31–7.23 (m, 1H), 7.15–7.10 (m, 1H), 7.09–7.04 (m, 1H), 4.04 (d, *J* = 10.5 Hz, 1H), 3.70 (d, *J* = 10.5 Hz, 1H), 3.40 (d, *J* = 10.0 Hz, 1H), 3.33 (d, *J* = 10.0 Hz, 1H), 2.24 (s, 3H), 1.52 (s, 3H). **$^{13}$C NMR** (126 MHz, $CDCl_3$) δ 168.6, 142.1, 135.0, 129.0, 123.9, 122.1, 117.4, 100.1, 62.3, 44.1, 26.4, 24.3, 19.1. **IR** (neat film, $cm^{-1}$) 3019, 2968, 2957, 2921, 2861, 1650, 1598, 1478, 1462, 1396, 1369, 1339, 1287, 1217, 1192, 1128. **HRMS** (ESI+) calculated 316.01983 m/z (found 316.01945 m/z for $C_{12}H_{15}$.INO).

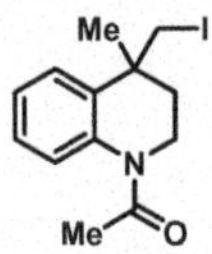


1-(4-(iodomethyl)-4-methyl-3,4-dihydroquinolin-1(2*H*)-yl)ethan-1-one (1.115r)

Prepared according to **GP10** using *N*-(2-iodophenyl)-*N*-(3-methylbut-3-en-1-yl)acetamide (65.8 mg, 0.2 mmol) for 24 h. The indoline was purified by flash column chromatography using pentanes:EtOAc (2:1 v:v) and was obtained as an off white solid (43.0 mg, 0.13 mmol, 65%, M.P.: 59–61 °C).

**$^1$H NMR** (500 MHz, $CDCl_3$) δ 7.32–7.27 (m, 1H), 7.25–7.20 (m, 1H), 7.20–7.15 (m, 1H), 4.16 (s, 1H), 3.48 (s, 1H), 3.44–3.29 (m, 2H), 2.25 (s, 3H), 2.15 (ddd, *J* = 14.0, 7.0, 5.5 Hz, 1H), 1.82–1.70 (m, 1H), 1.47 (s, 3H). **$^{13}$C NMR** (126 MHz, $CDCl_3$) δ 170.0, 138.3 (broad), 135.4 (broad), 127.0, 127.0, 126.2, 125.4, 124.9, 40.9

(broad), 37.2, 36.8, 28.2, 23.3, 20.6. **IR** (neat film, $cm^{-1}$) 3063, 2963, 2930, 2891, 1657, 1601, 1575, 1489, 1452, 1375, 1337, 1300, 1269, 1254, 1211, 1178. **HRMS** (ESI+) calculated 330.03548 m/z (found 330.03491 m/z for $C_{13}H_{17}$-INO).

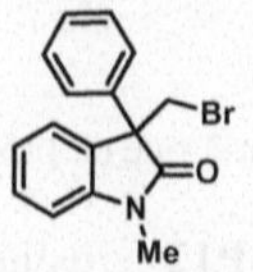


3-(bromomethyl)-1-methyl-3-phenylindolin-2-one (1.117)

Prepared according to **GP9** using *N*-(2-bromophenyl)-*N*-methyl-2-phenylacrylamide (1.6 g, 5.0 mmol) and $NiBr_2(PPh_3)_2$ as a precatalyst for 24 h. The oxindole was purified by flash column chromatography using pentanes:DCM:EtOAc (12:8:1 v:v:v) and was obtained as a white solid (126 mg, 0.4 mmol, 8%, M.P.: 111–112 °C).

**$^1$H NMR** (500 MHz, $CDCl_3$) δ 7.47–7.38 (m, 4H), 7.35–7.27 (m, 3H), 7.23–7.18 (m, 1H), 6.95 (d, $J$ = 8.0 Hz, 1H), 4.19 (d, $J$ = 10.0 Hz, 1H), 3.98 (d, $J$ = 10.0 Hz, 1H), 3.24 (s, 3H). **$^{13}$C NMR** (126 MHz, $CDCl_3$) δ 175.9, 144.2, 137.3, 129.7, 129.1, 128.8, 128.2, 127.1, 125.2, 122.7, 108.6, 57.1, 36.5, 26.6. **IR** (neat, $cm^{-1}$) 3058, 3027, 2967, 2932, 1707, 1609, 1492, 1467, 1443, 1372, 1348, 1244, 1222, 1126, 1078, 1023. **HRMS** (ESI+) calculated 316.03370 m/z (found 316.03325 m/z for $C_{16}H_{15}$-BrNO).

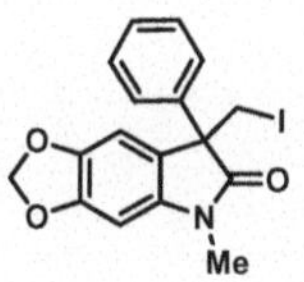


7-(iodomethyl)-5-methyl-7-phenyl-5,7-dihydro-6*H*-[1, 3]dioxolo[4,5-*f*]indol-6-one (1.115t)

Prepared according to a modification of **GP11** using *N*-(6-bromobenzo[*d*] [1, 3] dioxol-5-yl)-*N*-methyl-2-phenylacrylamide (72.0 mg, 0.2 mmol) and $NiI_2$ and $P(O\textit{i}Pr)_3$ as the precatalyst. The oxindole was purified by flash column chromatography using pentanes:DCM:EtOAc (12:8:1 v:v:v) and was obtained as a white solid (68.4 mg, 0.17 mmol, 84%, M.P.: 146–147 °C).

**$^1$H NMR** (500 MHz, $CDCl_3$) δ 7.46–7.40 (m, 2H), 7.36–7.27 (m, 3H), 6.88 (s, 1H), 6.54 (s, 1H), 6.01 (s, 2H), 4.02 (d, $J$ = 10.0 Hz, 1H), 3.70 (d, $J$ = 10.0 Hz, 1H), 3.19 (s, 3H). **$^{13}$C NMR** (126 MHz, $CDCl_3$) δ 176.3, 148.2, 143.4, 138.5, 137.8, 128.8, 128.1, 127.1, 122.5, 106.3, 101.4, 92.4, 56.9, 26.7, 10.9. **IR** (neat, $cm^{-1}$) 3058, 2909, 1698, 1626, 1614, 1494, 1422, 1389, 1326, 1197, 1129, 1073, 1037, 937, 880. **HRMS** (ESI+) calculated 408.00966 m/z (found 408.00938 m/z for $C_{17}H_{15}$-$INO_3$).

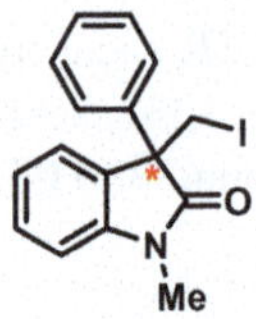


73-(iodomethyl)-1-methyl-3-phenylindolin-2-one (1.115a)

Prepared according to **GP12** using *N*-(2-iodophenyl)-*N*-methyl-2-phenylacrylamide (72.6 mg, 0.2 mmol) at 100 °C for 24 h. The oxindole was purified by flash column chromatography using pentanes:DCM:EtOAc (12:8:0.5 v:v:v) and was obtained as a white solid (36.3 mg, 0.10 mmol, 50%, 89:11 e.r.).

The enantiomeric ratio was determined by HPLC: Chiracel IA column, *n*-hexanes/*i*-PrOH = 90:10, flow rate: 1 mL/min. $[\alpha]_D^{20} = -116.55$ (c = 0.525, $CHCl_3$).

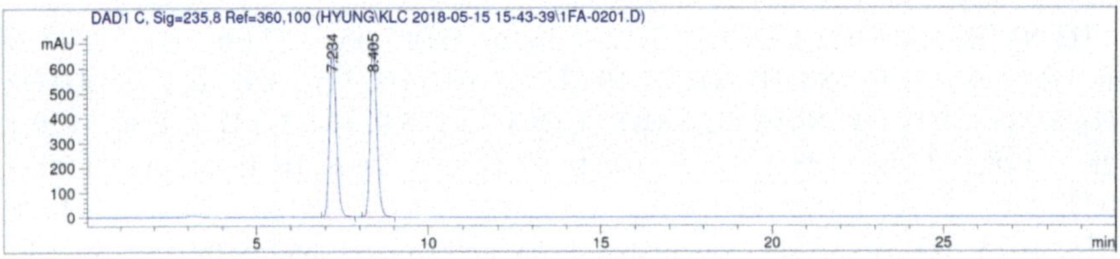


Signal 3: DAD1 C, Sig=235,8 Ref=360,100

| Peak # | RetTime [min] | Type | Width [min] | Area [mAU*s] | Height [mAU] | Area % |
|---|---|---|---|---|---|---|
| 1 | 7.234 | BB | 0.2114 | 9483.87012 | 694.12805 | 50.0291 |
| 2 | 8.405 | BB | 0.2070 | 9472.82324 | 704.00323 | 49.9709 |
| Totals : | | | | 1.89567e4 | 1398.13129 | |

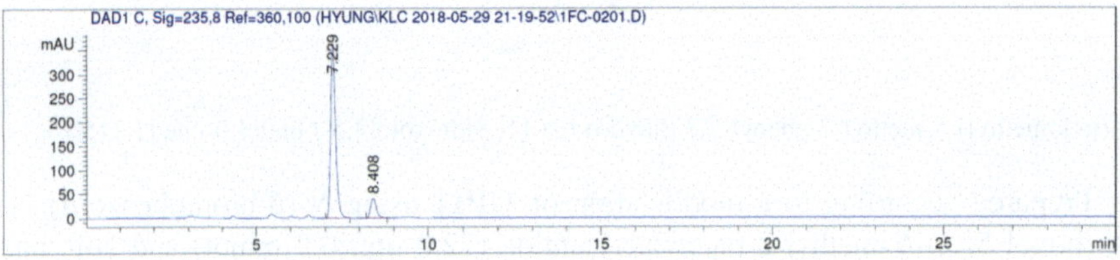


Signal 3: DAD1 C, Sig=235,8 Ref=360,100

| Peak # | RetTime [min] | Type | Width [min] | Area [mAU*s] | Height [mAU] | Area % |
|---|---|---|---|---|---|---|
| 1 | 7.229 | BB | 0.1493 | 3604.59595 | 364.59708 | 88.8408 |
| 2 | 8.408 | BB | 0.1718 | 452.76868 | 40.69995 | 11.1592 |
| Totals : | | | | 4057.36462 | 405.29702 | |

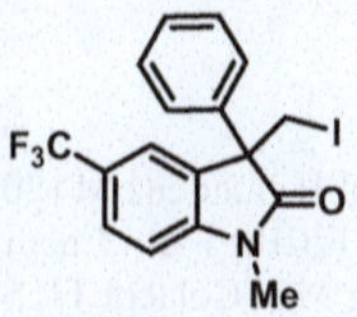


3-(iodomethyl)-1-methyl-3-phenyl-5-(trifluoromethyl)indolin-2-one (1.115f)

Prepared according to **GP12** using *N*-(2-iodo-4-(trifluoromethyl)phenyl)-*N*-methyl-2-phenylacrylamide (86.3 mg, 0.2 mmol) at 100 °C for 24 h. The oxindole was purified by flash column chromatography using pentanes:EtOAc (5:1 v:v) and was obtained as an off white solid (72.5 mg, 0.17 mmol, 84%, 78:22 e.r.).

The enantiomeric ratio was determined by HPLC: Chiracel IA column, *n*-hexanes/*i*-PrOH = 90:10, flow rate: 1 mL/min. $[\alpha]_D^{20} = -73.98$ (c = 0.50, $CHCl_3$).

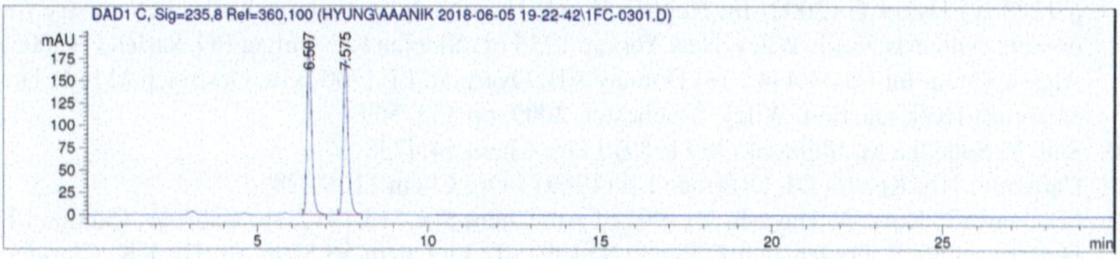


Signal 3: DAD1 C, Sig=235,8 Ref=360,100

| Peak # | RetTime [min] | Type | Width [min] | Area [mAU*s] | Height [mAU] | Area % |
|---|---|---|---|---|---|---|
| 1 | 6.507 | BB | 0.1384 | 1784.13220 | 195.71381 | 49.9478 |
| 2 | 7.575 | BB | 0.1565 | 1787.86084 | 173.13635 | 50.0522 |
| Totals : | | | | 3571.99304 | 368.85016 | |

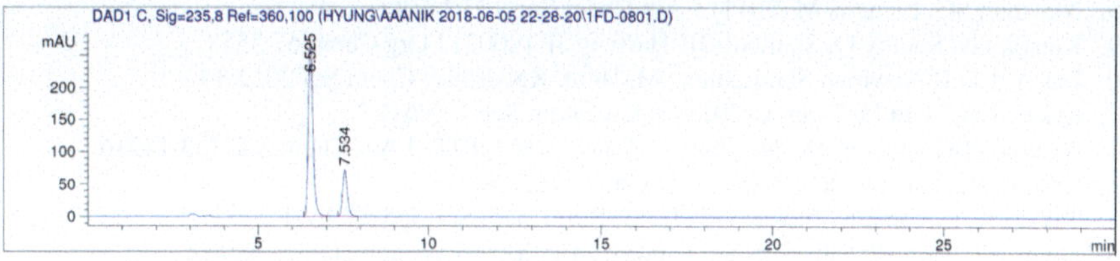


Signal 3: DAD1 C, Sig=235,8 Ref=360,100

| Peak # | RetTime [min] | Type | Width [min] | Area [mAU*s] | Height [mAU] | Area % |
|---|---|---|---|---|---|---|
| 1 | 6.525 | BB | 0.1410 | 2476.06812 | 270.06305 | 77.9391 |
| 2 | 7.534 | BB | 0.1495 | 700.85999 | 72.02408 | 22.0609 |
| Totals : | | | | 3176.92810 | 342.08713 | |

## References


1. (a) Petrone DA, Yoon H, Weinstabl H, Lautens M (2014) Angew Chem Int Ed 53:7908. (b) Yoon H, Marchese AD, Lautens M (2018) J Am Chem Soc 140:10950
2. Johansson Seechurn CCC, Kitching MO, Colacot TJ, Snieckus V (2012) Angew Chem Int Ed 51:5062
3. (a) Amatore C, Jutand A (2000) Acc Chem Res 33:14. (b) Beletskaya IP, Cheprakov AV (2000) Chem Rev 100:3009
4. von Schenck H, Akermark B, Svensson M (2003) J Am Chem Soc 125:3503
5. Hartwig JF (2010) Organotransition metal chemistry: from bonding to catalysis, 1st edn. University Science Books, Mill Valley
6. de Meijere A, Diederich F (eds) (2004) Metal-catalyzed cross coupling reactions, 2nd edn. Wiley-VCH, Weinheim, Germany
7. Xue L, Lin Z (2010) Chem Soc Rev 39:1692
8. Pérez-Rodríguez M, Braga AAC, García-Melchor M, Pérez-Temprano MH, Casares JA, Ujaque G, de Lera AR, Álvarez R, Maseras F, Espinet P (2009) J Am Chem Soc 131:3650
9. (a) Overman LE (1994) Pure Appl Chem 66:1423. (b) Bräse S, De Meijere A (2002) In: Negishi E (ed) Handbook of organopalladium chemistry for organic synthesis, vol 1. Wiley, New York, p 1223 (c) Dyker G (2002) In: Negishi E (ed) Handbook of organopalladium chemistry for organic synthesis, vol 1. Wiley, New York, p 1255 (d) Nicolau KC, Bulger PG, Sarlah D (2005) Angew Chem Int Ed 44:4442. (e) Dounay AB, Overman LE (2009) In: Oestreich M (ed) The Mizoroki-Heck reaction. Wiley, Chichester, **2009**, pp 533–569
10. Sato Y, Sodeoka M, Shibasaki M (1989) J Org Chem 54:4738
11. Carpenter NE, Kucera DJ, Overman LE (1989) J Org Chem 110:2328
12. (a) Ozawa F, Kubo A, Hayashi T (1991) J Am Chem Soc 113:1417. (b) Cabri W, Candiani I, DeBernardinis S, Francalanci F, Penco S (1991) J Org Chem 56:5796. (c) Hii KK, Claridge TD, Brown JM (2001) Helv Chim Acta 84:3043
13. Lapierre AJB, Geib SJ, Curran DP (2007) J Am Chem Soc 129:494
14. Roy AH, Hartwig JF (2001) J Am Chem Soc 123:1232
15. Roy AH, Hartwig JF (2003) J Am Chem Soc 125:13944
16. Shen X, Hyde AM, Buchwald SL (2010) J Am Chem Soc 132:14072
17. Newman SG, Lautens M (2010) J Am Chem Soc 132:11416
18. Newman SG, Lautens M (2011) J Am Chem Soc 133:1778
19. Kataoka N, Shelby Q, Stambuli JP, Hartwig JF (2002) J Org Chem 67:5553
20. Lan Y, Liu P, Newman SG, Lautens M, Houk KN (1987) Chem Sci 2012:3
21. Liu H, Li C, Qiu D, Tong X (2011) J Am Chem Soc 133:6187
22. Newman SG, Howell JK, Nicolaus N, Lautens M (2011) J Am Chem Soc 133:14916
23. Cramer R, Coulson DR (1975) J Org Chem 40:2267
24. Takagi K, Hayama N, Inokawa S (1980) Bull Chem Soc Jpn 53:3691
25. Higgs AT, Zinn PJ, Simmons SJ, Sanford MS (2009) Organometallics 28:6142
26. Zheng B, Tang F, Luo J, Schultz JW, Rath NP, Mirica LM (2014) J Am Chem Soc 136:6499
27. Nakao Y, Oda S, Hiyama T (2004) J Am Chem Soc 126:13904
28. Nakao Y, Yada A, Ebata S, Hiyama T (2006) J Am Chem Soc 129:2428
29. Nakao Y, Ebata S, Yada A, Hiyama T, Ikawa M, Ogoshi S (2008) J Am Chem Soc 130:12874
30. Watson MP, Jacobsen EN (2008) J Am Chem Soc 130:12594
31. Ninomiya I, Yamamoto O, Naito T (1976) J Chem Soc Chem Commun 437
32. Rana NK, Singh VK (2011) Org Lett 13:6520
33. Isrow D, Captain B (2011) Inorg Chem 50:5864
34. Low DW, Pattison G, Wieczsysty MD, Churchill GH, Lam HW (2012) Org Lett 14:2548

# Chapter 2
# Diastereoselective Pd-Catalyzed Aryl Cyanation and Aryl Borylation

**Abstract** In Chap. 2, a Pd-catalyzed diastereoselective anion capture cascade reaction is discussed. Building on the Pd-catalyzed carboiodination reaction en route to (+)-corynoline, a Pd-catalyzed aryl cyanation is discussed. The alkyl nitrile was synthesized in high yield and d.r. The second method generates borylated chromans as a single diastereomer in good yield. The alkyl boronate was further functionalized to showcase its utility.

## 2.1 Introduction

Palladium-catalysis has become ubiquitous in organic synthesis to accessing reactive synthetic handles from the respective halide counterpart. Chapter 1 focused on the regeneration of the carbon–halogen bond in a domino process. This chapter outlines the development of the aryl cyanation and aryl borylation transformations. These methods have been previously developed but the lack of examples demonstrating stereoselective transformations prompted our interest in this field.

## 2.2 Pd-Catalyzed Cyanation

Nitriles are an essential functionality in dyes, agrochemical, and the pharmaceutical industries. Additionally, this motif may serve as a precursor to a range of other significant functional groups such as aldehydes, ketones, amines, and amides. The traditional methods to install a nitrile group are the Sandmeyer or Rosenmund–von Braun reaction (Scheme 2.1) [2]. The Sandmeyer reaction uses an in situ generated diazonium salt with stoichiometric quantities of CuCN to install the nitrile group.

Parts of this chapter have been reproduced with permission from Yoon et al. [1].



H. Yoon, *Palladium and Nickel Catalyzed Transformations Forming Functionalized Heterocycles*, Springer Theses, https://doi.org/10.1007/978-3-030-54077-7_2

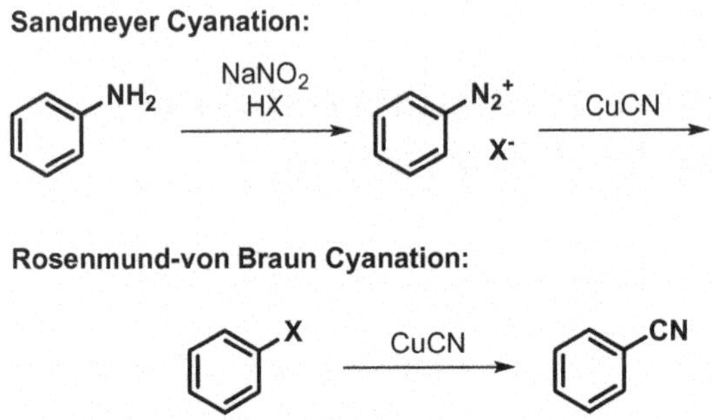


**Scheme 2.1** Traditional methods to forming aryl nitriles

Alternatively, the Rosenmund–von Braun reaction employs aryl halides in the presence of CuCN to give the respective aryl nitrile. Limitations with these methodologies include the necessity to use stoichiometric amounts of CuCN and elevated temperatures for aryl bromides and chlorides for the latter reaction.

In 2003, the Buchwald group alleviated some of the serious limitations associated with this methodology including the need for significantly higher reaction temperatures (150–280 °C) with aryl bromides while also employing catalytic amounts of copper [3]. Using a previously reported Cu/diamine catalyst system, the aryl nitriles were generated in excellent yields via the in situ generated aryl iodides from the respective aryl bromides (Scheme 2.2).

Although Cu-catalyzed cyanations have been most frequently reported, Pd-catalyzed cyanation provides complimentary routes to aryl and heteroaryl nitriles that are otherwise difficult to access [4]. Takagi et al. disclosed the first Pd-catalyzed cyanation using 2 mol% of $Pd(CN)_2$ and 2 equivalents of KCN to convert iodo- or bromobenzene to the corresponding benzonitrile (Scheme 2.3) [5].

Takagi and coworkers reported the preliminary mechanistic studies of the traditionally accepted Pd-catalyzed cyanation reaction and they also noted that in the

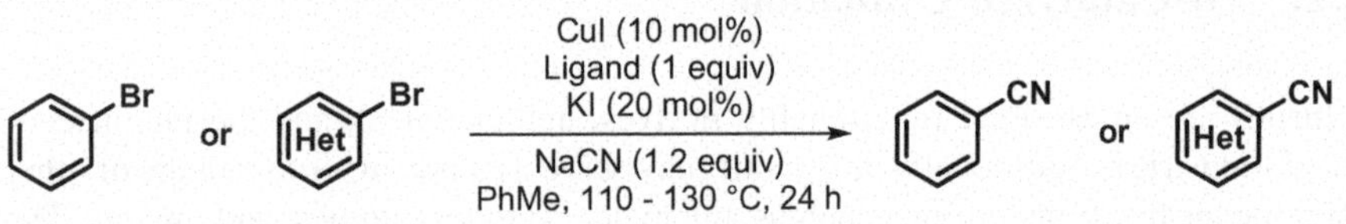


**Scheme 2.2** Cu-catalyzed Rosenmund-von Braun cyanation of aryl bromides via the in situ generated aryl iodides

**Scheme 2.3** Pd-catalyzed cyanation reported by the Sakakibara group

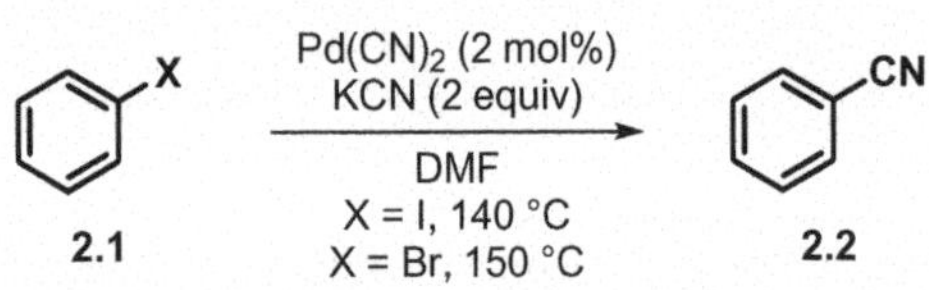

presence of excess cyanide, the reaction was inhibited (Scheme 2.4, catalytic cycle) [6]. Following these the studies, Grushin et al. were able to further probe the mechanism and isolate the complexes which were catalytic dead ends (Scheme 2.4, deactivation pathways) [7]. In the presence of water, the cyanide salts undergo hydrolysis to produce HCN which competes with the aryl halide for oxidative addition (**2.5**). In addition, catalyst inhibition was still observed in dry conditions via the generation of **2.6**. Thus the addition of HCN scavengers or reducing agents may greatly enhance the reactivity by regenerating the required $Pd^0$ catalyst. Inhibition was also observed among the on-cycle intermediates **2.3** and **2.4**. High concentration of $CN^-$ resulted in ligand exchange giving rise to the catalytically inactive species **2.7** [8].

Overcoming a major limitation associated with generating benzonitriles from aryl iodides or bromides, Jin and Confalone disclosed the first Pd-catalyzed cyanation of aryl chlorides (Scheme 2.5, A) [9]. Using a Pd/dppf catalyst and substoichiometric quantities of $Zn(CN)_2$, the authors were able to access electron rich and poor benzonitriles in excellent yields. In addition, Zn dust was found to be beneficial for the reaction as it restored the reactivity of the catalyst. Diverging away from toxic cyanide sources, the Beller group discovered that $K_4[Fe(CN)_6]$ was a competent transmetallating agent in the reaction [10]. $K_4[Fe(CN)_6]$ is considered to be a safe additive and is used in the food industry. Additionally, the salt is not hygroscopic and is considered to be easier to handle compared to other cyanide salts. Aryl bromide **2.10** was efficiently converted to the respective aryl nitrile **2.11** in excellent

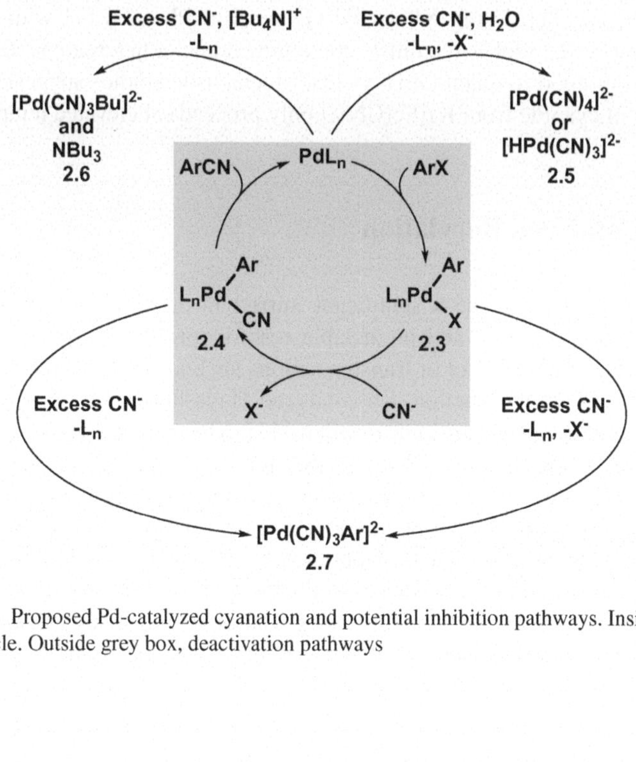


**Scheme 2.4** Proposed Pd-catalyzed cyanation and potential inhibition pathways. Inside grey box, catalytic cycle. Outside grey box, deactivation pathways

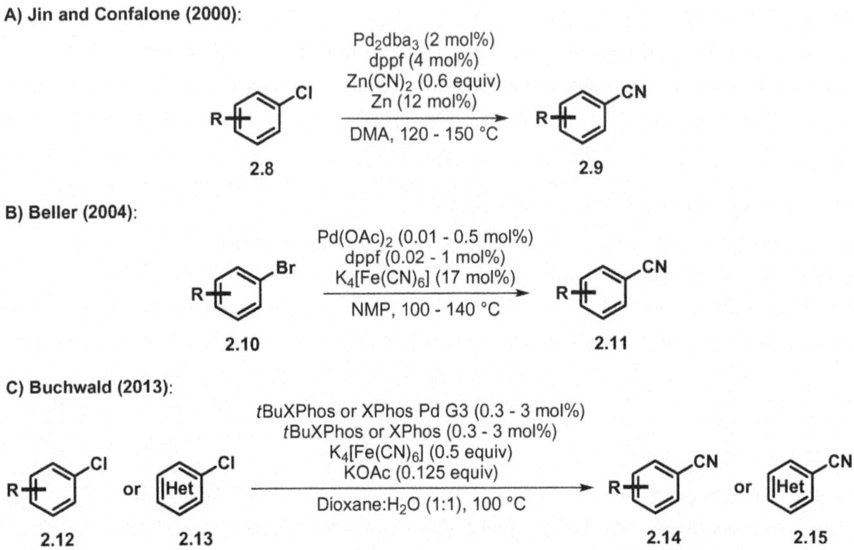


**Scheme 2.5** Advances in the Pd-catalyzed cyanation

yields. Complementary to their work, the Buchwald group were able to extend this methodology to reacting a range of aryl and heteroaryl chlorides with $K_4[Fe(CN)_6]$ (Scheme 2.5, C) [11]. Using XPhos Pd G3 or *t*BuXPhos Pd G3 with a catalyst loading as low as 0.2 mol%, the nitrile containing arene or heteroarene **2.14** or **2.15** were formed in good to quantitative yields. Mechanistic studies suggested that the dissociation of cyanide from $K_4[Fe(CN)_6]$ only proceeds at elevated temperatures.

## 2.3 Pd-Catalyzed Borylation

Organoboronates have made a significant impact in the field of transition metal catalysis due to their mild and predictable reactive profile. They have served as coupling partners in a myriad of transformations such as the Pd-catalyzed Suzuki-Miyaura cross coupling reaction, Rh-catalyzed Hayashi-Miyaura conjugate addition, and Cu-catalyzed Cham-Lam reaction [12]. The standard route to accessing organoboronates has been quenching aryllithium or arylmagnesium intermediates with a trialkyl borate (Scheme 2.6). However, this method poses many problems as it often requires cryogenic temperatures and it may be incompatible with various functional groups. Complimentary to this approach, the Pd-catalyzed borylation provides access to a range of complex boronates (Scheme 2.6).

The seminal report on the Pd-catalyzed borylation was made by Miyaura in 1995 where aryl bromides or iodides were successfully converted to the respective aryl

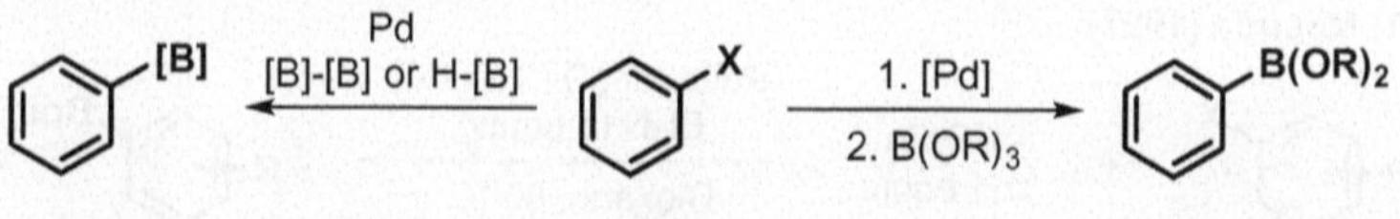


2.17 2.16 2.18

**Scheme 2.6** Routes to access aryl boronates

boronate **2.20** using bis(pinacolato)diboron ($B_2pin_2$) (Scheme 2.7, A) [13]. The mechanism of this transformation was later elucidated by Sakaki and coworkers via DFT studies (Scheme 2.7, B) [14]. The proposed mechanism proceeds via an oxidative addition and ligand exchange with KOAc to form the acetoxopalladium species **2.22**. The acetoxopalladium complex undergoes transmetallation with $B_2pin_2$ and subsequent reductive elimination to afford the aryl boronate **2.24**. The transmetallation of **2.22** with $B_2pin_2$ can be rationalized by the high reactivity of the Pd–O bond and the high oxophilicity of the boron center.

Concurrently, the Masuda group reported the Pd-catalyzed borylation using pinacolborane as the boron source (Scheme 2.8, A) [15]. The arylpalladium intermediate

**A) Miyaura (1995):**

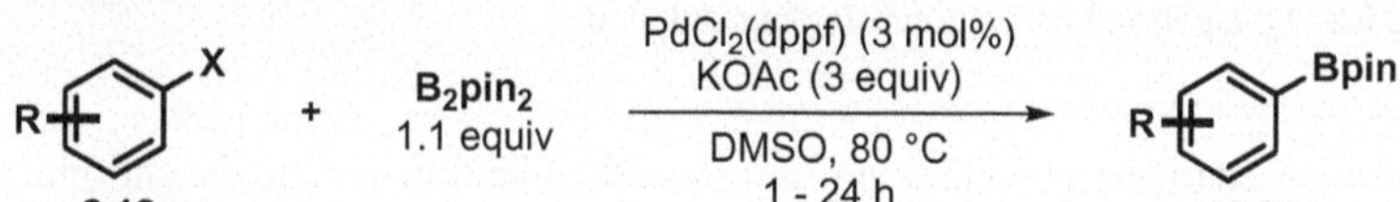


2.19
X = Br or I
2.20

**B) Proposed Mechanism:**

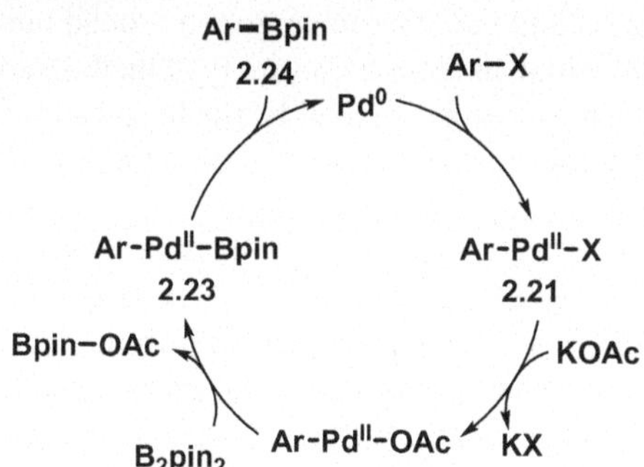


**Scheme 2.7** Pd-catalyzed Miyaura borylation and proposed mechanism

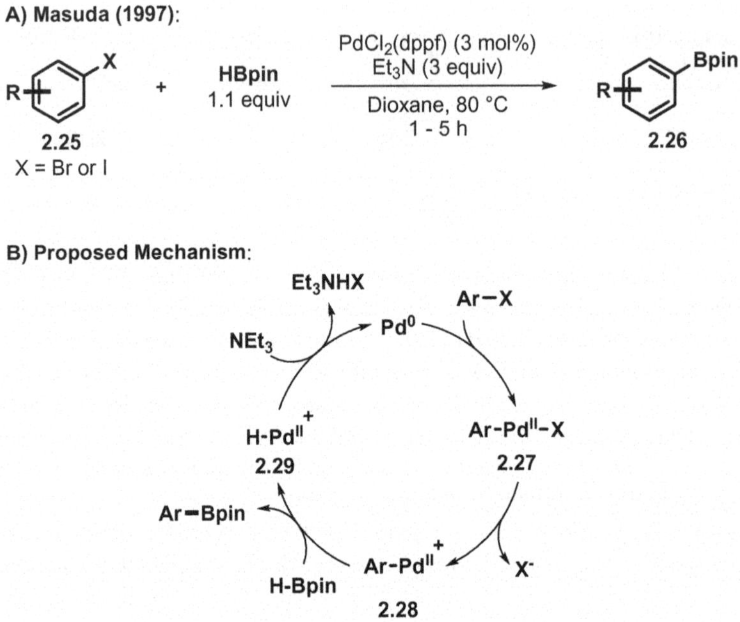


**Scheme 2.8** Pd-catalyzed borylation using pinacolborane

**2.27** is proposed to proceed via an amine assisted ionization to yield the cationic palladium species **2.28**. A σ-bond metathesis yields the aryl boronate and the $Et_3N$ regenerates the catalyst (Scheme 2.8, B) [16]. Although achieving an identical transformation, the mechanism of the two borylation reactions differ significantly in the formation of the C–B bond wherein one pathway undergoes a sequential transmetallation and reductive elimination and the other undergoes a σ-bond metathesis.

Following these two initial discoveries, a variety of methodologies were described showcasing advancements in this field. Notably, Buchwald and coworkers have found that using XPhos or SPhos, the borylation of aryl chlorides with $B_2pin_2$ or HBpin could be achieved respectively (Scheme 2.9, A) [17]. A limitation in the borylation reactions involve the use of pinacol as it contributes to >90% of the mass of the boron reagent and is often unwanted waste in the reaction. Addressing this concern, the Molander group reported a two step procedure to give the aryl trifluoroborate salts originating from aryl chlorides and bis-boronic acids (Scheme 2.9, B) [18]. Using a similar catalyst system disclosed by Buchwald, they were able to borylate the aryl chlorides to generate the aryl boronic acid and a subsequent reaction with $KHF_2$ yielded **2.35**.

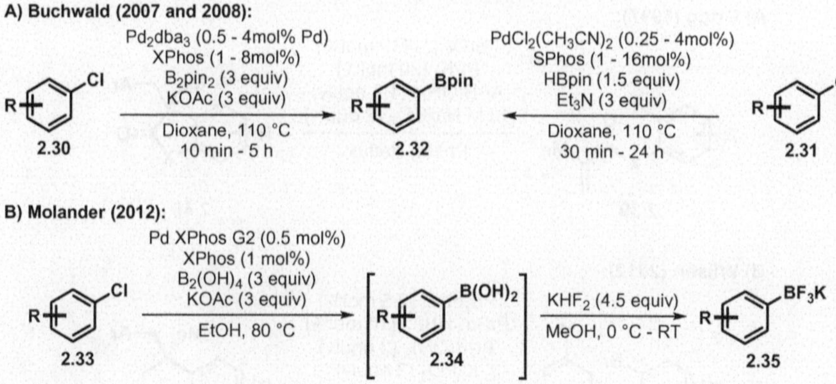


**Scheme 2.9** Recent advancements in the Pd-catalyzed borylation

## 2.4 Pd-Catalyzed Domino-Heck Anion Capture Cascade Reaction

As mentioned in Chap. 1, the intramolecular Pd-catalyzed Heck reaction is a powerful transformation which allows the formation of complex heterocycles in a singular step. Typically, aryl halide **2.36** undergoes oxidative addition and carbopalladation to form the alkylpalladium intermediate **2.37**. In the absence of β-hydrogens and in the presence of a compatible trapping agent, the Pd-catalyzed Heck reaction progresses to become the Pd-catalyzed anion capture cascade reaction (Scheme 2.10) [19].

Since the pioneering work by Grigg and coworkers, the anion capture cascade reaction has been used to furnish complex heterocycles while terminating with a variety of nucleophiles [20].

### 2.4.1 *Pd-Catalyzed Domino-Heck Arylation*

Using boronic acids as a compatible coupling partner, the Pd-catalyzed domino-Heck arylation was first disclosed by Grigg in 1997 (Scheme 2.11, A) [21]. Reacting

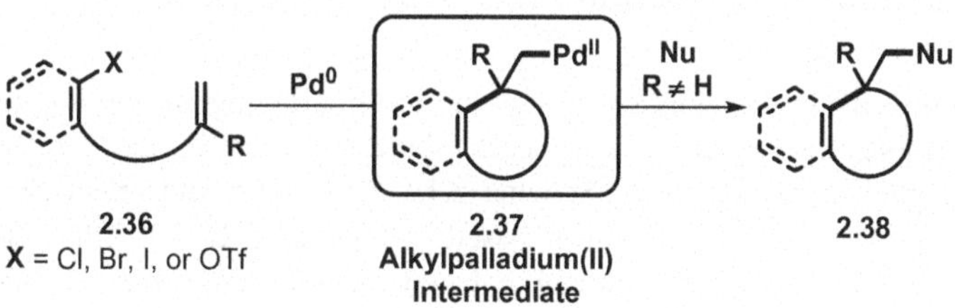


**Scheme 2.10** Pd-catalyzed domino-Heck anion capture cascade reaction

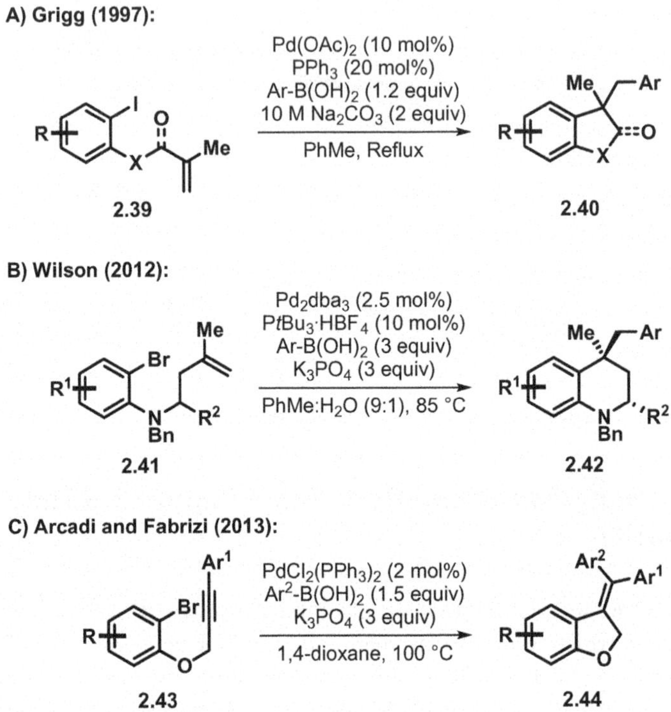


**Scheme 2.11** Pd-catalyzed domino-Heck arylation sequences

iodoaryl ethers or iodoacrylamides with aryl and heteroaryl boronic acids yielded the benzofurans and oxindoles. Similarly, Wilson from Merck reported the synthesis of arylated tetrahydroquinolines **2.42** as single diastereomer in excellent yields (Scheme 2.11, B) [22]. Complimentary to these methods, Arcadi and Fabrizi reported a *syn*-selective Pd-catalyzed cyclization of bromoaryl propargyl ether **2.43** and boronic acids. Utilizing $PdCl_2(PPh_3)_2$ in the transformation, the authors were able to generate **2.44** as a single stereoisomer (Scheme 2.11, C) [23].

### *2.4.2 Pd-Catalyzed Domino-Heck Direct Arylation*

As opposed to using organoboronates as the terminating agent, the Fagnou group reported a more atom economical process where the direct arylation of the alkyl-palladium species was achieved (Scheme 2.12, A) [24]. A site selective arylation was achieved when employing sulfur based heterocycles. In 2015, the Zhu group extended this methodology by publishing the enantioselective variant (Scheme 2.12, B) [25]. They showed that by using a variant of the Pfaltz ligands and starting with the aryl triflate **2.45**, arylated oxindole **2.48** was formed in good yield and excellent

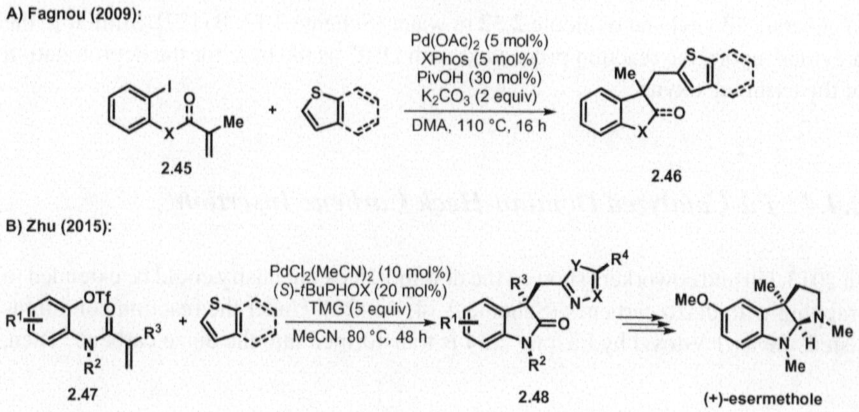


**Scheme 2.12** Pd-catalyzed domino-Heck direct arylation sequences

ee. Using this method, they were able to achieve an asymmetric total synthesis of (+)-esermethole.

## 2.4.3 *Pd-Catalyzed Domino-Heck Sonogashira Reaction*

In 2002, Stará and Starý reported the Pd-catalyzed domino-Heck Sonogashira reaction of iodoaryl propargyl ethers (Scheme 2.13, A) [26]. The authors found that in the presence of copper, the direct Sonogashira coupling occurred. The domino reaction proceeded in the absence of copper and minimized steric congestion on the tethered alkyne. Alternatively, in 2016, Guo and coworkers discovered a Pd-catalyzed method

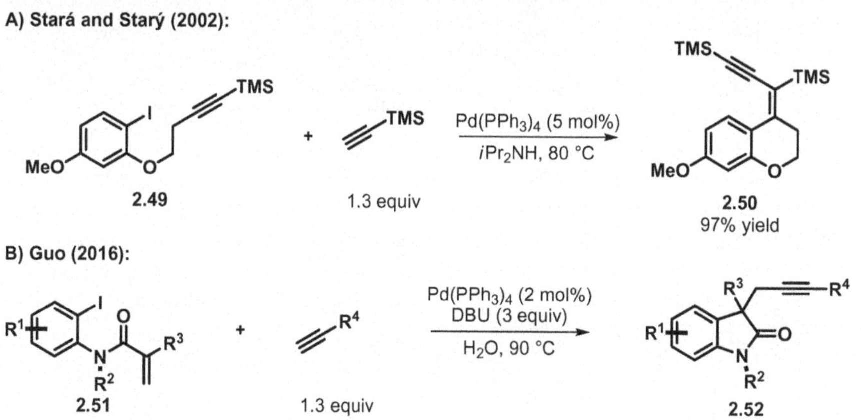


**Scheme 2.13** Pd-catalyzed domino-Heck Sonogashira reactions

to generate alkynylated oxindole **2.52** in water (Scheme 2.13, B) [27]. Similar to the previous report, the reaction proceeded with DBU as the base for the deprotonation of the terminal alkyne.

### 2.4.4 *Pd-Catalyzed Domino-Heck Carbene Insertion*

In 2013, Gu and coworkers showed the domino-Heck chemistry could be extended to trapping with diazo carbenes (Scheme 2.14, A) [28]. Under the reaction conditions listed, the aryl *N*-tosyl hydrazone **2.54** is transformed into the diazo carbene. Then,

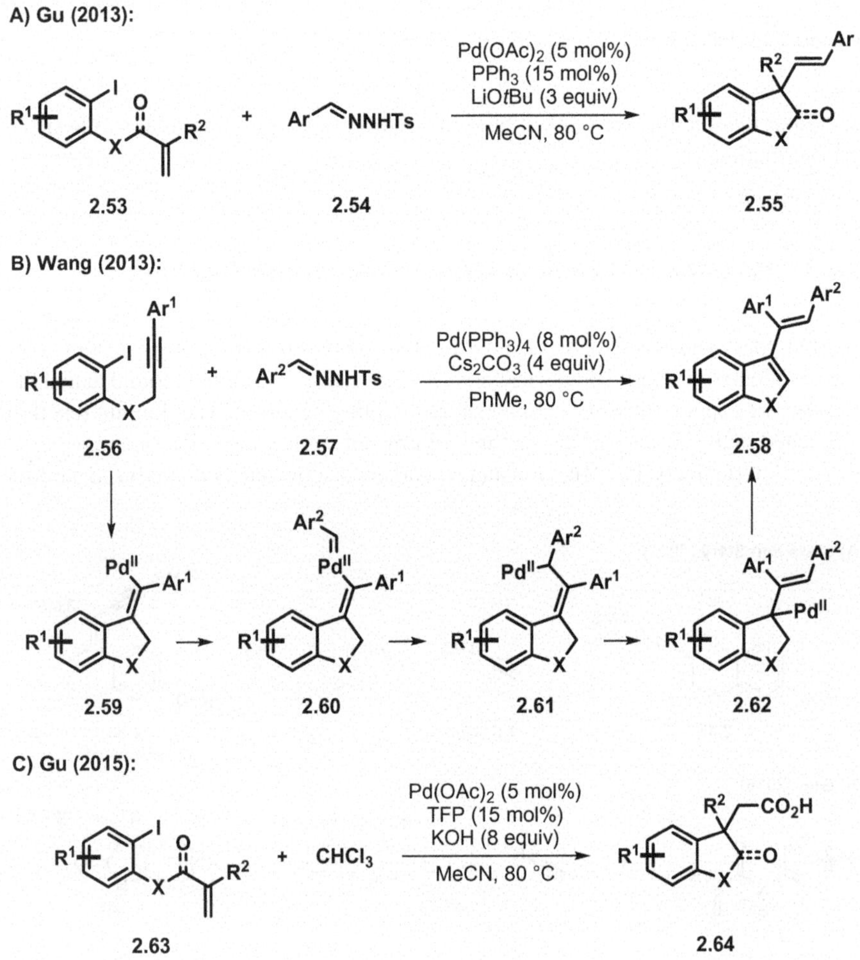


**Scheme 2.14** Pd-catalyzed domino-Heck carbene insertion reactions

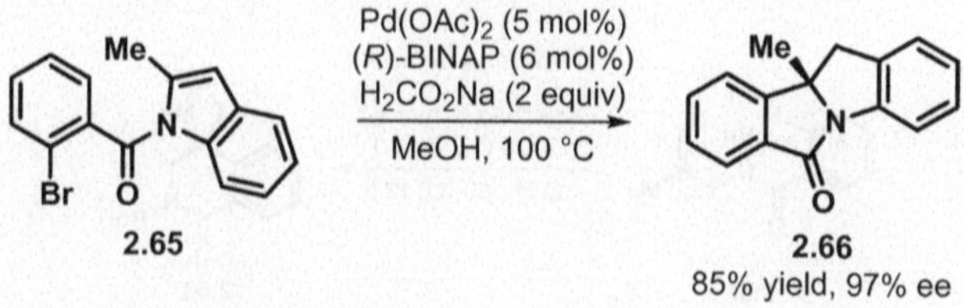


**Scheme 2.15** Asymmetric Pd-catalyzed dearomative reductive Heck reaction

nitrogen is released and the alkylpalladium intermediate undergoes a 1,1 insertion and β-hydride elimination to give **2.55**. Concurrently, Zhang and Wang reported the synthesis of 3-vinyl benzofurans and indoles (Scheme 2.14, B) [29]. The mechanism proceeds through a 1,1 insertion of the carbene into the vinylpalladium intermediate **2.59**. The generated π-allyl palladium species **2.61** isomerizes and undergoes β-hydride elimination to form heterocycle **2.58**. As opposed to using hydrazones as carbene surrogate, the Gu group reported the synthesis of neopentyl carboxylic acids **2.64** originating from chloroform (Scheme 2.14, C) [30]. This method provided a safer and alternative route to installing carboxylic acids.

### *2.4.5 Pd-Catalyzed Dearomative Heck Reaction*

Although the early discoveries were performed by Grigg and coworkers, the Jia group rejuvenated interest into dearomative Pd-catalyzed cyclizations by enantioselectively forming indolines via a reductive Heck process (Scheme 2.15) [31]. Following this work, our group reported the highly diastereoselective dearomative cyanation [32] and arylation [33] while the Jia group reported the enantioselective alkynylation [34].

### *2.4.6 Pd-Catalyzed Domino Cyanation Reactions*

Since the original report by Torii in 1992, the Pd-catalyzed domino cyanation reaction gained significant interest. In the following year, Grigg and coworkers showed that the reaction is applicable in an intramolecular sequence (Scheme 2.16, A) [35]. In the presence of 18-crown-6 and KCN, alkyl nitrile **2.68** was formed in moderate yield. The Zhu group further advanced the domino-Heck cyanation chemistry by employing $K_4[Fe(CN)_6]$ to generate oxindole **2.70** and found Difluorophos gave moderate enantioselectivity (Scheme 2.16, B) [36]. Using this method as the key transformation, (±)-physostigmine and (±)-horsfiline were successfully synthesized.

Alternative methods to introduce nitrile moieties while avoiding the use of toxic cyanide sources have been explored. In 2006, Takemoto and coworkers described the intramolecular Pd-catalyzed cyanoamidation to form lactams or oxindoles starting

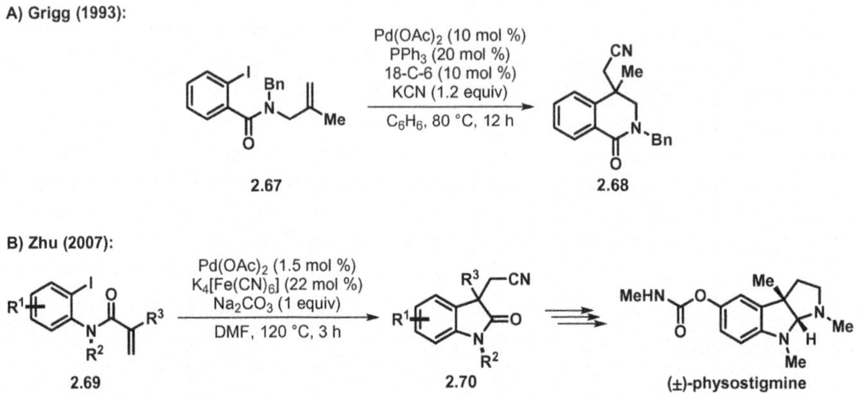


**Scheme 2.16** Pd-catalyzed domino-Heck cyantion reactions

from cyanoformamide **2.71** (Scheme 2.17, A) [37]. Employing this method, a mixture of both the *E* and *Z* isomers were detected. Later in 2008, the Takemoto group found that using a combination of the Feringa ligand with added DMPU, oxindole **2.74** was synthesized quantitative yield and good e.r. (Scheme 2.17, B) [38]. Although the mentioned methods generate similar products as the traditional anion capture cascade reaction, these two approaches represent a perfectly atom economical process eliminating the necessity of starting from an aryl halide.

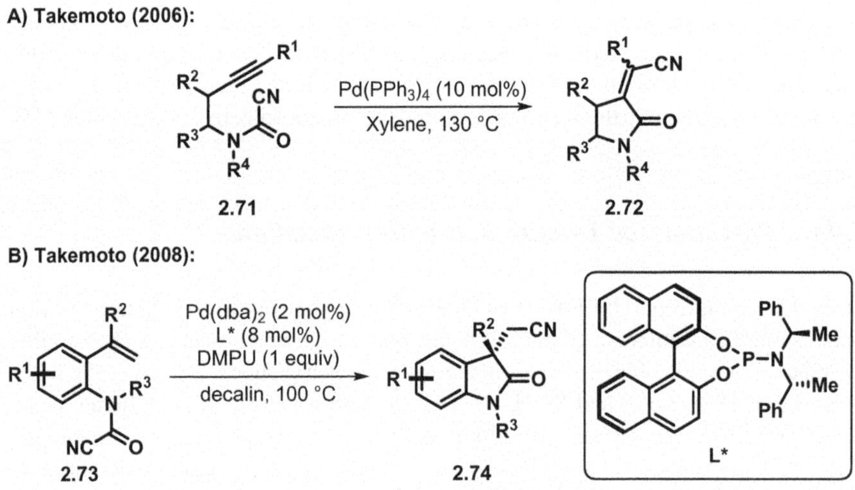


**Scheme 2.17** Intramolecular Pd-catalyzed cyanoformidation reactions

### 2.4.7 Pd-Catalyzed Borylation Reactions

Since the initial Pd-catalyzed borylation reaction in 1995, borylation has been studied and applied in a range of domino reactions generating carbo and heterocycles. Cárdenas reported a Pd-catalyzed borylative cyclization of 1,6-enyne **2.75** (Scheme 2.18, A) [39]. The method was proposed to proceed through a cyclometallation or a stepwise hydropalladation and carbopalladation. Interestingly, the authors proposed that the generated exocyclic double bond coordinates to the Pd-center to prevent β-hydride elimination and allow reductive elimination of the C–B bond. Later in 2011, the Bäckvall group focused their attention on the oxidative carbocyclization–borylation reaction of enallenes (Scheme 2.18, B) [40]. Using benzoquinone as the external oxidant, they were able also able to achieve borylation of carbon atoms bearing β-hydrogens. They further realized that chiral Brønsted acids were crucial to attaining high enantioselectivity [41]. Although the Pd-catalyzed borylation reaction was realized in 1995, the Pd-catalyzed domino-Heck variant did not appear until 2015 when the Van der Eycken group reported the synthesis of borylated oxindoles **2.80** (Scheme 2.18, C) [42].

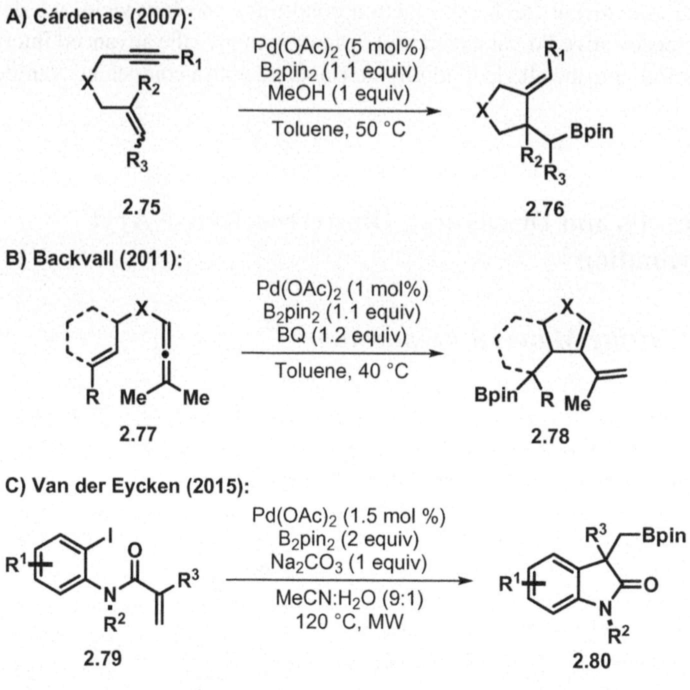


**Scheme 2.18** Pd-catalyzed domino borylation reactions

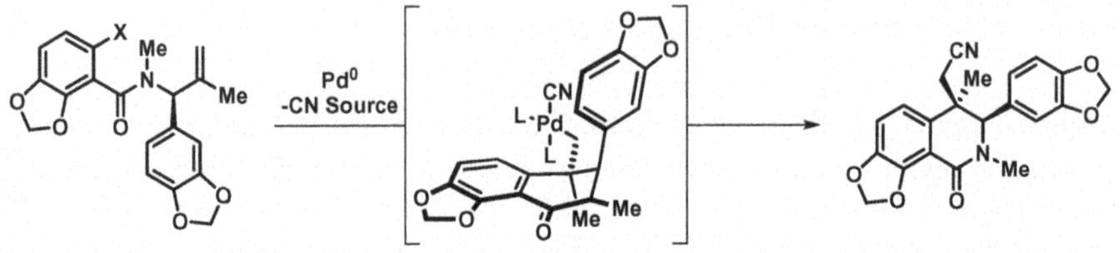


2.81 2.82 2.95j

**Scheme 2.19** Proposed diastereoselective Pd-catalyzed aryl cyanation

## 2.5 Research Goal: Part 1

Previously, we disclosed the formal synthesis of (+)-corynoline that involved carboiodination as the key step. In an effort to improve our approach, we sought to address some of the shortcomings of the synthesis. A major issue arises from the nucleophilic displacement of the alkyl iodide to generate the respective nitrile. The reaction was moderate yielding and required high temperature, long reaction times, super stoichiometric quantities of KCN and 18-crown-6 (56% yield over 2 steps). As a method to circumvent the harsh reaction conditions and low yield, we developed the diastereoselective Pd-catalyzed aryl cyanation to give the advanced intermediate **1.106g** by trapping the alkylpalladium intermediate with a competent cyanide source (Scheme 2.19).

## 2.6 Results and Discussion: Diastereoselective Aryl Cyanation

### *2.6.1 Starting Material Preparation*

The enantioenriched carboxamides were prepared in a similar route described in Chap. 1. The key difference in the two routes is the use of 2-bromobenzoic acids in the Schotten-Baumann amide coupling (Scheme 2.20).

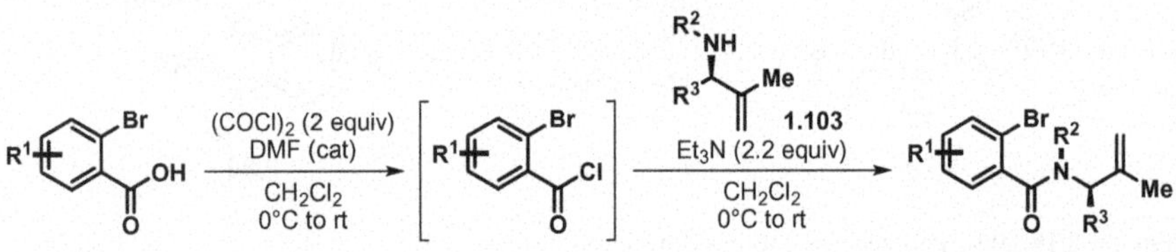


2.83 2.84 2.85

**Scheme 2.20** General route to the enantioenriched 2-bromoaryl carboxamides

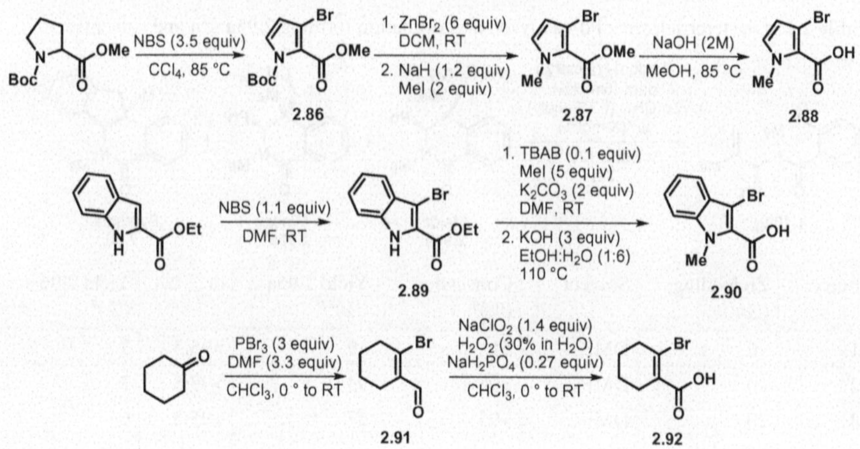


**Scheme 2.21** Synthetic routes to access bromoheterocycles and the vinyl bromide

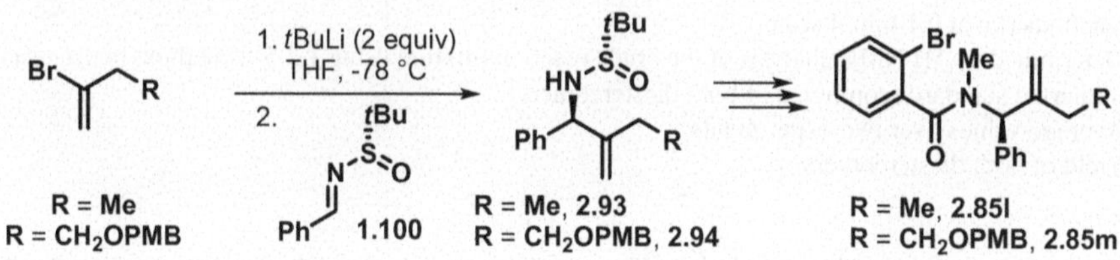


**Scheme 2.22** Synthetic route to access carboxamides with varied vinyl substituents

Some of the heterocycles and the vinyl bromide **2.92** were accessed in a straightforward fashion (Scheme 2.21). Sequential bromination, –Boc deprotection, methylation and hydrolysis of Boc-proline-methyl ester generated pyrrole **2.88**. Similarly, **2.90** was acquired from bromination, methylation and hydrolysis of indole methyl ester. The vinylbromide **2.92** was accessed from a formylation, bromination and Pinnick oxidation.

Additionally, the substrates with substituents other than $–CH_3$ were prepared via lithiation of the respective vinylbromides and diastereoselective addition at −78 °C. The addition products (**2.93** and **2.94**) underwent the subsequent general procedures to form carboxamides **2.85l** and **2.85m** (Scheme 2.22).

## *2.6.2 Optimization*

Based on the literature precedence for the Pd-catalyzed cyanation of aryl halides, the enantioenriched amide **1.106g** was subjected to $Pd(dba)_2$, dppf, and $Zn(CN)_2$ in DMF at 120 °C to give the alkyl nitrile **2.95a** in moderate yield and excellent dr

**Table 2.1** Diastereoselective Pd-catalyzed aryl cyanation forming **2.95a**: Zn and solvents

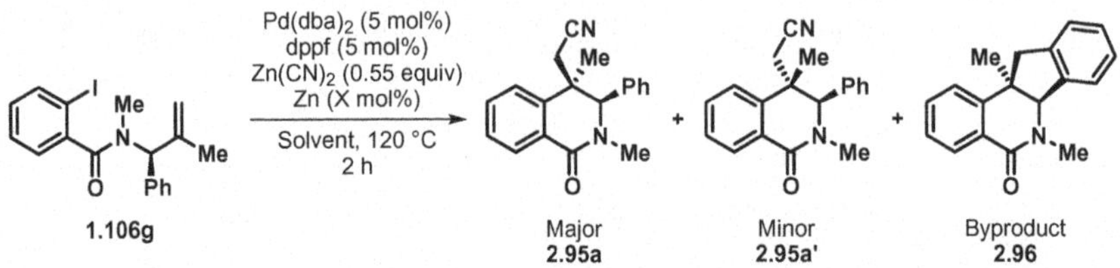


| Entry | Zn loading (%) | Solvent | Conversion (%)[a] | Yield 2.95a (%)[a,c] | dr | Yield 2.96 (%)[a] |
|---|---|---|---|---|---|---|
| 1 | 0 | DMF | 51 | 49 | >95:5 | 7 |
| 2[b] | 10 | DMF | >95 | 93 | >95:5 | 2 |
| 3 | 20 | DMF | >95 | 87 | >95:5 | 6 |
| 4 | 10 | PhMe | – | – | – | – |
| 5 | 10 | Dioxane | <5 | – | – | – |
| 6 | 10 | DMA | 7 | – | – | – |

Reactions run at 0.1 mmol scale
[a]Determined by $^{1}$H NMR analysis of the crude reaction mixture using 1,3,5-trimethoxybenzene as an internal standard (combined of both diastereomers)
[b]Average values over two experiments
[c]Yield of both diastereomers

(Table 2.1, 49% yield, > 20:1 dr). The reaction proceeded in moderate conversion and byproduct **2.96** was found in low yield. The formation of **2.96** is attributed to the direct C–H functionalization of the major diastereomer of the alkylpalladium intermediate which may lead to artificially lowered dr. Catalytic amounts of zinc powder were added and gratifyingly, the desired product was obtained in excellent yield and dr (entry 2 and 3). Performing the reaction in PhMe, dioxane and DMA led to trace conversion or no product formation (entry 4–6).

Afterwards, a range of ligands and Pd-sources were screened (Table 2.2). During this time, we found that performing the reaction on 0.1 mmol scale gave inconsistent results and scaling up to 0.3 mmol scale alleviated the issue. Upon the scale up, dppf was found to be nonoptimal for the transformation as a significant amount of byproduct **2.96** was present (Entries 1–3). Preformed catalysts were also screened and $Pd(PtBu_3)_2$ generated **2.95a** in the highest yield albeit in slightly lower dr (84% yield, 88:12 dr, entry 6). Employing the less reactive aryl bromide **2.85a** furnished the desired product in higher dr (entry 7).

The reaction was further optimized to give **2.95a** in 90% yield and 95:5 dr (Table 2.3, entry 1). In addition, individual parameters were manipulated from the optimized conditions to gain further insight into the reaction. The addition of zinc powder was found to give comparable yields but lower dr (entry 2). Other bulky Pd sources such as $Pd(QPhos)_2$ or *t*BuXPhos Pd G2 led to lower yield and dr (entries 3 and 4). Dioxane and PhMe were inferior solvents in the reaction. This may be due to the lower solubility of $Zn(CN)_2$ in these solvents (entries 5 and 6). Replacing $Zn(CN)_2$

**Table 2.2** Diastereoselective Pd-catalyzed aryl cyanation forming **2.95a**: Catalysts

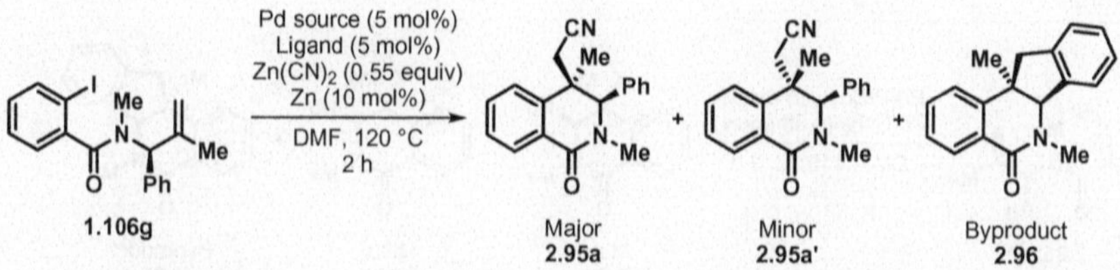


| Entry | Pd source (5 mol% Pd) | Ligand | Conversion (%)[a] | Yield 2.95a (%)[a,b] | dr | Yield 2.96 (%)[a] |
|---|---|---|---|---|---|---|
| 1 | $Pd(dba)_2$ | dppf | >95 | 63 | >95:5 | 16 |
| 2 | $Pd_2(dba)_3$ | dppf | >95 | 74 | >95:5 | 15 |
| 3 | $Pd_2(dba)_3{\cdot}CH_3Cl$ | dppf | >95 | 61 | >95:5 | 15 |
| 4 | $Pd(dppf)Cl_2$ | – | >95 | 36 | >95:5 | – |
| 5 | $Pd(PtBu_2Ph)_2$ | – | >95 | 66 | 89:11 | – |
| 6 | $Pd(PtBu_3)_2$ | – | >95 | 84 | 88:12 | – |
| 7[c] | $Pd(PtBu_3)_2$ | – | >95 | 84 | 92:8 | – |

Reactions run at 0.1 mmol scale
[a]Determined by $^1$H NMR analysis of the crude reaction mixture using 1,3,5-trimethoxybenzene as an internal standard (combined of both diastereomers)
[b]Yield of both diastereomers
[c]Aryl bromide **2.85a** was used the reaction

with the less toxic equivalent $K_4[Fe(CN)_6]$ generated **2.96** exclusively (entry 7). A plausible explanation is that the rate of cyanide transfer using $K_4[Fe(CN)_6]$ is slower and the presence of $Na_2CO_3$ accelerates the direct arylation. Aryl iodides and chlorides generated **2.95a** in lower yield; however, the addition of Zn powder restored the reactivity of the aryl iodide (entries 8–10). Lastly, performing the reaction at 90 °C resulted in slower formation of **2.95a** while having marginal effects on the dr (entry 11).

### 2.6.3 *Substrate Scope*

Having found the optimal reaction conditions, the scope of the aryl- and heteroaryl cyanation was explored. Of note, all of the enantioenriched alkylnitriles were obtained with no erosion of er (98:2 to > 99:1). *Ortho*- substituted substrate **2.85b** was tolerated under the standard reaction conditions, yielding the desired product **2.95b** in excellent yield and dr, suggesting that the catalytic system is not sensitive to sterically encumbered aryl halides. Dihalogenated carboxamide **2.85c** underwent the cyclization affording the desied chlorodihydroisoquinolinone **2.95c** in 89% yield

**Table 2.3** Diastereoselective Pd-catalyzed aryl cyanation forming **2.95a**: variation from optimized conditions

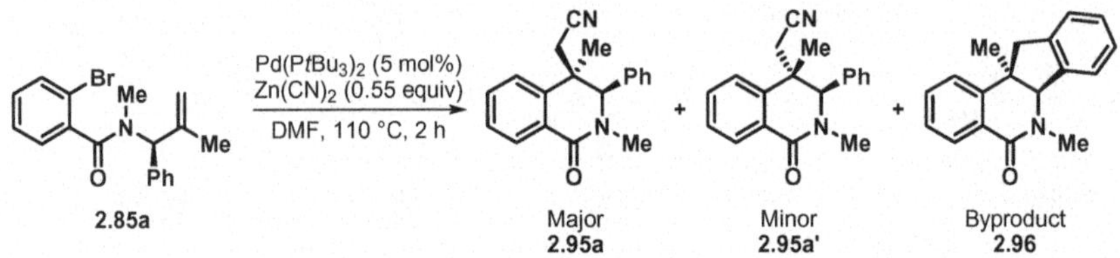


| Entry | Variation from "optimized conditions" | Yield 2.95a (%)[a,b] | dr | Yield 2.96 (%)[a] |
|---|---|---|---|---|
| 1[c] | None | (90) | 95:5 | – |
| 2 | Zn powder (5 mol%) | 89 | 93:7 | – |
| 3 | Pd(QPhos)-2 instead of Pd($PtBu_3$)$_2$ | 69 | 90:10 | – |
| 4 | *t*BuXPhos Pd G2 instead of Pd($PtBu_3$)$_2$ | 4 | 94:6 | – |
| 5 | Dioxane instead of DMF | 11 | 75:25 | – |
| 6 | PhMe instead of DMF | – | – | – |
| 7 | $K_4[Fe(CN)_6]$ instead of $Zn(CN)_2$ | – | – | 32 |
| 8 | ArI instead of ArBr | 55 | 93:7 | – |
| 9[d] | ArI instead of ArBr (Zn power added) | 77 | 88:12 | – |
| 10 | ArCl instead of ArBr | 57 | 94:6 | – |
| 11 | 90 °C instead of 110 °C | 73 | 96:4 | – |

Reactions were run on 0.3 mmol scale
[a]Determined by $^1$H NMR analysis of the crude reaction mixture using 1,3,5-trimethoxybenzene as an internal standard (combined of both diastereomers)
[b]Yield of both diastereomers
[c]Yield in parentheses are isolated yields
[d]10 mol% of Zn power was added

and 94:6 dr, suggesting that a selective di-functionalization process is viable. Additionaly, in the excess of $Zn(CN)_2$, a tandem aryl cyanation and direct cyanation of both arylhalides was observed (**2.95c'**). Changing from the methyl to benzyl *N*-protecting group resulted slightly diminished selectivity (**2.95d**, 91:9 dr). Chloro- and trifluoromethyl-substituted aryl bromides **2.85e** and **2.85f** furnished the desired products in 70 and 78% yields with 91:9 and 89:11 dr, respectively, demonstrating the tolerance of electron deficient functional groups. Substituents on the allylic aromatic group (**2.85g–2.85i**) were tolerated and the corresponding dihydroisoquinolinones (**2.95g–2.95i**) were obtained in high yields with good diastereoselectivity. (+)-Corynoline precursor **2.85j** was cyclized in 77% yield with 92:8 dr under the standard conditions. Gratifyingly, on gram scale (4.75 mmol), **2.95j** was realized in 79%

**Table 2.4** Diastereoselective Pd-catalyzed aryl cyanation

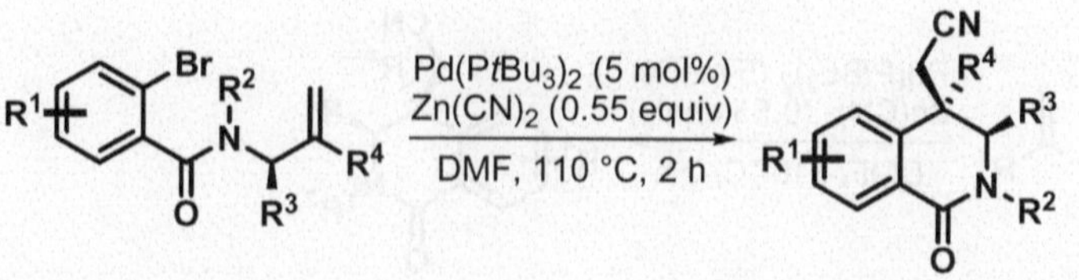


2.85 2.95

| Entry | Substrate | Product | Yield (%)[a,b,c] |
|---|---|---|---|
| 1 | 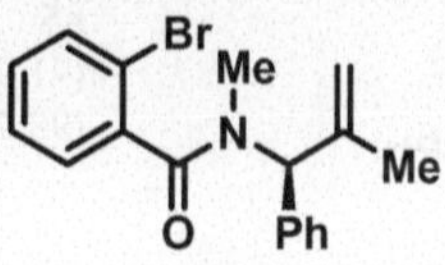 2.85a | 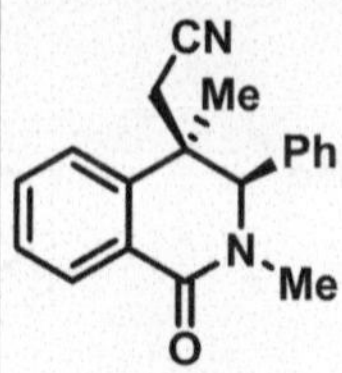 2.95a | 90 (95:5 dr) (>99:1 er) |
| 2 | 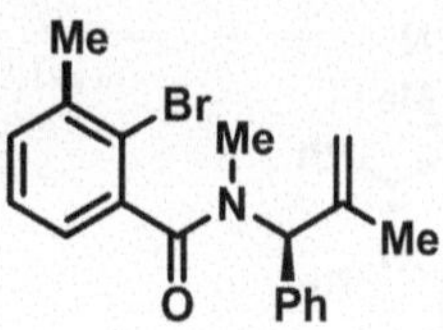 2.85b | 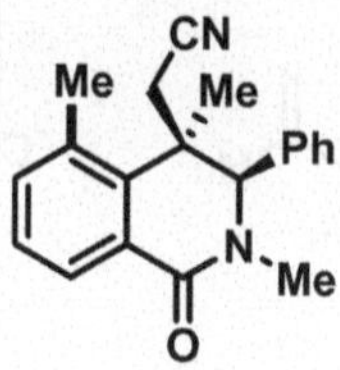 2.95b | 89 (>95:5 dr) (>99:1 er) |
| 3 | 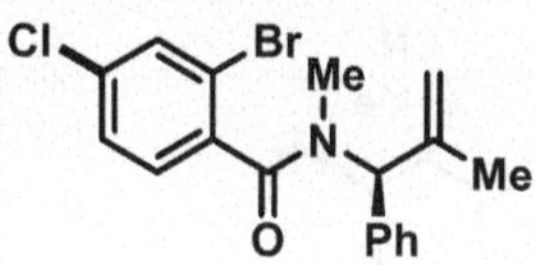 2.85c | 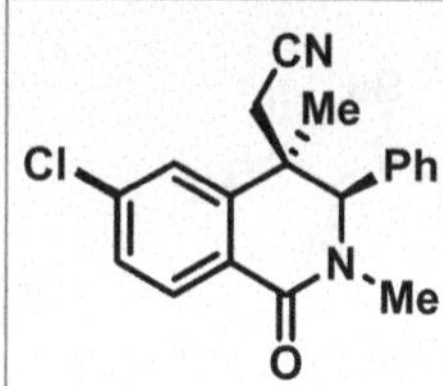 2.95c | 89 (94:6 dr) (99:1 er) |
| 4[d] | 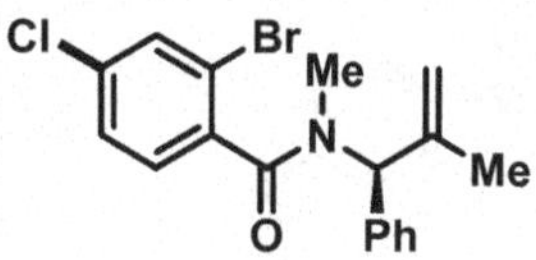 2.85c | 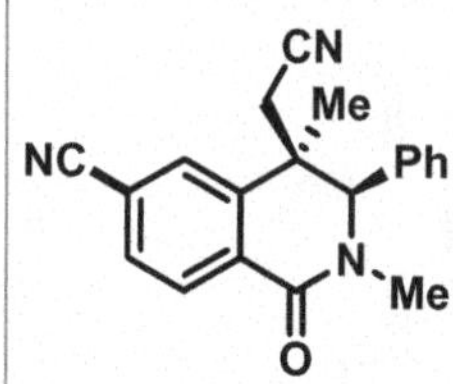 2.95c' | 67 (95:5 dr) (99:1 er) |

(continued)

**Table 2.4** (continued)

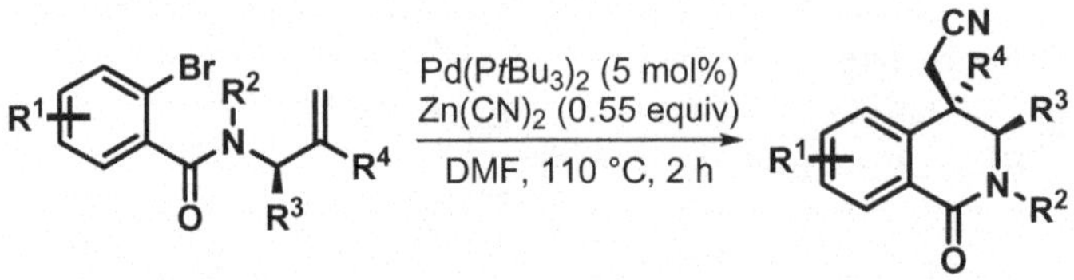


**2.85** **2.95**

| Entry | Substrate | Product | Yield (%)[a,b,c] |
|---|---|---|---|
| 5 | 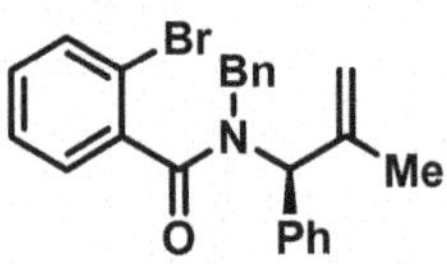 **2.85d** | 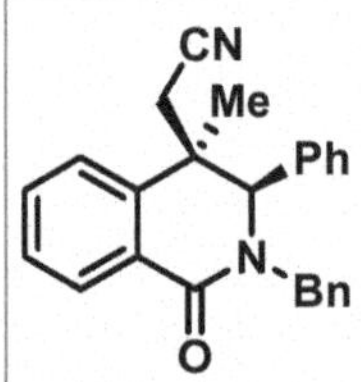 **2.95d** | 85 (91:9 dr) (98:2 er) |
| 6 | 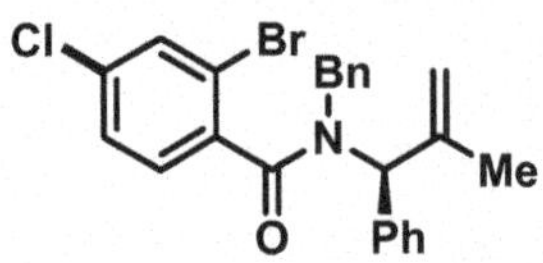 **2.85e** | 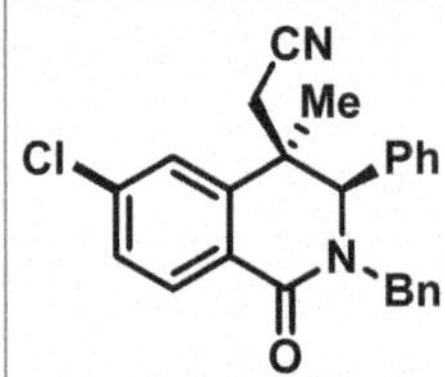 **2.95e** | 70 (91:9 dr) (>99:1 er) |
| 7 | 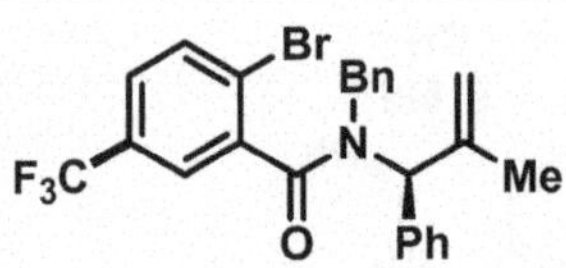 **2.85f** | 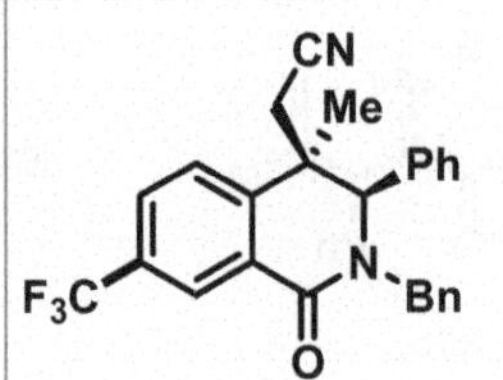 **2.95f** | 78 (89:11 dr) (99:1 er) |
| 8 | 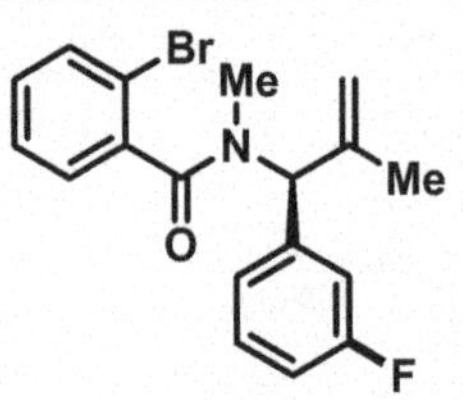 **2.85g**[f] | 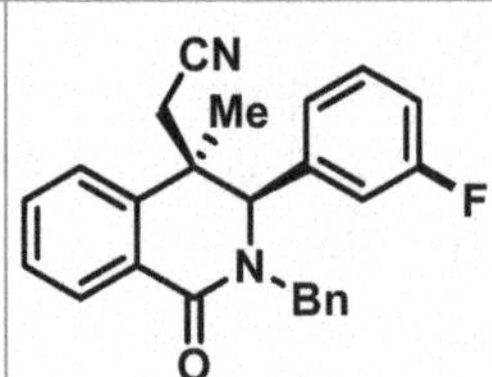 **2.95g**[f] | 93 (90:10 dr) (99:1 er) |

(continued)

**Table 2.4** (continued)

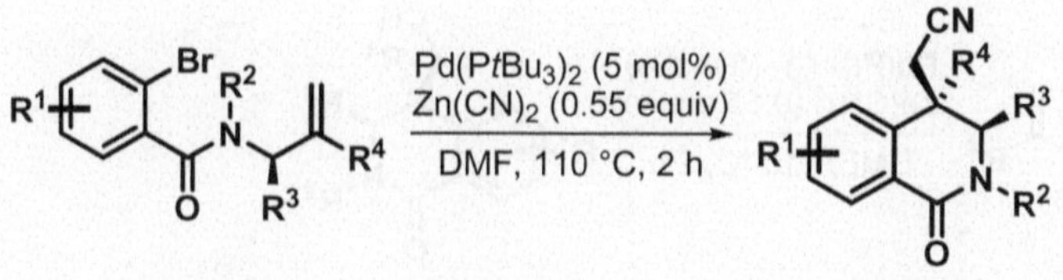


**2.85** **2.95**

| Entry | Substrate | Product | Yield (%)[a,b,c] |
|---|---|---|---|
| 9 | 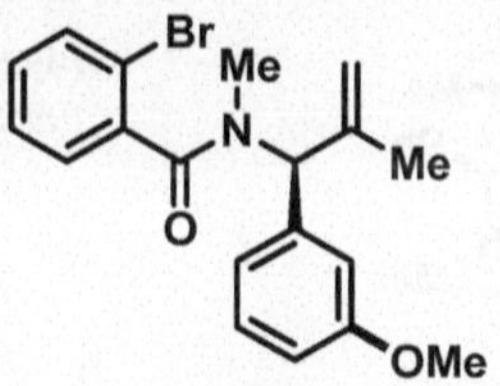 **2.85h** | 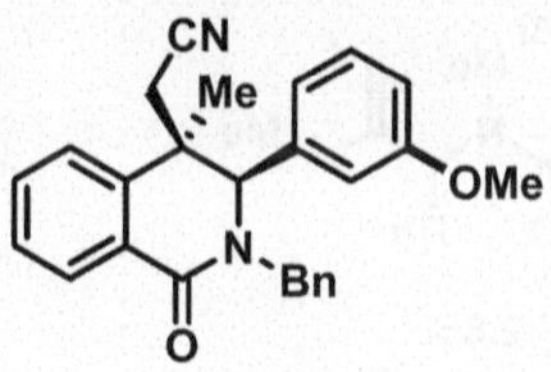 **2.95h**[f] | 86 (90:10 dr) (>99:1 er) |
| 10 | 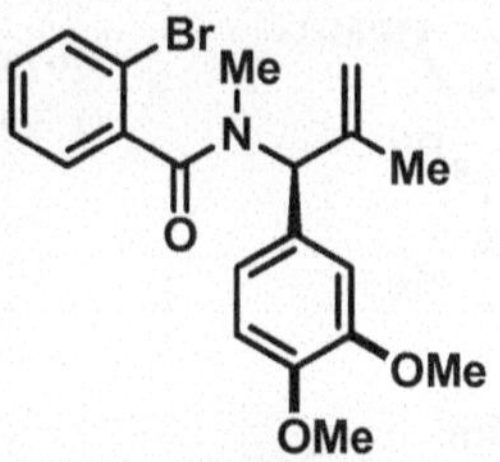 **2.85i**[f] | 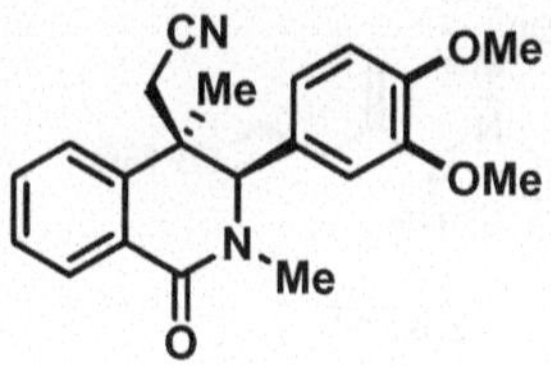 **2.95i**[f] | 94 (94:6 dr) (99:1 er) |
| 11 | 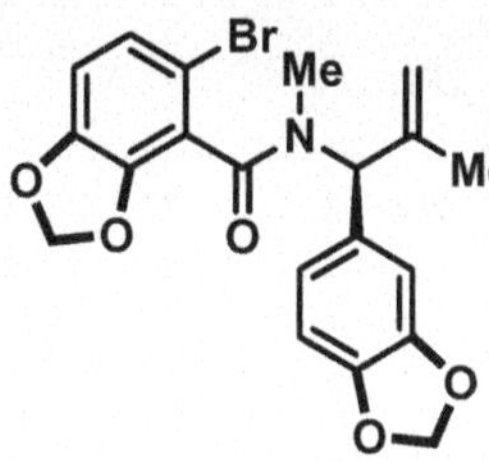 **2.85j**[f] | 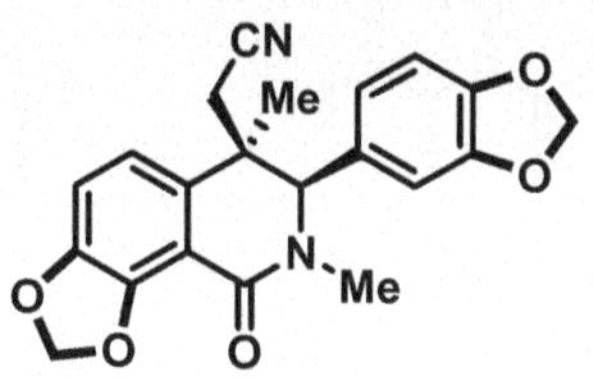 **2.95j**[f] | 5 mol% Pd 77 (92:8 dr) (>99:1 er) 2.5 mol% Pd[e] 79 (>95:5 dr) (>99:1 er) |
| 12 | 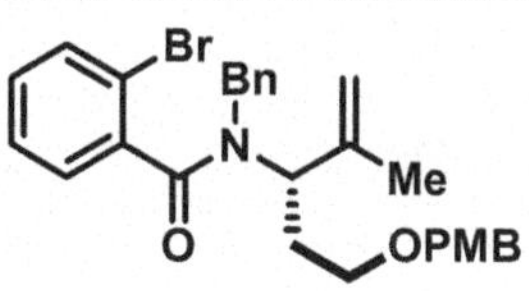 **2.85k** | 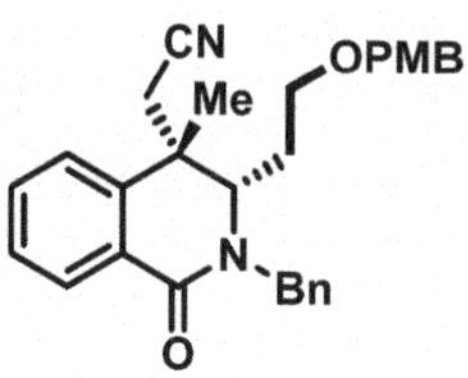 | 66 (53:47 dr) (>99:1 er) |

(continued)

**Table 2.4** (continued)

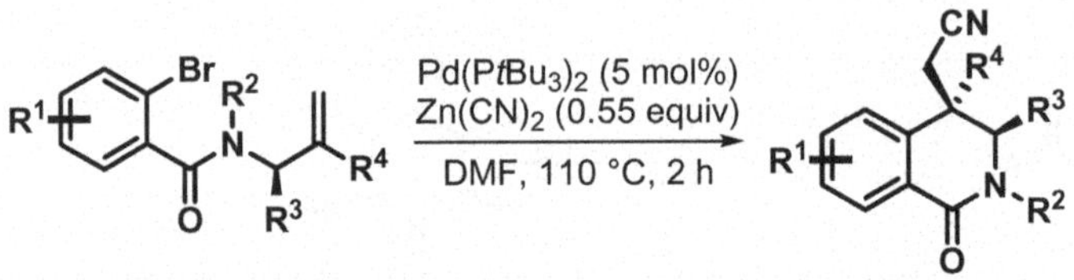


2.85 2.95

| Entry | Substrate | Product | Yield (%)[a,b,c] |
|---|---|---|---|
| 13 | 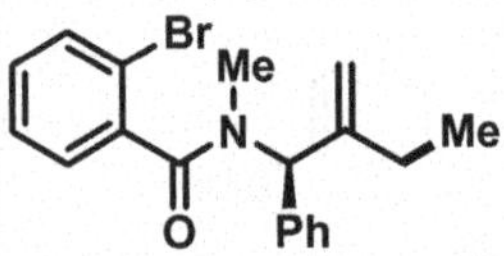 2.85l[f] | 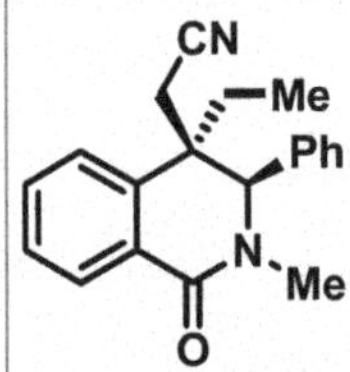 2.95l[f] | 73 (95:5 dr) (99:1 er) |
| 14 | 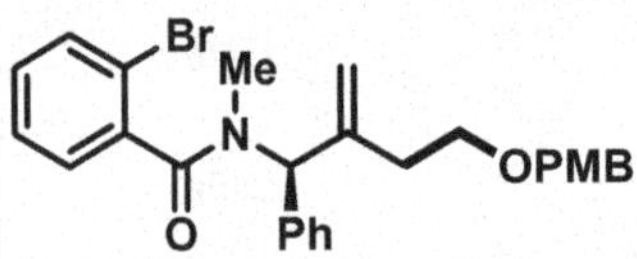 2.85m[f] | 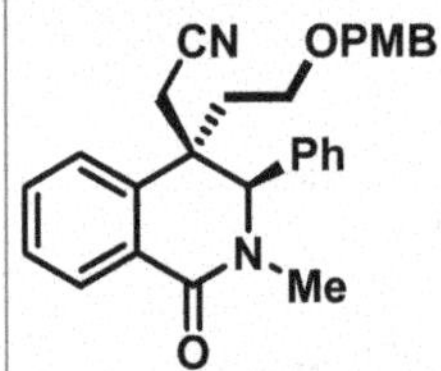 2.95m[f] | 77 (92:8 dr) (>99:1 er) |
| 15[f] | 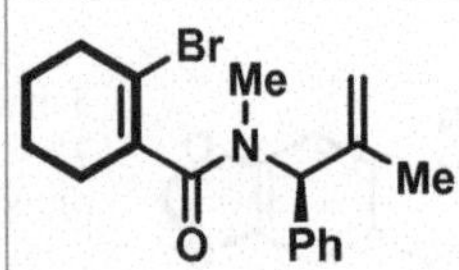 2.85n[f] | 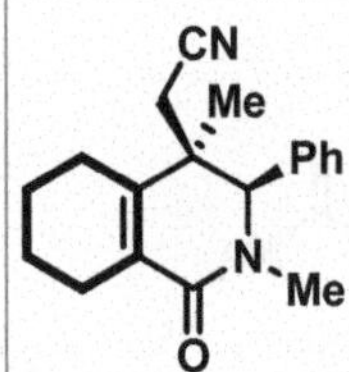 2.95n[f] | 67 (87:13 dr) (99:1 er) |
| 16 | 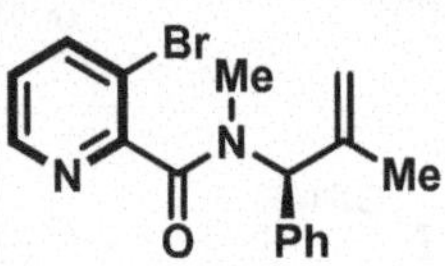 2.85o | 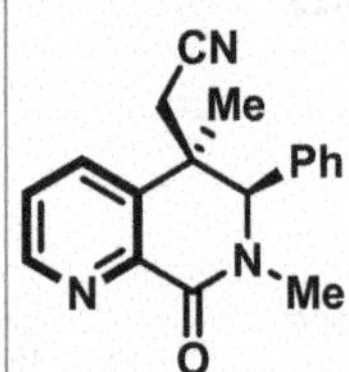 2.95o | 70 (93:7 dr) (98:2 er) |

(continued)

**Table 2.4** (continued)

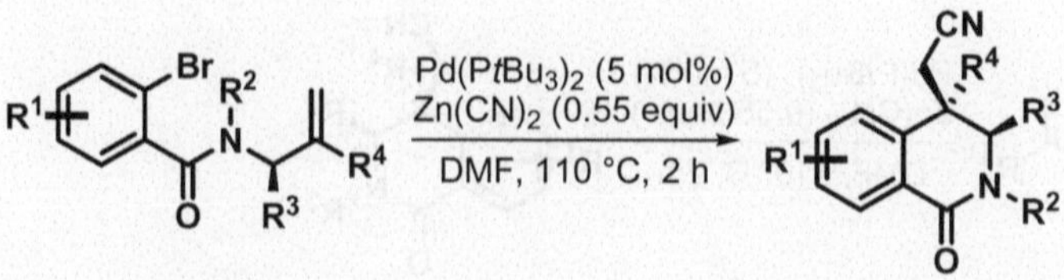


**2.85** **2.95**

| Entry | Substrate | Product | Yield (%)[a,b,c] |
|---|---|---|---|
| 17 | 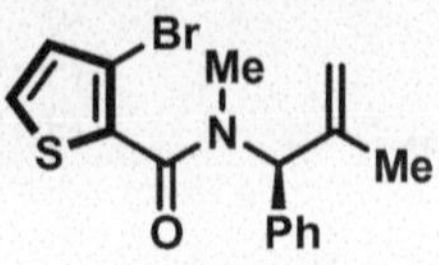 **2.85p**[f] | 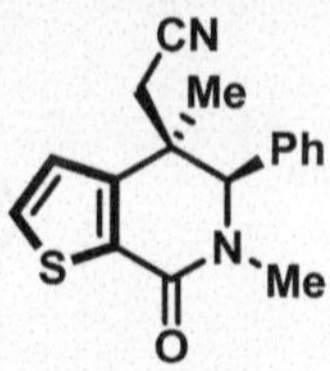 **2.95p**[f] | 87 (93:7 dr) (>99:1 er) |
| 18 | 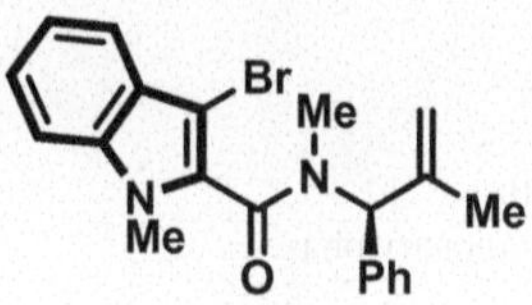 **2.85q** | 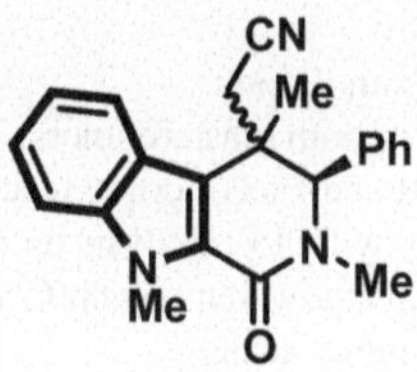 **2.95q** | 77 (40:60 dr) (>99:1 er) |
| 19 | 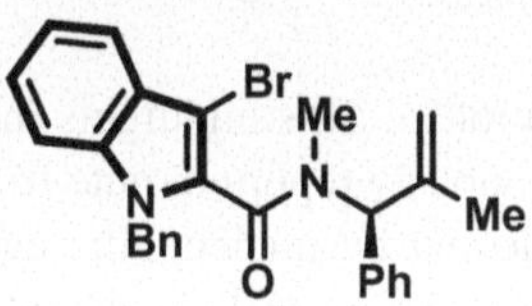 **2.85r** | 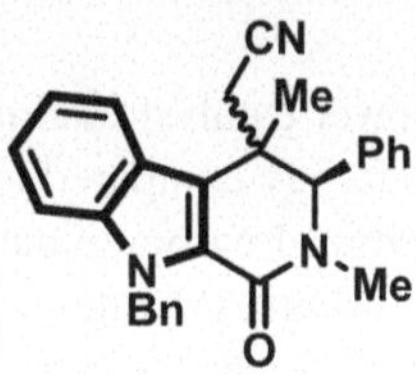 **2.95r**[f] | 85 (50:50 dr) (99:1 er) |

(continued)

**Table 2.4** (continued)

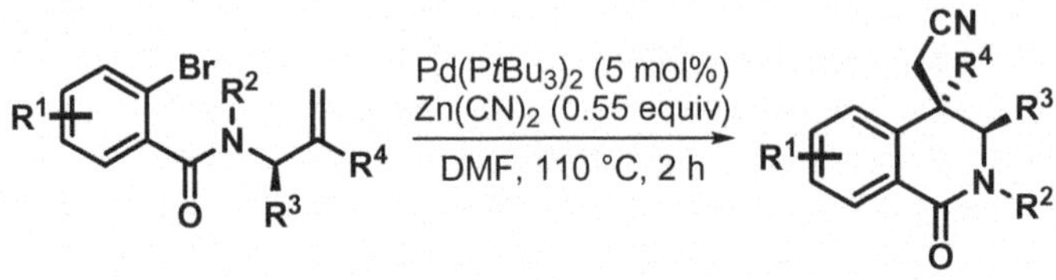


2.85 2.95

| Entry | Substrate | Product | Yield (%)[a,b,c] |
|---|---|---|---|
| 20 | 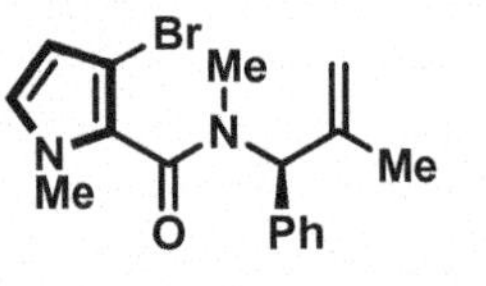 2.85s | 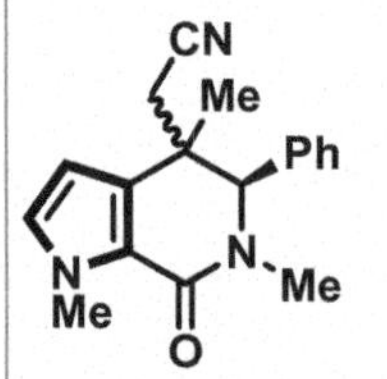 2.95s | 92 (50:50 dr) (99:1 er) |

Substrate scope
Reactions were run on 0.3 mmol scale
[a]Combined isolated yields of both diastereomers
[b]dr determined by $^1$H NMR analysis of crude reaction mixture
[c]er values were determined by HPLC analysis on a chiral stationary phase
[d]Reaction was run with 1.05 equivalents of $Zn(CN)_2$
[e]Reaction was run on 4.75 mmol scale
[f]Reaction was run in the presence of 50 mol% PMP
[g]This compound was prepared by David Petrone

yield > 95:5 dr under lower catalyst loading (2.5 mol%). This highlights increased efficiency of this methodology compared to the previously reported route from 56% over two steps (Pd-catalyzed carboiodination followed by nucleophilic cyanation using KCN) to 79% in a single Pd-catalyzed transformation. Alkyl substrate **2.85k** containing a PMB protected ether was transformed to **2.95k** in 66% yield with no selectivity (53:47 dr). This finding is consistent with previous models suggesting that diastereoselectivity is governed by the $A^{1,2}$ strain in this class of substrates. Other vinyl substituted amides (**2.85l** and **2.85m**) cyclized to form the desired products **2.95l** and **2.95m** in 73 and 77% yield and 95:5 and 92:8 dr, respectively. Product **2.95n** was isolated in 67% yield with an 87:13 dr from the corresponding vinyl bromide **2.85n** with the aid of the amine base PMP. Trace amounts of cyclopropanation product were observed, which is thought to arise from the intramolecular carbopalladation of the activated olefin by the key neopentyl Pd(II) intermediate. 3-Bromopicolinic acid derivative **2.85o** was transformed to the corresponding product **2.95o** in 70% yield with 93:7 dr. Bromothiophene **2.85p** afforded heterocycle **2.95p** in 87% yield and 93:7 dr. Indole derivatives **2.85q** and **2.85r** successfully underwent heteroaryl cyanation, affording **2.95q** and **2.95r** in 77 and 85% yield, albeit with negligible diastereoselectivity. In the case of *N*-Me protected **2.85q**, a switch in stereochemistry of the major diastereomer was observed, and the opposite product was found to be in slight excess. Lastly, pyrrole **2.85s** converted to the desired product in 92% yield. In correlation to the findings from **2.85q–2.85s**, the electron rich pyrrole moiety is thought to be the key factor affecting the selectivity of the cyclization (Table 2.4).

### 2.6.4 *Derivatization of Alkyl Nitriles*

Product derivatization studies were performed to showcase the synthetic utility of the enantioenriched dihydroisoquinolinones. Pure diastereomers of **2.95a** and **2.95c** were chosen to explore the reactivity of the key functional groups (Scheme 2.23). A Mizoroki-Heck reaction of the aryl chloride was accomplished using a variation to Fu's conditions, which furnished the *trans* alkene product **2.97** in 91% yield [43]. Primary amide **2.98** was obtained in 83% yield via nitrile hydrolysis under basic conditions. The respective acid derivative could not be obtained, owing to the severe steric crowding around the function group. The global reduction was achieved using an excess of $LiAlH_4$ in refluxing THF, producing diamine **2.99** in 85% yield.

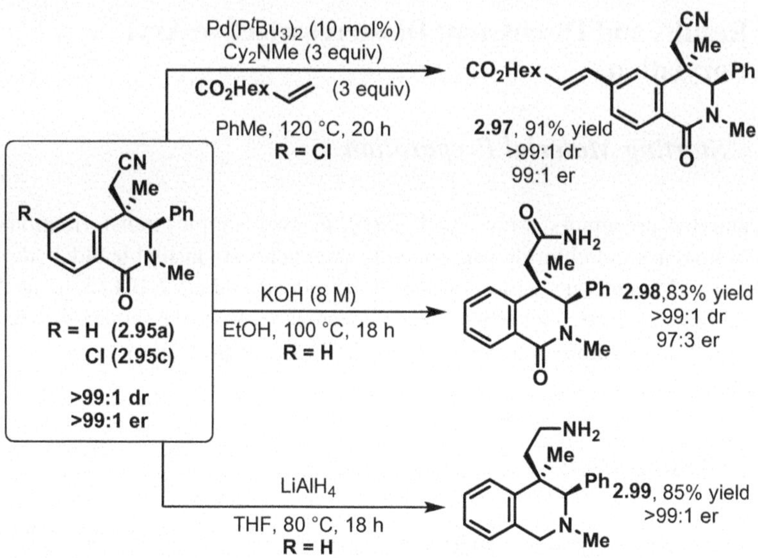


**Scheme 2.23** Derivatization studies on the alkyl nitrile **2.95a** and **2.95c**

## 2.7 Research Goal: Part 2

With the successful realization of the diastereoselective intramolecular Pd-catalyzed aryl cyanation, we sought to explore other valuable coupling moieties. Scouting the literature for reagents capable of undergoing transmetallation, we became interested in the Pd-catalyzed borylation. The initial report was made by Miyaura in 1995 wherein aryl halides were transformed into their respective aryl boronates in the presence of a palladium source, diboron species and an acetate salt (Scheme 2.7). A range of follow up reactions have been disclosed on the borylation and most notably in 2007, Cardénas reported a borylative cyclization 1,6-enynes (Scheme 2.18, A). The transformation used MeOH as the hydride source and completed the catalytic cycle with the borylation of the alkyl palladium species. Encouraged by these findings, we envisioned an intramolecular Pd-catalyzed domino-Heck borylation sequence to access sterically encumbered alkyl boronates in a catalytic fashion. During the development of this project, the Van der Eycken reported a microwave assisted Pd-catalyzed domino-Heck aryl borylation sequence of *o*-iodoanilides to form indolinone-3-methyl boronates (Scheme 2.18, C). Complimentary to their findings, we report the first highly diastereoselective Pd-catalyzed aryl borylation sequence of tethered aryl iodides to furnish the borylated chromans.

## 2.8 Results and Discussion: Diastereoselective Aryl Borylation

### *2.8.1 Starting Material Preparation*

The catalytic precursors were synthesized in two simple transformations. 2-Methallylmagnesium chloride was added to commercially available aldehydes and a subsequent Mitsunobu reaction furnished the iodoaryl ethers **2.101** (Scheme 2.24) [44]. Alternatively, the iodoaniline **2.104** was prepared from the sequential addition of 2-methallymagnesium chloride to imine **2.102** and *N*-protection.

### *2.8.2 Optimization*

Subjecting the iodoaryl ether **2.101a** to $Pd_2(dba)_3$, KOAc, and $B_2Pin_2$ in DMF at 80 °C for 4 h was found to be the optimal conditions (Table 2.5). The alkyl boronate **2.105a** was formed in 82% yield and >20:1 dr and the relative stereochemistry was unambiguously confirmed by spectroscopic analysis and single crystal X-ray crystallography. A series of variation experiments were performed to examine the effects of changes to the optimal conditions (Table 2.5). It should be noted that in all cases, no direct borylation occurred to form the respective aryl boronate. Less polar solvents such as PhMe and 1,4-dioxane did not yield any product (entries

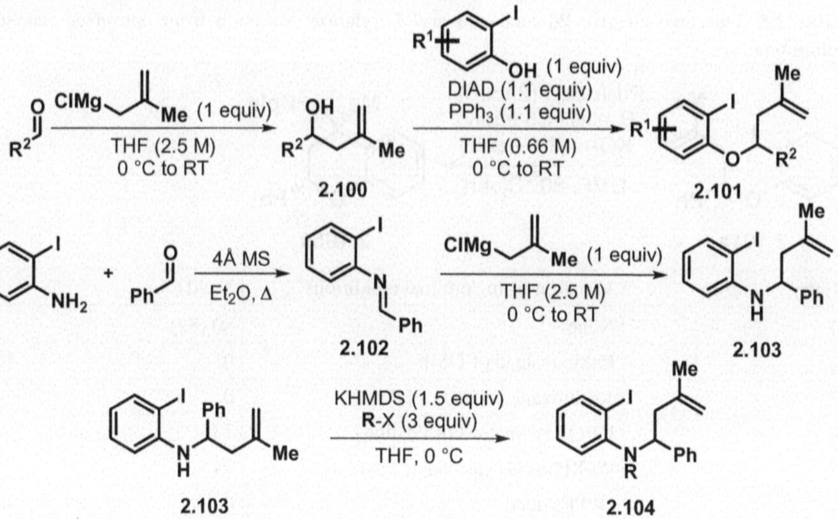


**Scheme 2.24** General synthetic route to accessing the iodoaryl ethers

2 and 3). Other Pd sources including those reported by Cárdenas and Buchwald, produced the desired product in lower yields but with consistent dr's (entries 4 and 5 respectively). The incorporation of a bidentate ligand inhibited formation of **2.105a** (entry 6). Acetate bases were found to be optimal as opposed to carbonates and alkyl amines (entries 7–9). Attempts to lower the equivalents of KOAc and $B_2pin_2$ led to full conversion but in moderate yields (entries 10 and 11). Lastly, aryl bromides and aryl chlorides gave incomplete conversion to **2.105a** (entries 12 and 13).

### *2.8.3 Substrate Scope*

Having determined the ideal reaction conditions to afford the alkyl boronates, we then explored the scope of the reaction (Table 2.6). It should be noted **2.105** was formed as the exclusive product (>20:1 dr). Trifluoromethyl bearing aryl iodide **2.101b** afforded **2.105b** in excellent yield (88% yield). Electron-rich aryl iodide **2.101c** cyclized in good yield with prolonged reaction time (**2.105c**, 62% yield, 12 h). Multi-substituted aryl iodides **2.101d** and **2.101e** produced the desired product **2.105d** and **2.105e** in 75 and 72% yield respectively with higher catalyst loading. Electron-rich, electron-poor and *ortho*-substituted aryl groups were tolerated in good to excellent yields (**2.101f**–**2.101h**, 74-88% yield). Substituting the tethered phenyl substituent to a thienyl group was also tolerated and afforded the final product in 73% yield (**2.105i**). Additionally, aryl iodides with tethered aliphatic groups **2.101j** and **2.101k** cyclized in moderate to good yield to produce the respective alkyl boronates **2.105j** and **2.105k** in excellent dr (56 and 70% yield respectively, >20:1 dr). An iodopyridyl group was tolerated

**Table 2.5** Diastereoselective Pd-catalyzed aryl borylation: variation from optimized reaction conditions

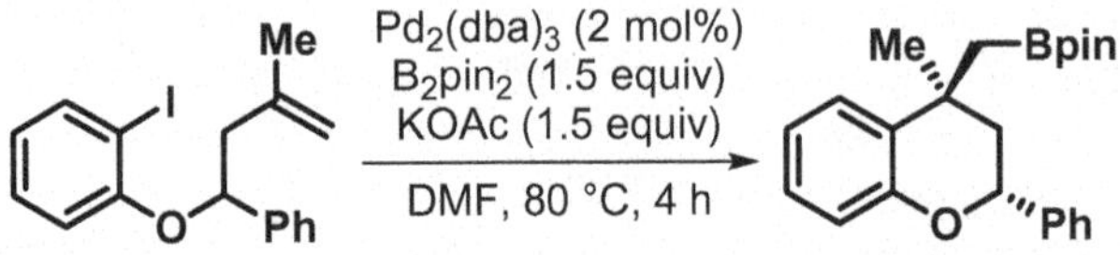


2.101a 2.105a

| Entry | Variation from "optimal conditions" | Yield (%) |
|---|---|---|
| 1 | None | 80 (82) |
| 2 | PhMe instead of DMF | 0 |
| 3 | 1,4-dioxane instead of DMF | 0 |
| 4 | $Pd(OAc)_2$ instead of $Pd_2dba_3$ | 73 |
| 5 | Pd XPhos G1 instead of $Pd_2dba_3$ | 74 |
| 6 | DPPF added | 0 |
| 7 | NaOAc instead of KOAc | 52 |
| 8 | $K_2CO_3$ instead of KOAc | 45 |
| 9 | $Et_3N$ instead of KOAc | 9 |
| 10 | 1.1 equiv instead of 1.5 equiv of KOAc | 62 |
| 11 | 1.1 equiv instead of 1.5 equiv of $B_2pin_2$ | 47 |
| 12 | ArBr instead of ArI | 43 |
| 13 | ArCl instead of ArI | 16 |

Reactions were run on 0.2 mmol scale
[a]Determined by $^1H$ NMR analysis of the crude reaction mixtures using 1,3,5-trimethoxybenzene as internal standard
[b]Yield in parentheses are isolated yields. [c]Reaction was run with 4 mol% of DPPF
[c]Reaction was run with 4 mol% of DPPF

in modest yields with prolonged reaction time while, iodoquinolinols were tolerated under the standard reaction conditions in good yield (**2,105l** and **2.105m**, 44% and 78% yield respectively).

### *2.8.4 Limitations*

Substrate **2.101n** which contains a 1-iodonaphthyl group was subjected to the reaction conditions but failed to produce **2.105n**. Quantitative recovery of starting material was observed, which can be attributed to the steric hindrance around the aryl iodide. Under the optimal reaction conditions, aryl iodides bearing an amine linker

**Table 2.6** Diastereoselective Pd-catalyzed aryl borylation

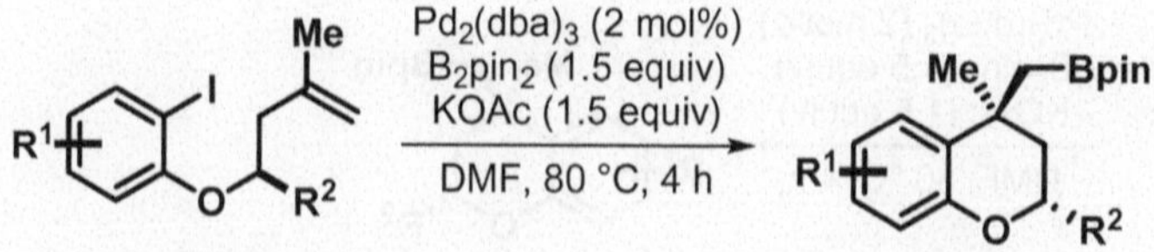


2.101 → 2.105

| Entry | Substrate | Product | Yield (%)[a,b,c] |
|---|---|---|---|
| 1 | 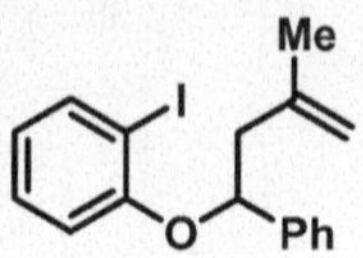 2.101a | 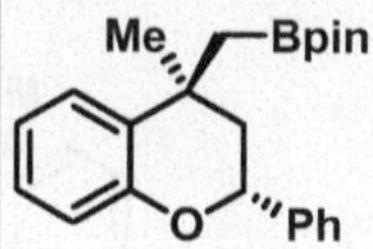 2.105a | 82 |
| 2 | 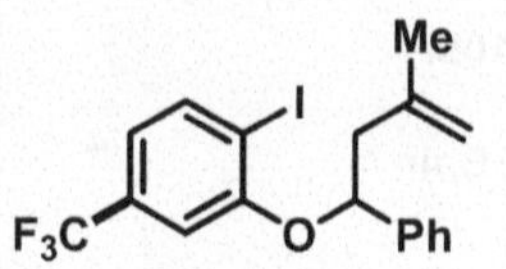 2.101b | 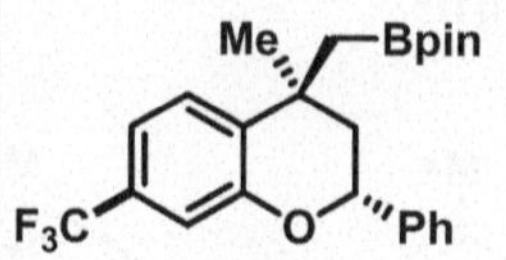 2.105b | 88 |
| 3[b] | 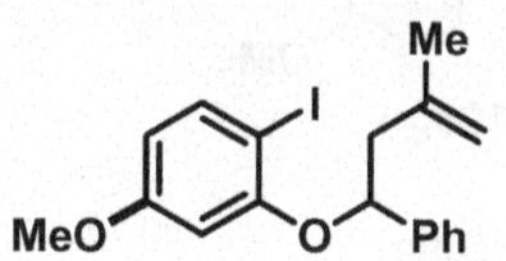 2.101c | 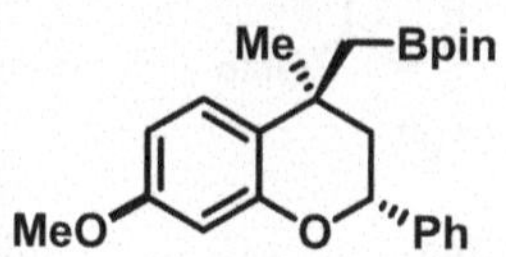 2.105c | 62 |
| 4[c] | 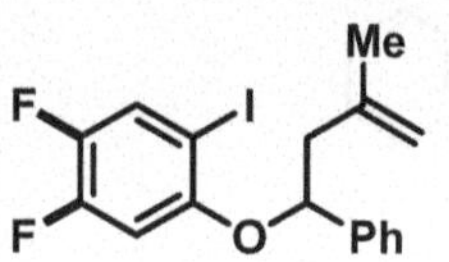 2.101d[d] | 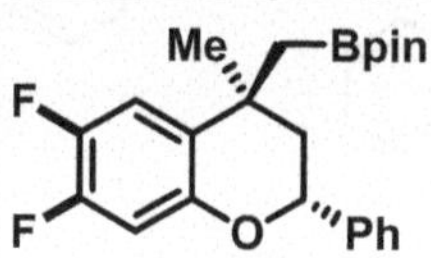 2.105d[d] | 75 |
| 5[c] | 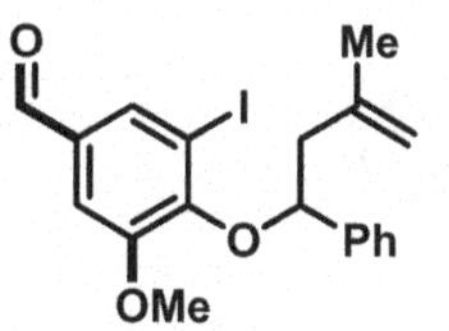 2.101e | 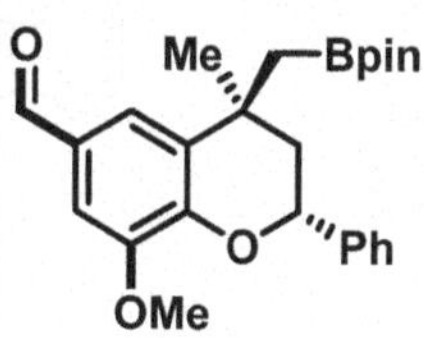 2.105e | 72 |

(continued)

**Table 2.6** (continued)

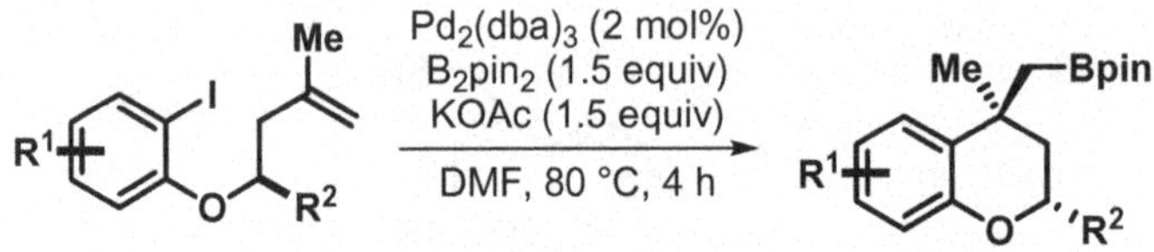


**2.101** **2.105**

| Entry | Substrate | Product | Yield (%)[a,b,c] |
|---|---|---|---|
| 6 | 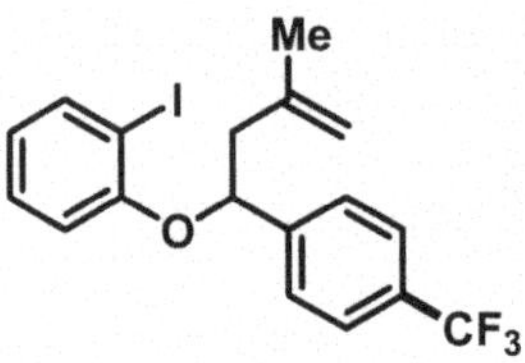  **2.101f**[d] | **2.105e** 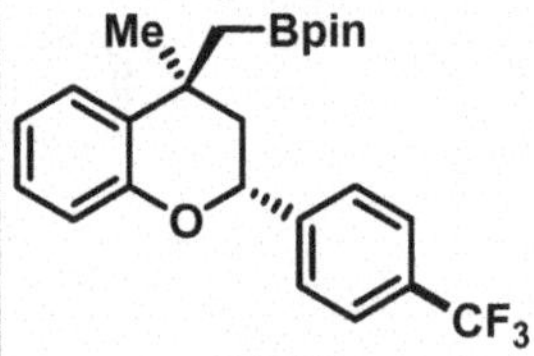  **2.105f** | 88 |
| 7 | 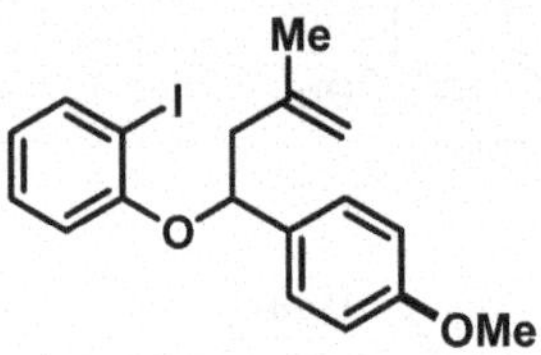  **2.101g**[d] | 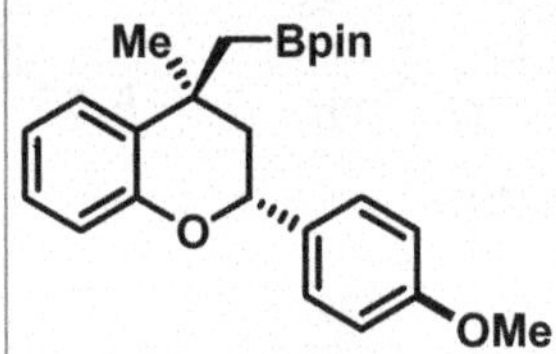  **2.105g** | 74 |
| 8 | 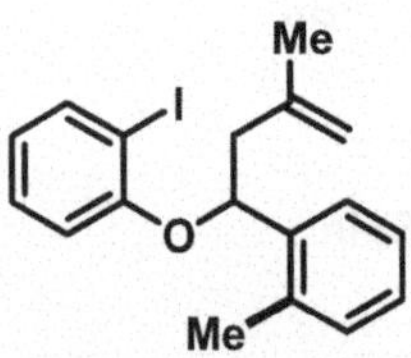  **2.101h** | 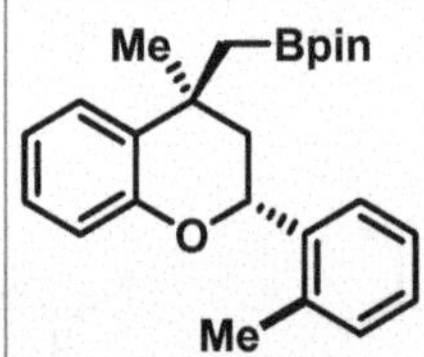  **2.105h** | 75 |
| 9 | 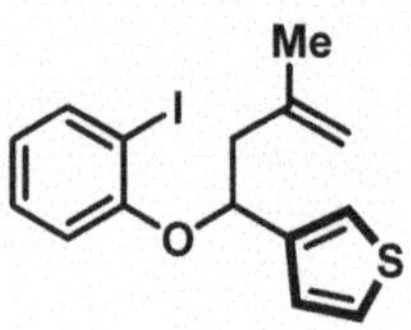  **2.101i** | 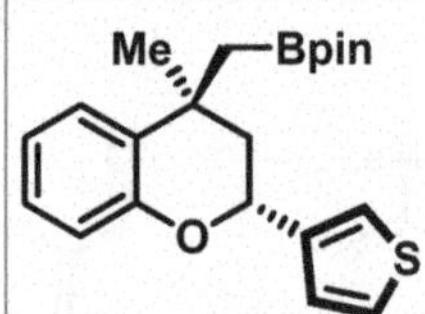  **2.105i** | 73 |
| 10 | 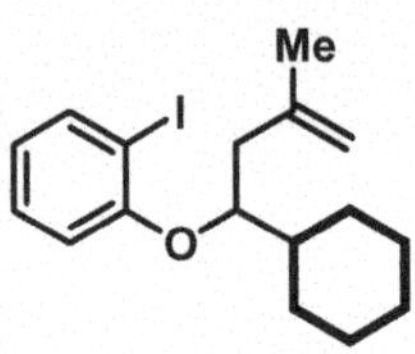  **2.101j** | 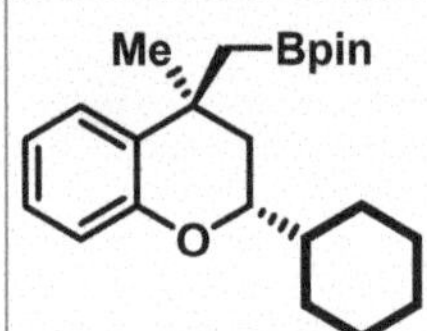  **2.105j** | 56 |

(continued)

**Table 2.6** (continued)

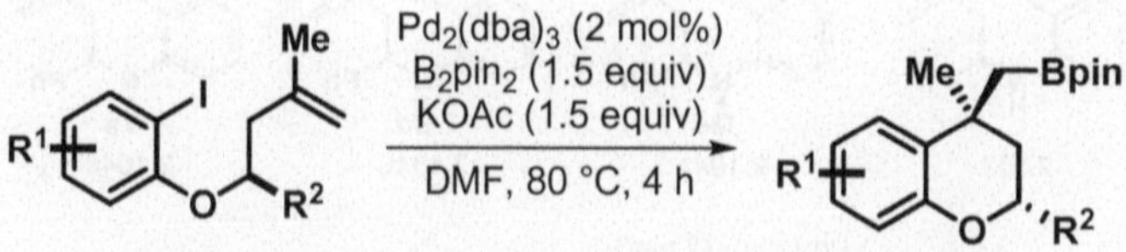

2.101 → 2.105

| Entry | Substrate | Product | Yield (%)[a,b,c] |
|---|---|---|---|
| 11 | 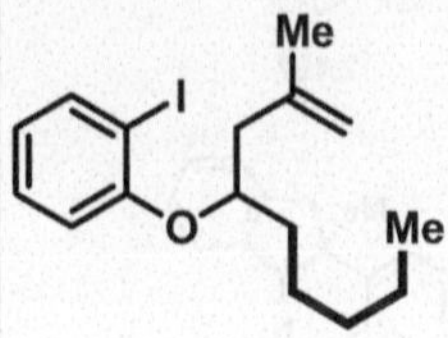 2.101k[d] | 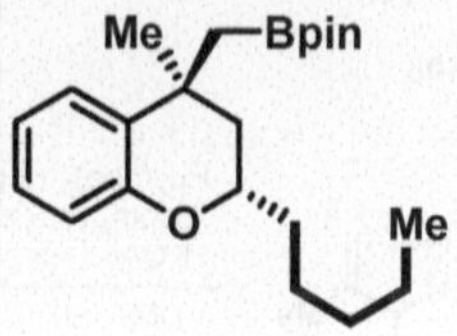 2.105k | 70 |
| 12[b] | 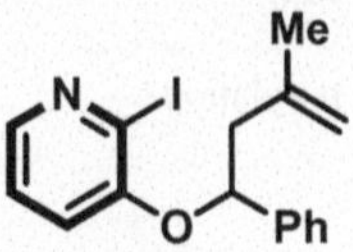 2.101l | 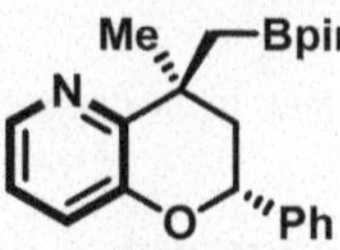 2.105l | 44 |
| 13 | 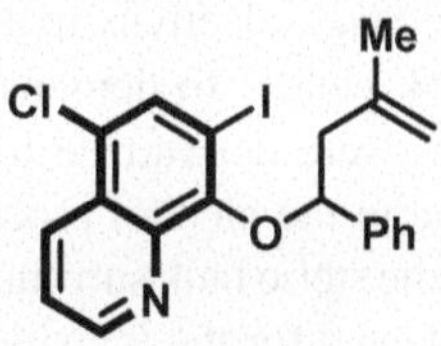 2.101m[d] | 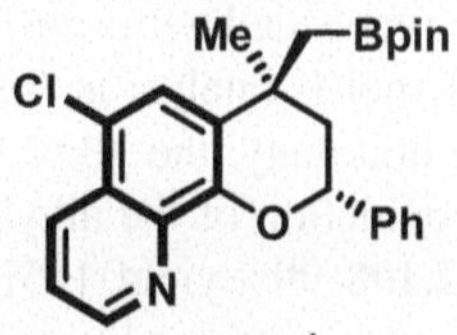 2.105m[d] | 78 |

Substrate scope
Reactions were run on 0.2 mmol scale. All reactions gave > 20:1 dr
[a]Isolated yields
[b]Reaction was run for 12 h
[c]Reaction was run with 4 mol% catalyst
[d]This compound was prepared by Alvin Jang

did not react to form the respective borylated tetraisoquinolines (**2.103**, **2.104a–2.104c**). Additionally, attempts to deviate from the chroman scaffolds led to unsatisfactory results. Employing the isochroman precursor **2.106** or the enantioenriched carboxamide **1.106a** exclusively formed side products arising from the C–H functionalization of the tethered aromatic group (Scheme 2.25).

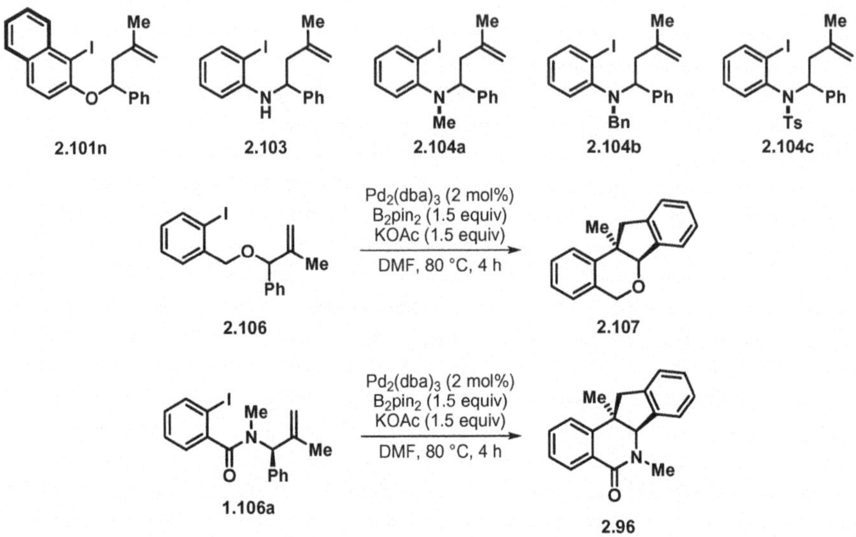


**Scheme 2.25** Unsuccessful substrates in the diastereoselective aryl borylation

## *2.8.5 Derivatization of Alkyl Boronates*

The gram-scale reaction proceeded with consistent yield and selectivity to the model substrate **2.105a** with 1 mol% catalyst at 0.2 M (80% yield). To demonstrate the applicability of this methodology, the alkyl boronates were derivatized to various functional groups. The oxidation of the alkyl boronate to the alcohol proceeded in near quantitative yield (**2.108**, 99% yield) [45]. Due to the steric bulk surrounding the alkyl boronate, particular reaction conditions were selected for the transition-metal catalyzed cross coupling reactions. A Cu(I)-catalyzed Chan-Lam coupling outlined by Watson yielded the desired amide in moderate yield (**2.109**, 45% yield) [46]. Additionally, a Pd-catalyzed Suzuki reaction of **2.105a** was accomplished using a variation of a method outlined by Molander [47] and Capretta [48] producing the desired product **2.110** in moderate yield (44% yield) (Scheme 2.26).

## *2.8.6 Postulated Mechanism*

A plausible mechanism, in line with those previously postulated for both the Pd-catalyzed domino-Heck cascade reactions and Miyaura borylation is illustrated (Scheme 2.27). The aryl iodide **2.101a** undergoes oxidative addition and carbopalladation forming the alkylpalladium(II) intermediate **2.112**. The high diastereoselectivity in the carbopalladation forming **B** was consistent with previous literature. The *trans*- adduct was thought to form exclusively from the minimized axial-axial

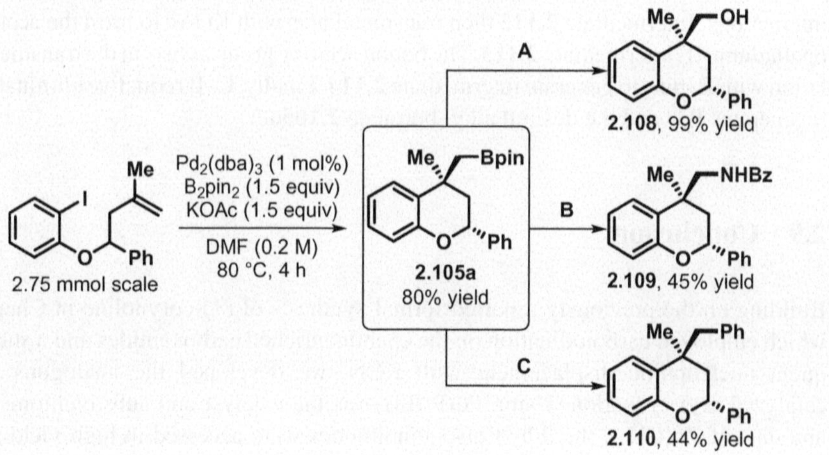


**Scheme 2.26** Gram scale diastereoselective Pd-catalyzed domino-Heck aryl borylation and derivatization. Reaction conditions: **A** 30% $H_2O_2$, 3 M NaOH, EtOH, 0 °C. **B** Benzamide, CuBr, Di-tert-butyl peroxide, NaOTMS, *t*BuOH, 75 °C. **C** PhI, $Pd(OAc)_2$, 1,3,5,7-tetramethyl-6-phenyl-2,4,8-trioxa-6-phosphadamantane, K*t*BuO, PhMe, 100 °C

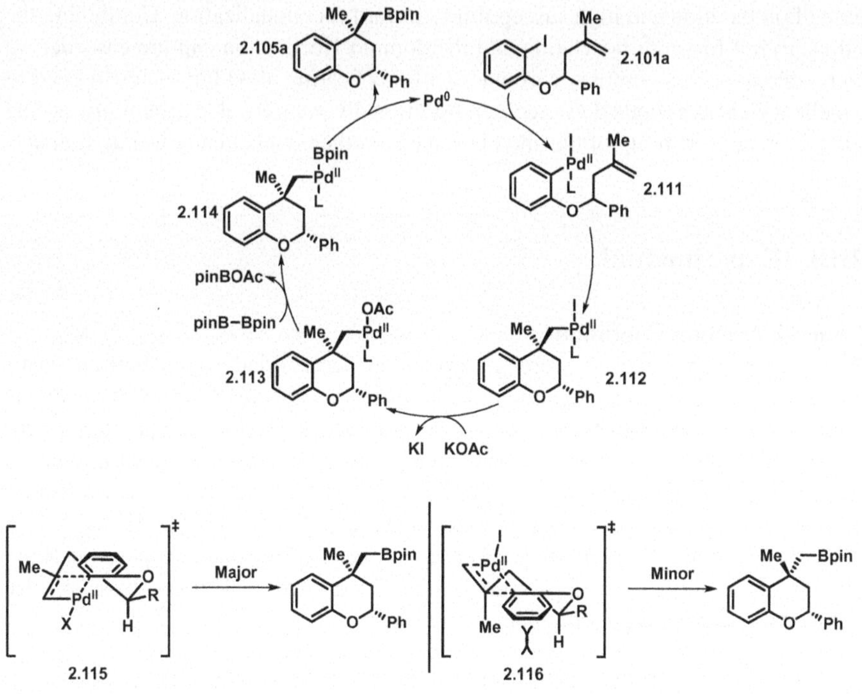


**Scheme 2.27** Proposed mechanism of diastereoselective Pd-catalyzed domino-Heck aryl borylation

interactions. Intermediate **2.112** then transmetallates with KOAc to form the acetoxopalladium(II) intermediate **2.113**. The bound acetoxy group assists in the transmetalation with $B_2pin_2$ to generate intermediate **2.114**. Finally, C–B reductive elimination regenerates $Pd^0$ and the desired alkyl boronate **2.105a**.

## 2.9 Conclusion

Building on the previously reported formal synthesis of (+)-corynoline in Chap. 1 which employed carboiodination on the enantioenriched carboxamides and a subsequent nucleophilic displacement with KCN, we developed the analogous Pd-catalyzed aryl cyanation. Using $Pd(PtBu_3)_2$ as the catalyst and substoichiometric amounts of $Zn(CN)_2$, the dihydroisoquinolinones were accessed in high yield and excellent diastereoselectivity. This transformation also provided a more efficient process to accessing the (+)-corynoline precursor in higher yield and dr than the previously reported two step procedure (79% yield and > 95:5 dr vs. 56% yield over two steps and 91:9 dr).

Organoboronates have been a valuable reagent in organic synthesis due to their ease of preparation and high susceptibility toward functionalization. Combining the Pd-catalyzed Miyaura borylation and the domino-Heck anion capture cascade, we were able to access complex chroman scaffolds bearing alkyl boronates in good to excellent yield as a single diastereomer. Additionally, we were able to demonstrate the functional convertibility of the alkyl boronate to other synthetically useful moieties.

## 2.10 Experimental

**General Reaction Conditions**

All non-aqueous reactions were performed in flame dried round bottom flasks sealed with a fitted rubber septum under an inert atmosphere of argon unless otherwise stated. All reactions were magnetically stirred and elevated temperatures were reported as the temperature of the surrounding oil bath. Reactions were monitored by thin layer chromatography (TLC) or by crude $^1$H-NMR analysis of a worked up aliquot. TLC visualization was performed under a UV lamp or $KMnO_4$/CAM stain developed with heat. Solvent evaporation was conducted by rotary evaporation at the appropriate temperature and pressure. All reported yields reflect spectroscopically ($^1$H-NMR) pure material unless otherwise stated.

**Materials**

Unless stated otherwise, all reagents were used as received and the following reaction solvents were distilled under anhydrous conditions over the appropriate drying agent and transferred under argon via a syringe. Dichloromethane was distilled over $CaH_2$, tetrahydrofuran was distilled over Na (1% w:v) and benzophenone (1% w:v), 1,4-dioxane was distilled over Na (1% w:v) and benzophenone (1% w:v), and triethylamine was distilled over KOH. Dimethylformamide was distilled over 5Å molecular sieves and stored over 5 Å molecular sieves (water content was kept lower than 50 ppm). 1,3,5-trimethoxybenzene was crushed into a fine powder by a mortar and pestle, dried overnight in vacuo and stored in a desiccator.

**Analysis**

$^1$H-NMR and $^{13}$C-NMR spectra of catalytic starting material and products were obtained on the Agilent DD2 500 equipped with a 5 mm Xsens Cold Probe. $^1$H-NMR and $^{13}$C-NMR spectra of the catalytic starting material precursors were obtained from one of the following spectrometers: Varian NMR system 400, Bruker Avance III 400, Varian Mercury 400 or Varian Mercury 300. All $^{19}$F-NMR spectra were obtained on the Varian Mercury 300 and Varian Mercury 400. Measurements were carried out at 23 °C and chemical shifts (δ) are reported as parts per million (ppm). The solvent resonance was used as the internal standard for $^1$H-NMR (Chloroform at 7.26 ppm) and $^{13}$C-NMR (Chloroform at 77.0 ppm). The *J* values are reported in hertz (Hz) and are rounded off to the nearest 0.5 Hz. Multiplicities are abbreviated as follows: singlet (s), doublet (d), triplet (t), quartet (q), multiplet (m) and broad (br). All accurate mass values were obtained from the following spectrometers: Agilent 6538 Q-TOF (ESI) and JEOL AccuTOF-DART. Melting points were obtained on a Fisher-Johns Melting Point Apparatus and uncorrected. Infrared (IR) spectra were obtained as a neat film or dissolved in $CHCl_3$ on a NaCl disk using a Shimadzu FTIR-8400S FT-IR spectrometer.

**General Procedures**

**General Procedure 1**:

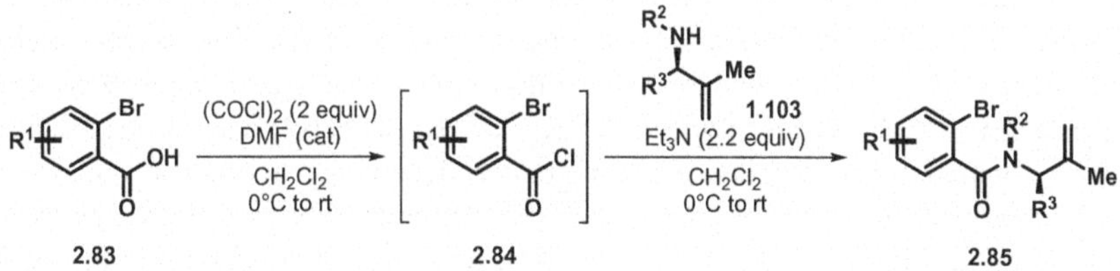


2.83 2.84 2.85

A flame dried flask was charged with the *o*-bromobenzoic acid derivative (1 equiv) and was purged with argon for 10 min. The contents of the flask were taken up in dry

DCM (10 mL), and anhydrous DMF (2 drops) was added. The resulting suspension was cooled to 0 °C and oxalyl chloride (2 equiv) was added drop wise over 5 min. Once gas evolution had ceased, the reaction mixture was warmed to room temperature, and stirred vigorously. Once the slow gas evolution ceased at this temperature, the reaction mixture was concentrated to dryness on a rotary evaporator, followed by high-vacuum to remove any excess oxalyl chloride. The flask containing the crude acid chloride was fitted with a septum and purged with argon for 10 min. The contents of the flask were taken up in dry DCM (10 mL) and cooled to 0 °C. To this solution was added dry $Et_3N$ (2 equiv) which immediately resulted in the appearance of a persistent orange-to -red colour. A solution (2.5 mL) of chiral allylic amine (**1.103**, 1.1 equiv) and dry $NEt_3$ (2 equiv) in DCM was added drop wise over 10 min. The reaction was stirred at 0 °C for 10 min and then warmed to room temperature where it was stirred for an additional 10 min. The reaction was quenched by doubling the reaction volume with a concentrated aqueous solution of $NaHCO_3$. The layers were separated and the aqueous layer was extracted with DCM (3x). The combined organic layers were washed with brine, dried over $Na_2SO_4$, filtered, and concentrated *in vacuo*. The pure chiral *N*-allyl carboxamides **2.85** were purified using Si gel flash column chromatography using the indicated mobile phase.

**General Procedure 2**:

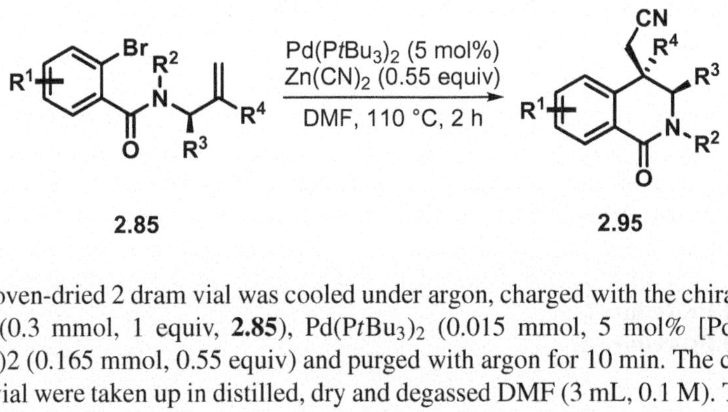


**2.85** **2.95**

An oven-dried 2 dram vial was cooled under argon, charged with the chiral linear amide (0.3 mmol, 1 equiv, **2.85**), $Pd(PtBu_3)_2$ (0.015 mmol, 5 mol% [Pd]), and Zn(CN)2 (0.165 mmol, 0.55 equiv) and purged with argon for 10 min. The contents of the vial were taken up in distilled, dry and degassed DMF (3 mL, 0.1 M). The vial was fitted with a Teflon lined screw cap under a stream of argon passed through a large inverted glass funnel, sealed using Teflon tape, and placed in a pre-heated oil bath at 110 °C for 2 h. At this time either TLC analysis or $^1H$ NMR analysis of a small aliquot indicated full conversion of starting material, and the reaction vial was cooled and $H_2O$ and EtOAc or DCM was added. The combined layers were washed with water (4x) and brine (2x), dried over $Na_2SO_4$, filtered, and concentrated in vacuo. The combined organic layers were dried filtered through a 2 cm plug of silica gel in a pipette eluting with 100% EtOAc. The crude dihydroisoquinolinones were purified via silica gel flash column chromatography using the indicated mobile phase.

**General Procedure 3:**

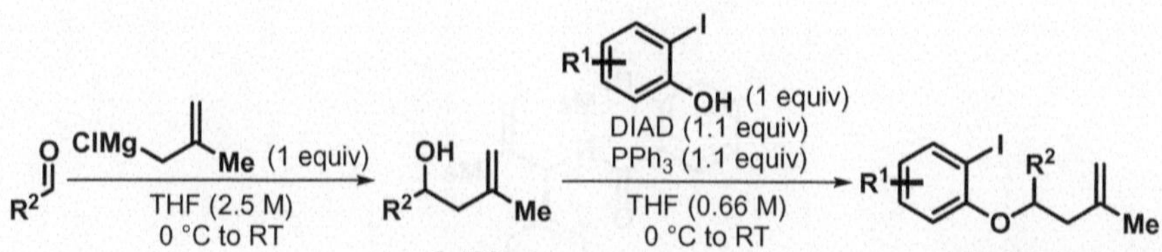


2.100

2.101

The aldehyde (1 equiv) was added to a flask and was purged with argon for 10 min. The aldehyde was then charged with THF (2.5 M) and the resulting solution was brought to 0 °C and stirred for 5 min. At this time, 2-methallylmagnesium chloride (1 equiv, 0.5 M in THF) was added dropwise and was stirred for 10 min at 0 °C until the reaction was warmed to room temperature. The reaction was monitored by TLC analysis and once full conversion was observed, the reaction was brought to 0 °C and quenched with concentrated ammonium chloride. The mixture was diluted with water and was transferred to a separatory funnel and extracted with EtOAc (3x). The combined organic layers were washed with brine, dried over $Na_2SO_4$ and concentrated in vacuo. **2.100** (1 equiv), iodophenol (1 equiv) and $PPh_3$ (1.1 equiv) were added to a flask and was purged with argon for 10 min. The contents were then dissolved in anhydrous THF (0.66 M) and the resultant solution was cooled to 0 °C. At this time, DIAD (1.1 equiv) was added dropwise and then the reaction mixture was warmed to room temperature for 12 h. After the indicated time, the reaction was quenched with $H_2O$ and extracted with EtOAc (3x). The combined organic layers were washed with brine, dried over $Na_2SO_4$ and concentrated *in vacuo*. The pure aryl ether was purified via Si gel flash chromatography using the indicated mobile phase.

**General Procedure 4:**

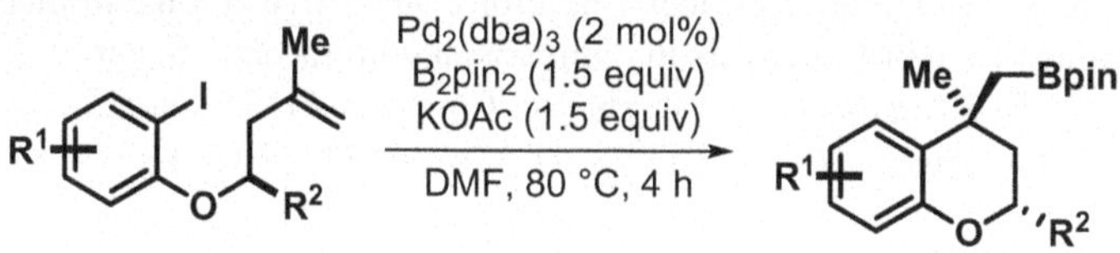


2.101

2.105

To a flame dried or oven dried 2 dram vial cooled under argon, **2.101** (1 equiv), $Pd_2(dba)_3$ (4 mol% Pd), bis(pinacolato)diboron (1.5 equiv), and KOAc (1.5 equiv) were added and allowed to purge for 10 min. DMF (0.1 M) was added. A Teflon line screw cap was fitted on the two dram vial, sealed with Teflon tape and place in a preheated oil bath at 80 °C for 4 h. Once TLC or $^1H$-NMR analysis confirmed full conversion of the starting material, the reaction mixture was cooled to room temperature. Once cooled, the reaction was extracted with EtOAc/$H_2O$ (3x), dried over $Na_2SO_4$ and passed through a 2 cm plug of silica gel in a pipette using EtOAc.

The pure chromans were obtained via Si gel flash column chromatography using the indicated mobile phase.

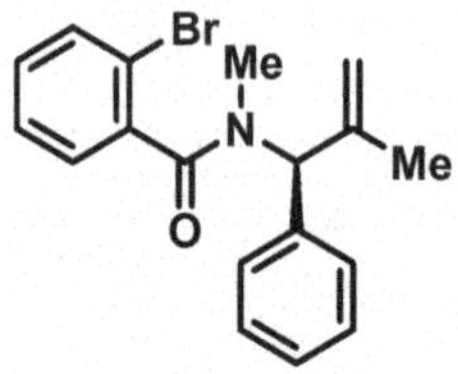

S)-2-bromo-*N*-methyl-*N*-(2-methyl-1-phenylallyl)benzamide (2.85a)

Prepared according to **GP1** using 2-bromobenzoic acid (388 mg, 1.93 mmol) and (*S*)-*N*,2-dimethyl-1-phenylprop-2-en-1-amine (342 mg, 2.13 mmol). The carboxamide was purified by flash column chromatography using hexanes: EtOAc (3:1 v:v) and was obtained as an extremely viscous, clear and colourless oil (586 mg, 1.7 mmol, 88%).

*$^1$H NMR Analysis Showed the Desired Compound to Be Present as a Complex Mixture of 4 Rotamers, and Only Approximate Integration Values Are Displayed.*

**$^1$H NMR** (500 MHz, $CDCl_3$) δ 7.64–7.50 (m, 1H), 7.40–7.13 (m, ~7.5H), 7.07 (td, *J* = 7.5, 1.0 Hz, ~0.2H), 7.04–7.00 (m, ~0.3H), 6.75 (dd, *J* = 7.5, 1.5 Hz, ~0.2H), 6.39 (s, ~0.5H), 5.24–4.58 (m, ~2.5H), 3.19–2.48 (m, 3H), 1.92–1.44 (m, 3H). **$^{13}$C NMR** (125 MHz, $CDCl_3$) δ 170.4, 169.7, 169.4, 143.1, 142.7, 141.8, 138.9, 138.8, 138.7, 138.2, 137.9, 137.4, 136.4, 135.6, 133.0, 132.9, 132.8, 132.7, 130.2, 130.1, 130.1, 129.7, 129.0, 128.6, 128.5, 128.4, 128.4, 128.1, 127.8, 127.7, 127.7, 127.6, 127.5, 127.5, 127.1, 127.1, 126.8, 119.2, 119.1, 118.8 overlapping 118.8, 118.0, 114.8, 114.1, 113.0, 67.8, 67.8, 61.9 overlapping 61.8, 33.1, 33.0, 30.4, 30.2, 22.4, 21.9, 21.4, 21.1. **IR** (neat, $cm^{-1}$) 3086, 3059, 3028, 2972, 2943, 2918, 2852, 1616, 1476, 1387, 1327, 1250, 1180, 1121, 1076, 1032. **HRMS** (DART) calculated 344.06500 m/z (found 344.06504 m/z for $C_{18}H_{19}BrNO$). Enantiomeric purity was determined by HPLC analysis in comparison with racemic material $T_R$ = 10.0 (minor) $T_R$ = 14.8 (major) (>99:1 er shown; Chiralcel AD-H column, 10% $^i$PrOH in hexanes 1.0 mL/min, 220 nm). $[\alpha]_D^{20}$ −81.3 (c = 0.74, $CHCl_3$).

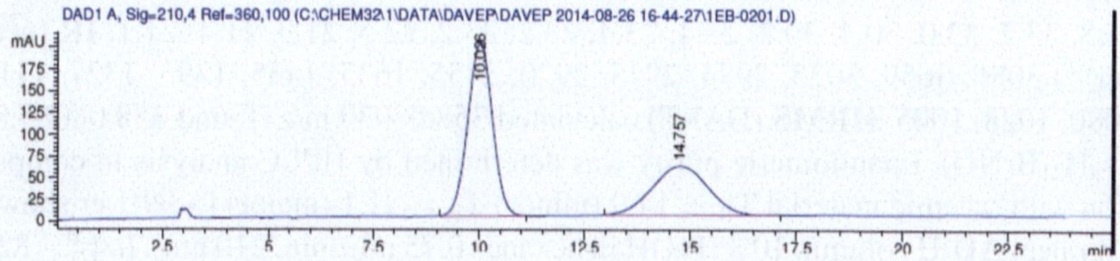


Signal 1: DAD1 A, Sig=210,4 Ref=360,100

| Peak # | RetTime [min] | Type | Width [min] | Area [mAU*s] | Height [mAU] | Area % |
|---|---|---|---|---|---|---|
| 1 | 10.026 | BB | 0.6345 | 8126.56494 | 199.80232 | 61.1147 |
| 2 | 14.757 | BB | 1.2963 | 5170.66650 | 56.78761 | 38.8853 |

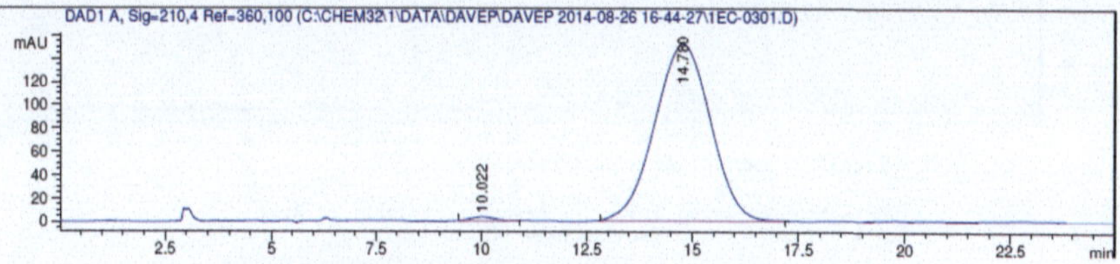


Signal 1: DAD1 A, Sig=210,4 Ref=360,100

| Peak # | RetTime [min] | Type | Width [min] | Area [mAU*s] | Height [mAU] | Area % |
|---|---|---|---|---|---|---|
| 1 | 10.022 | BB | 0.4809 | 121.58823 | 3.16009 | 0.8691 |
| 2 | 14.780 | BB | 1.4023 | 1.38679e4 | 151.45363 | 99.1309 |

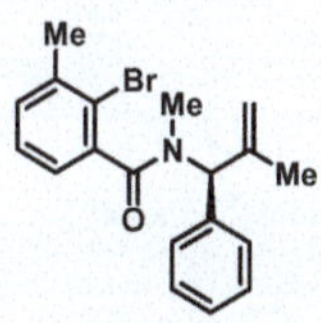


(*S*)-2-bromo-*N*,3-dimethyl-*N*-(2-methyl-1-phenylallyl)benzamide (2.85b)

Prepared according to **GP1** using 2-bromo-3-methylbenzoic acid (215 mg, 1 mmol) and (*S*)-*N*,2-dimethyl-1-phenylprop-2-en-1-amine (177 mg, 1.1 mmol). The carboxamide was purified by flash column chromatography using hexanes: EtOAc (3:1 v:v) and was obtained as clear and colourless oil (268 mg, 0.75 mmol, 75%).

*$^1$H NMR Analysis Showed the Desired Compound to Be Present as a Complex Mixture of 4 Rotamers, and Only Approximate Integration Values Are Displayed.*

**$^1$H NMR** (600 MHz, $CDCl_3$) δ 7.43–7.14 (m, 7H), 7.17–7.07 (m, ~0.5H), 7.03–6.92 (m, ~0.7H), 6.57–6.52 (m, ~0.1H), 6.44–6.37 (m, ~0.6H), 5.23–5.10 (m, 1H), 5.07–4.89 (m, 1H), 3.15 (s, ~0.4H), 2.83 (s, 0.7H), 2.59 (s, ~0.7H), 2.51 (s, ~0.8H), 2.45 (s, ~1.1H), 2.42 (s, ~0.8H), 2.39 (s, ~0.8H), 1.90–1.84 (m, 1H), 1.82 (s, ~0.8H), 1.49–1.42 (m, ~0.7H). **$^{13}$C NMR** (151 MHz, $CDCl_3$) δ 170.9, 170.2, 170.0, 169.9, 143.1, 142.9, 142.0, 141.9, 139.5, 139.4, 139.1, 139.0, 138.9, 138.7, 138.4, 137.5, 136.6, 135.8, 131.0, 130.9, 130.7, 129.8, 129.7, 129.0, 128.5, 128.5, 128.4, 128.3, 128.1, 127.7, 127.7, 127.6, 127.6, 127.6, 127.4, 127.0, 126.8, 125.0, 124.8, 124.8, 124.6, 121.5, 121.4, 121.2, 121.1, 118.0, 114.8, 114.1, 113.0, 67.9, 67.8, 61.9,

61.8, 33.2, 33.0, 30.4, 30.2, 23.4, 23.4, 23.2, 23.2, 22.5, 21.9, 21.4, 21.1. **IR** (neat, $cm^{-1}$) 3086, 3059, 3025, 2974, 2945, 2920, 2855, 1635, 1435, 1393, 1327, 1110, 1080, 1028, 1005. **HRMS** (DART) calculated 358.08039 m/z (found 358.08065 for $C_{19}H_{21}BrNO$). Enantiomeric purity was determined by HPLC analysis in comparison with racemic material $T_R = 14.9$ (minor) $T_R = 21.1$ (major) (> 99:1 er shown; Chiralcel AD-H column, 10% $^{i}$PrOH in hexanes 0.75 mL/min, 210 nm). $[\alpha]_D^{20}$ −82.0 (c = 0.59, $CHCl_3$).

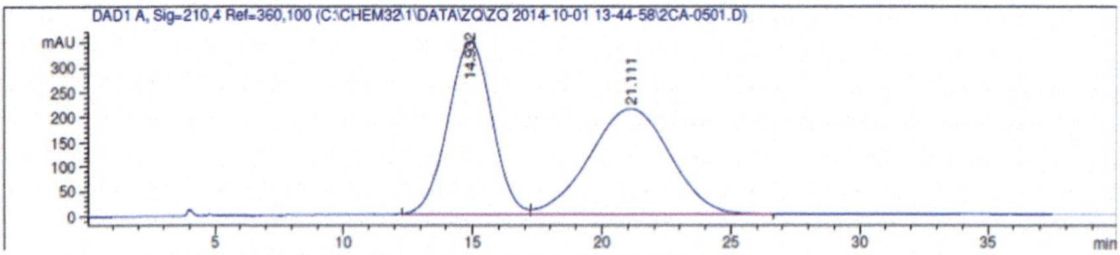


Signal 1: DAD1 A, Sig=210,4 Ref=360,100

| Peak # | RetTime [min] | Type | Width [min] | Area [mAU*s] | Height [mAU] | Area % |
|---|---|---|---|---|---|---|
| 1 | 14.932 | BV | 1.7668 | 4.03063e4 | 347.59616 | 46.9314 |
| 2 | 21.111 | VB | 2.5522 | 4.55771e4 | 212.67628 | 53.0686 |

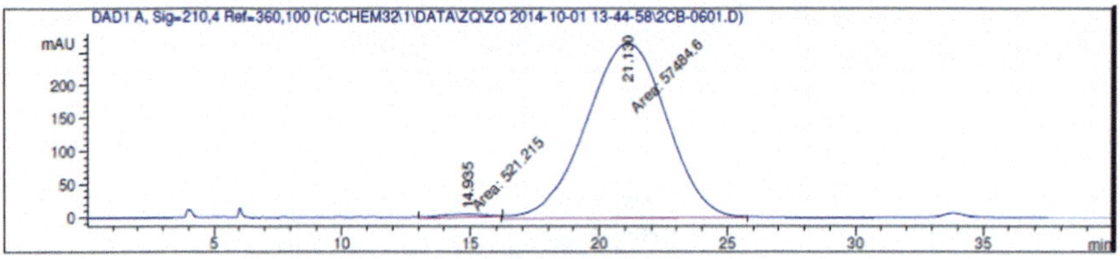


Signal 1: DAD1 A, Sig=210,4 Ref=360,100

| Peak # | RetTime [min] | Type | Width [min] | Area [mAU*s] | Height [mAU] | Area % |
|---|---|---|---|---|---|---|
| 1 | 14.935 | MM | 1.4719 | 521.21490 | 4.18988 | 0.8986 |
| 2 | 21.130 | MM | 3.6123 | 5.74846e4 | 265.22968 | 99.1014 |

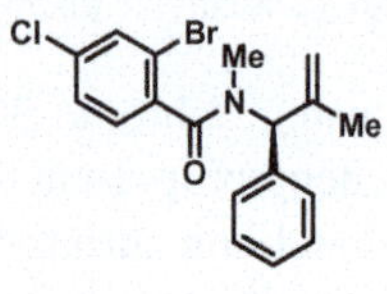


(*S*)-2-bromo-4-chloro-*N*-methyl-*N*-(2-methyl-1-phenylallyl)benzamide (2.85c)

Prepared according to **GP1** using 2-bromo-4-chlorobenzoic acid (353 mg, 1.5 mmol) and (*S*)-*N*,2-dimethyl-1-phenylprop-2-en-1-amine (266 mg, 1.65 mmol). The carboxamide was purified by flash column chromatography using hexanes: EtOAc (4:1 v:v) and was obtained as a clear and colourless oil (503 mg, 1.3 mmol, 89%).

*$^{1}$H NMR analysis showed the desired compound to be present as a complex mixture of at least 4 rotamers, and only approximate integration values are displayed.*

**$^{1}$H NMR** (500 MHz, $CDCl_3$) δ 7.69–7.53 (m, ~0.8H), 7.42–7.09 (m, 6H), 7.09–6.97 (m, ~0.5H), 6.67 (d, *J* = 8.2 Hz, ~0.2H), 6.36 (s, ~0.5H), 5.26–5.13 (m, 1H),

5.07–4.85 (m, ~1.25H), 4.59 (s, ~0.2H), 3.23–2.47 (m, 3H), 1.93–1.49 (m, 3H). **$^{13}$C NMR** (126 MHz, $CDCl_3$) δ 169.6, 168.8, 168.6, 143.0, 142.5, 141.7, 138.6, 137.3, 137.1, 136.7, 136.4, 136.2, 132.8, 132.7, 132.6, 132.5, 129.7, 129.0, 128.7, 128.5, 128.4, 128.3, 128.1, 127.9, 127.7, 127.5, 127.4, 127.2, 119.9, 119.8, 119.4, 118.3, 114.8, 114.3, 113.0, 68.0, 67.9, 62.1, 33.2, 33.0, 30.5, 30.3, 22.5, 21.9, 21.4, 21.1. **IR** (neat, cm$^{-1}$) 3086, 3028, 2972, 2918, 1636, 1586, 1549, 1450, 1373, 1248, 1099, 1074, 1030, 912. **HRMS** (DART) calculated 378.02603 m/z (found 378.02738 for $C_{18}H_{18}BrClNO$). Enantiomeric purity was determined by HPLC analysis in comparison with racemic material $T_R = 12.8$ (minor) $T_R = 22.5$ (major) (99:1 er shown; Chiralcel AD-H column, 10% $^i$PrOH in hexanes 1.0 mL/min, 210 nm). $[\alpha]_D^{20}$ −82.4 (c = 1.0, $CHCl_3$).

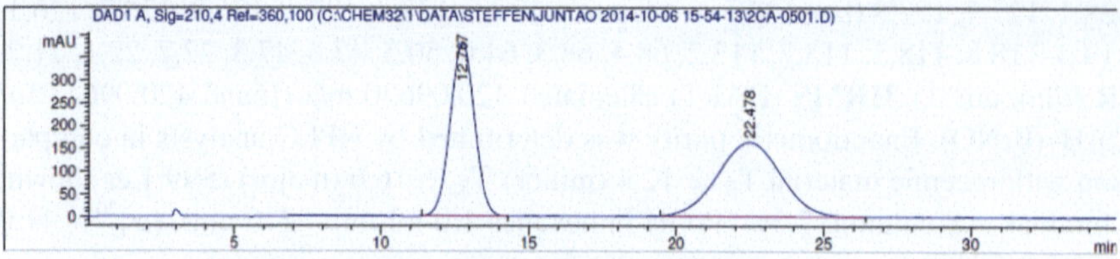


Signal 1: DAD1 A, Sig=210,4 Ref=360,100

| Peak # | RetTime [min] | Type | Width [min] | Area [mAU*s] | Height [mAU] | Area % |
|---|---|---|---|---|---|---|
| 1 | 12.767 | BB | 0.8921 | 2.18141e4 | 383.19662 | 47.0185 |
| 2 | 22.478 | BB | 1.9534 | 2.45805e4 | 164.11627 | 52.9815 |

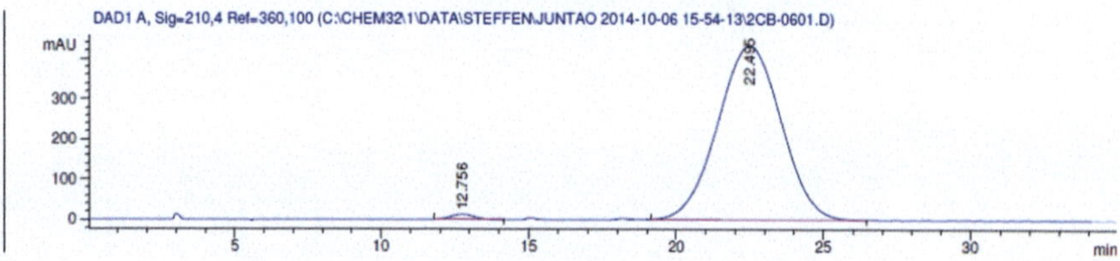


Signal 1: DAD1 A, Sig=210,4 Ref=360,100

| Peak # | RetTime [min] | Type | Width [min] | Area [mAU*s] | Height [mAU] | Area % |
|---|---|---|---|---|---|---|
| 1 | 12.756 | BB | 0.7317 | 723.77069 | 13.11717 | 1.1018 |
| 2 | 22.495 | BB | 2.2289 | 6.49642e4 | 432.86398 | 98.8982 |

(*S*)-*N*-benzyl-2-bromo-*N*-(2-methyl-1-phenylallyl)benzamide (2.85d)

Prepared according to **GP1** using 2-bromobenzoic acid (258 mg, 0.96 mmol) and (*S*)-*N*-benzyl-2-methyl-1-phenylprop-2-en-1-amine (250 mg, 1.05 mmol). The carboxamide was purified by flash column chromatography using hexanes: EtOAc (7:2 v:v) and was obtained as a white solid (290 mg, 0.7 mmol, 72%).

*$^{1}H$ NMR analysis showed the desired compound to be present as a complex mixture of 4 rotamers, and only approximate integration values are displayed.*

**$^{1}$H NMR** (500 MHz, $CDCl_3$) δ 7.67–7.55 (m, ~0.8H), 7.52–6.80 (m, ~12.5H), 6.72–5.97 (m, ~0.5H), 5.42–4.00 (m, ~5H), 1.84 – 1.78 (m, ~0.5H), 1.61 (s, ~0.3H), 1.57–1.54 (m, ~0.6H), 1.36–1.32 (m, ~1.7H). **$^{13}$C NMR** (125 MHz, $CDCl_3$) δ 171.2, 170.4, 143.9, 143.3, 142.5, 138.6, 138.3, 138.1 overlapping 138.0, 137.2, 134.7, 133.1, 132.8, 132.4, 130.6 overlapping 130.5 (br), 130.2 overlapping 130.1 (br), 129.8 (br), 128.8 (br), 128.5, 128.4, 128.3 overlapping 128.3 overlapping 128.2, 128.1, 127.9, 127.8 (br), 127.5 overlapping 127.5, 127.4, 127.1, 126.9, 126.7, 126.1, 119.3, 118.8, 118.5, 113.7, 113.2, 68.4, 68.3, 64.0, 50.3, 47.6, 47.3, 22.7, 22.1, 21.2. **IR** (film, $cm^{-1}$). **HRMS** (DART) calculated 420.09630 m/z (found 420.09643 for $C_{24}H_{23}BrNO$). Enantiomeric purity was determined by HPLC analysis in comparison with racemic material $T_R$ = 12.4 (minor) $T_R$ = 16.6 (major) (>99:1 er shown; Chiralpak AS column, 7.5% $^{i}$PrOH in hexanes 1.0 mL/min, 210 nm). $[\alpha]_D^{20}$ −41.0 (c = 1.0, $CHCl_3$).

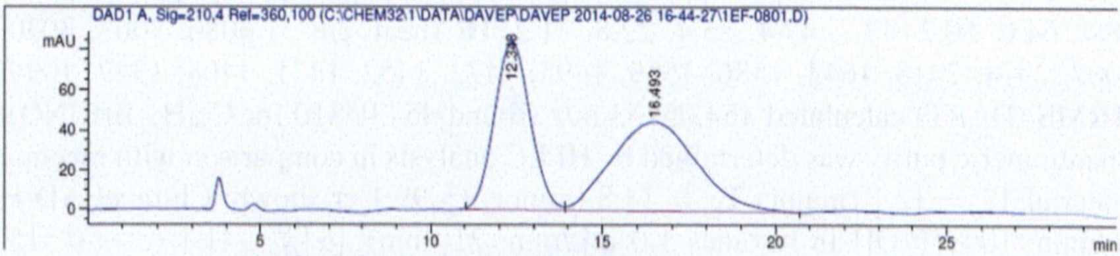


Signal 1: DAD1 A, Sig=210,4 Ref=360,100

| Peak # | RetTime [min] | Type | Width [min] | Area [mAU*s] | Height [mAU] | Area % |
|---|---|---|---|---|---|---|
| 1 | 12.348 | BV | 1.1071 | 6066.58447 | 83.95277 | 44.4628 |
| 2 | 16.493 | VB | 1.9994 | 7577.59229 | 44.70697 | 55.5372 |

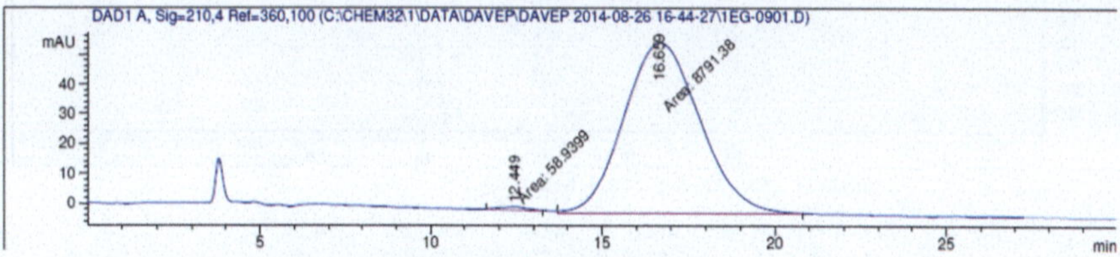


Signal 1: DAD1 A, Sig=210,4 Ref=360,100

| Peak # | RetTime [min] | Type | Width [min] | Area [mAU*s] | Height [mAU] | Area % |
|---|---|---|---|---|---|---|
| 1 | 12.449 | MM | 0.9483 | 58.93990 | 1.03593 | 0.6660 |
| 2 | 16.659 | MM | 2.5544 | 8791.37793 | 57.36147 | 99.3340 |

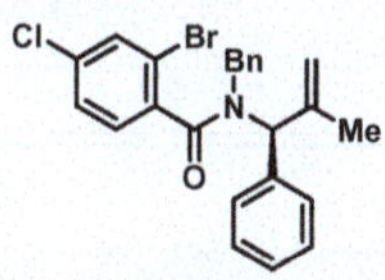


(*S*)-*N*-benzyl-2-bromo-4-chloro-*N*-(2-methyl-1-phenylallyl)benzamide (2.85e)

Prepared according to **GP1** using 2-bromo-4-chlorobenzoic acid (193 mg, 0.82 mmol) and (*S*)-*N*-benzyl-2-methyl-1-phenylprop-2-en-1-amine (210 mg, 0.9 mmol). The carboxamide was purified by flash column chromatography using hexanes: EtOAc (7:2 v:v) and was obtained as a clear crystalline solid (273 mg, 0.6 mmol, 73%, M.P.: 111–112 °C).

*$^1$H NMR analysis showed the desired compound to be present as a complex mixture of 4 rotamers, and only approximate integration values are displayed.*

**$^1$H NMR** (600 MHz, $CDCl_3$) δ 7.68–7.63 (m, ~0.5H), 7.60 (d, $J$ = 2.0 Hz, ~0.2H), 7.53–6.81 (m, ~11.5H), 6.76 (d, $J$ = 8.0 Hz, ~0.1H), 6.74–6.01 (m, ~0.6H), 5.32 (d, $J$ = 14.5 Hz, ~0.5H), 5.29–5.27 (m, ~0.6H), 5.19–5.12 (m, ~1.2H), 5.08 (s, ~0.2H), 4.90 (s, ~0.6H), 4.81 (s, ~0.2H), 4.67 (d, $J$ = 14.5 Hz, ~0.2H), 4.53 (m, ~0.1H), 4.39 (d, $J$ = 16.0 Hz, ~0.2H), 4.05 (d, $J$ = 14.5 Hz, ~0.6H), 1.84–1.32 (m, 3H). **$^{13}$C NMR** (126 MHz, $CDCl_3$) δ 170.4, 169.5, 143.9, 143.2, 142.4, 138.4, 138.1, 137.3, 137.0, 136.8, 136.6, 135.4, 135.3, 135.0, 134.4, 132.9, 132.5, 132.0, 130.6, 130.5, 128.7, 128.5, 128.4, 128.3, 128.2, 128.0, 127.9, 127.7, 127.5, 127.5, 127.3, 126.8, 126.7, 126.2, 119.9, 119.4, 118.8, 113.9, 113.3, 111.4, 68.5,

68.4, 64.0, 50.2, 47.7, 47.4, 25.4, 22.8, 21.2. **IR** (neat, $cm^{-1}$) 3086, 3063, 3030, 3007, 2974, 2918, 1643, 1586, 1559, 1495, 1472, 1452, 1431, 1408, 1327, 1099. **HRMS** (DART) calculated 454.05733 m/z (found 454.05810 for $C_{24}H_{22}BrClNO$). Enantiomeric purity was determined by HPLC analysis in comparison with racemic material $T_R$ = 12.1 (major) $T_R$ = 14.8 (minor) (> 99:1 er shown; Chiracel AD-H column, 10% $^{i}$PrOH in hexanes 1.0 mL/min, 210 nm). $[\alpha]_D^{20}$ −41.1 (c = 0.515, $CHCl_3$).

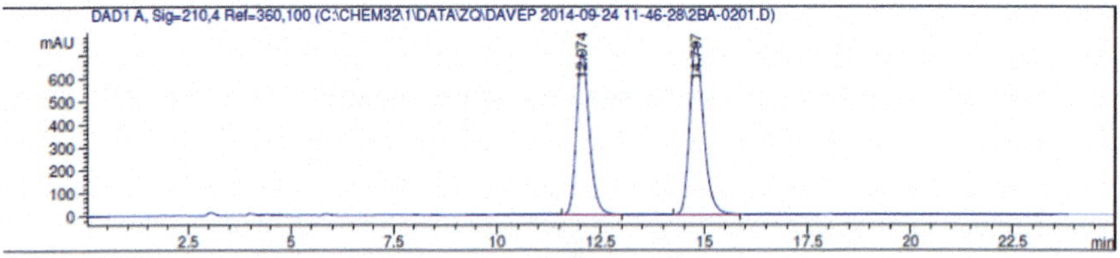


Signal 1: DAD1 A, Sig=210,4 Ref=360,100

| Peak # | RetTime [min] | Type | Width [min] | Area [mAU*s] | Height [mAU] | Area % |
|---|---|---|---|---|---|---|
| 1 | 12.074 | BB | 0.3363 | 1.53775e4 | 705.54602 | 46.6989 |
| 2 | 14.797 | BB | 0.3576 | 1.75515e4 | 753.94946 | 53.3011 |

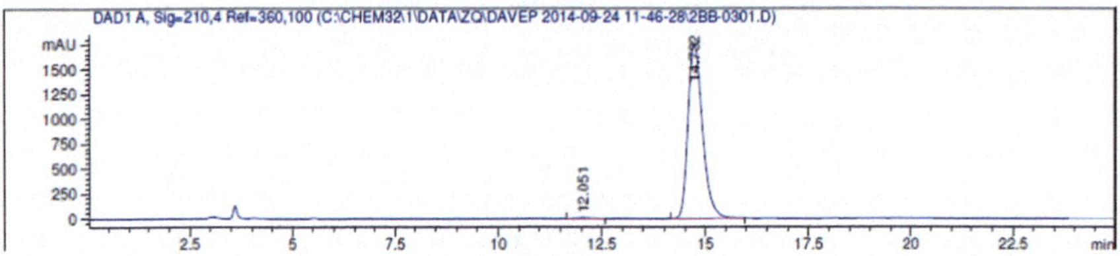


Signal 1: DAD1 A, Sig=210,4 Ref=360,100

| Peak # | RetTime [min] | Type | Width [min] | Area [mAU*s] | Height [mAU] | Area % |
|---|---|---|---|---|---|---|
| 1 | 12.051 | BB | 0.3251 | 174.14737 | 8.22233 | 0.4032 |
| 2 | 14.732 | BB | 0.3794 | 4.30188e4 | 1758.78772 | 99.5968 |

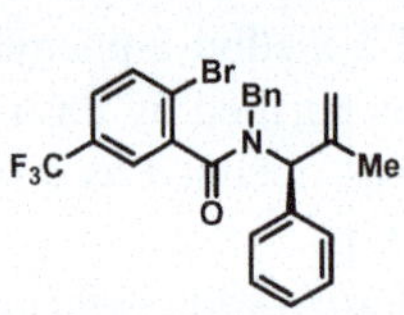


(*S*)-*N*-benzyl-2-bromo-*N*-(2-methyl-1-phenylallyl)-5-(trifluoromethyl)benzamide (2.85f)

Prepared according to **GP1** using 2-bromo-5-(trifluoromethyl)benzoic acid (258.24 mg, 0.96 mmol) and (*S*)-*N*-benzyl-2-methyl-1-phenylprop-2-en-1-amine (250 mg, 1.05 mmol). The carboxamide was purified by flash column chromatography using hexanes: EtOAc (15:1 v:v) and was obtained as white powder (314 mg, 0.64 mmol, 67%, M.P.: 89–91 °C).

*$^{1}H$ NMR analysis showed the desired compound to be present as a complex mixture of 4 rotamers, and only approximate integration values are displayed.*

**$^{1}$H NMR** (500 MHz, $CDCl_3$) δ 7.78 (d, *J* = 8.5 Hz, ~0.5H), 7.74–6.73 (m, 12H), 6.63–6.14 (m, ~0.5H), 5.38–5.14 (m, ~2.5H), 4.96 (s, ~0.2H), 4.82 (s, ~0.6H), 4.80 (s, ~0.2H), 4.69 (d, *J* = 14.5 Hz, ~0.1H), 4.61 (d, *J* = 17.0 Hz, ~0.2H), 4.45–4.30 (m, ~0.2H), 4.05 (d, *J* = 15.0 Hz, ~0.6H), 1.90–1.76 (m, ~0.6H), 1.55 (s, ~0.5H), 1.35–1.31 (m, ~1.8H). **$^{13}$C NMR** (126 MHz, $CDCl_3$) δ 169.9, 169.0, 144.0, 143.1, 142.3, 139.1, 138.9, 138.5, 138.0, 137.8, 136.9, 134.1, 133.6, 133.4, 132.8, 130.5, 130.1, 129.8, 129.6, 129.3, 128.9, 128.6, 128.5, 128.4, 128.1, 127.9, 127.8, 127.6, 127.5, 127.1, 127.0, 126.8, 126.7, 126.7, 126.7, 126.3, 124.7, 124.7, 124.7, 124.6, 124.6, 124.5, 123.1, 122.9, 122.9, 122.3, 120.2, 119.3, 114.1, 113.3, 68.8, 68.5, 63.8, 49.9, 47.9, 47.4, 22.8, 22.1, 20.9. **$^{19}$F NMR** (564 MHz, $CDCl_3$) δ −62.83, −63.03, −63.18, −63.29. **IR** (neat, $cm^{-1}$)3088, 3065, 3030, 2918, 1645, 1605, 1497, 1454, 1418, 1393, 1335, 1173, 1132, 1080, 1034. **HRMS** (DART) calculated 488.08369 m/z (found 488.08439 m/z for $C_{25}H_{22}BrF_3NO$). Enantiomeric purity was determined by HPLC analysis in comparison with racemic material $T_R$ = 10.05 (major) $T_R$ = 11.0 (minor) (99:1 er shown; Chiracel AD-H column, 5% $^{i}$PrOH in hexanes 1.0 mL/min, 210 nm). $[\alpha]_D^{20}$ −26.403 (c = 0.515, $CHCl_3$).

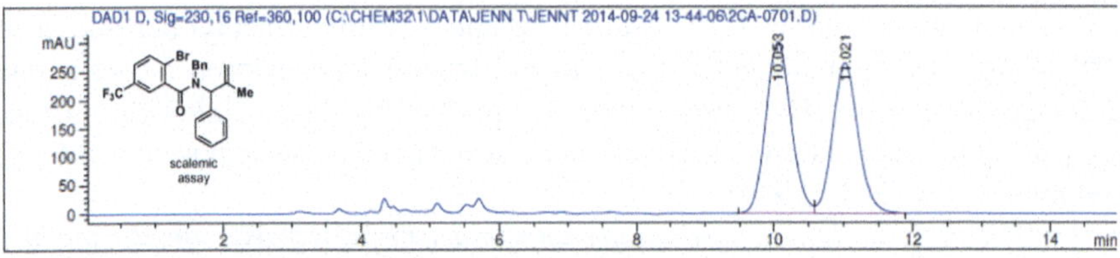


Signal 4: DAD1 D, Sig=230,16 Ref=360,100

| Peak # | RetTime [min] | Type | Width [min] | Area [mAU*s] | Height [mAU] | Area % |
|---|---|---|---|---|---|---|
| 1 | 10.053 | BV | 0.3933 | 7632.86865 | 301.52750 | 53.2043 |
| 2 | 11.021 | VB | 0.3919 | 6713.48242 | 264.72144 | 46.7957 |

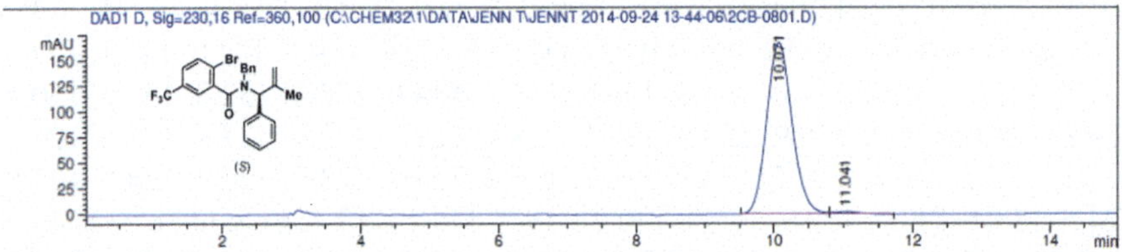


Signal 4: DAD1 D, Sig=230,16 Ref=360,100

| Peak # | RetTime [min] | Type | Width [min] | Area [mAU*s] | Height [mAU] | Area % |
|---|---|---|---|---|---|---|
| 1 | 10.071 | BB | 0.3950 | 4258.08105 | 167.24721 | 98.7482 |
| 2 | 11.041 | BB | 0.3734 | 53.97638 | 2.14646 | 1.2518 |

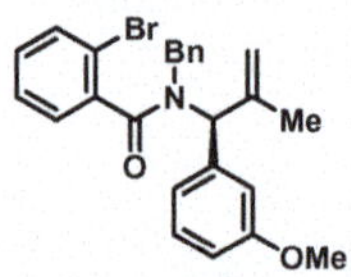

(*S*)-*N*-benzyl-2-bromo-*N*-(1-(3-methoxyphenyl)-2-methylallyl)benzamide (2.85h)

Prepared according to **GP1** using 2-bromobenzoic acid (124 mg, 0.62 mmol) and (*S*)-*N*-benzyl-1-(3-methoxyphenyl)-2-methylprop-2-en-1-amine (184 mg, 0.68 mmol). The carboxamide was purified by flash column chromatography using hexanes:DCM:EtOAc (10:2:1 v:v:v) and was obtained as a white solid (192 mg, 0.43 mmol, 70%, M.P.: 122–123 °C).

*$^1$H NMR analysis showed the desired compound to be present as a complex mixture of at least 3 rotamers, and only approximate integration values are displayed.*

**$^1$H NMR** (500 MHz, $CDCl_3$) δ 7.63 (m, ~0.6H), 7.37–6.78 (m, 11H), 6.68–6.51 (m, ~1H), 5.38 (d, $J$ = 14.5 Hz, ~0.6H), 5.29–5.25 (m, ~0.6H), 5.22–5.06 (m, ~1.3H), 4.89 (s, ~0.6H), 4.83 (s, ~0.2H), 4.74 (d, $J$ = 15.0 Hz, ~0.2H), 4.56–4.39 (m, ~0.1H), 4.01 (d, $J$ = 14.5 Hz, ~0.6H), 3.78–3.58 (m, 3H)1.80 (s, ~0.5H), 1.56–1.53 (m, 1H), 1.38–1.33 (m, 1H). **$^{13}$C NMR** (125 MHz, $CDCl_3$) δ 171.2, 170.4, 159.7, 159.6, 159.5, 143.8, 143.1, 140.2, 138.4, 138.3, 138.3, 138.0, 136.1, 133.1, 132.8, 130.2, 130.1, 129.5, 129.3, 129.2, 128.4, 128.0, 128.0, 127.6, 127.5, 127.4, 127.1, 126.9, 126.7, 126.2, 122.4, 120.4, 119.3, 118.7, 118.6, 116.3, 114.5, 114.0, 113.8, 113.3, 112.9, 68.5, 68.2, 55.3, 55.1 overlapping 55.1, 47.6, 47.4, 22.7, 21.3. **IR** (film, cm$^{-1}$) 3063, 2937, 1651, 1600, 1494, 1431, 1406, 1327, 1261, 1165, 1045. **HRMS** (ESI +) calculated 450.1063 m/z (found 450.1074 for $C_{25}H_{25}BrNO_2$). Enantiomeric purity was determined by HPLC analysis in comparison with racemic material $T_R$ = 13.1 (minor) $T_R$ = 15.7 (major) (99:1 er shown; Chiracel AD-H column, 10% $^i$PrOH in hexanes 1.0 mL/min, 210 nm). $[\alpha]_D^{20}$ −35.2 (c = 0.5, $CHCl_3$).

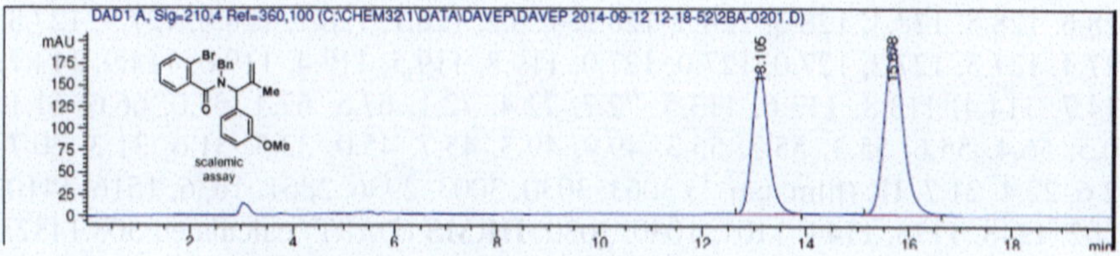


Signal 1: DAD1 A, Sig=210,4 Ref=360,100

| Peak # | RetTime [min] | Type | Width [min] | Area [mAU*s] | Height [mAU] | Area % |
|---|---|---|---|---|---|---|
| 1 | 13.105 | BB | 0.3310 | 3559.24121 | 164.10214 | 41.5027 |
| 2 | 15.678 | BB | 0.3954 | 5016.68506 | 194.19519 | 58.4973 |

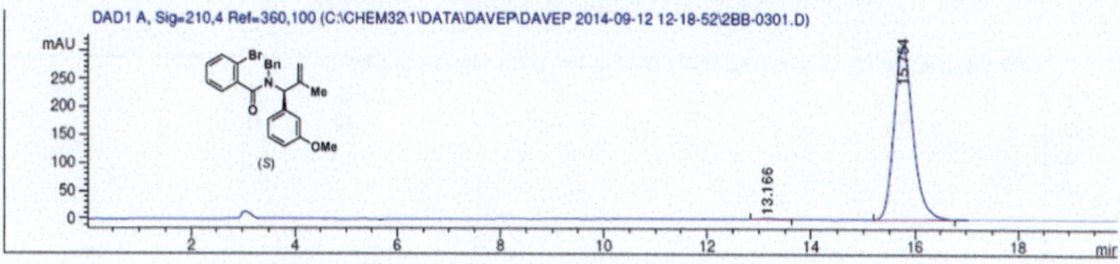


Signal 1: DAD1 A, Sig=210,4 Ref=360,100

| Peak # | RetTime [min] | Type | Width [min] | Area [mAU*s] | Height [mAU] | Area % |
|---|---|---|---|---|---|---|
| 1 | 13.166 | BB | 0.3092 | 37.95173 | 1.73510 | 0.4678 |
| 2 | 15.754 | BB | 0.3977 | 8074.25342 | 310.18015 | 99.5322 |

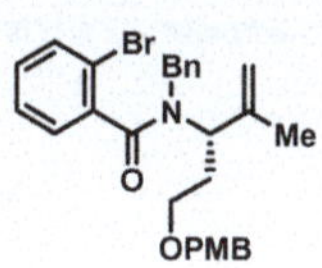


(*S*)-*N*-benzyl-2-bromo-*N*-(5-((4-methoxybenzyl)oxy)-2-methylpent-1-en-3-yl)benzamide (2.85k)

Prepared according to **GP1** using 2-bromobenzoic acid (136 mg, 0.68 mmol) and (*S*)-5-((4-methoxybenzyl)oxy)-*N*,2-dimethylpent-1-en-3-amine (244 mg, 0.752 mmol, 1.1 equiv). The carboxamide was purified by flash column chromatography using hexanes:EtOAc (3:1 v:v) and was obtained as a clear and colourless oil (219 mg, 0.43 mmol, 63%).

*$^1$H NMR analysis showed the desired compound to be present as a complex mixture of at least 3 rotamers, and only approximate integration values are displayed.*

**$^1$H NMR** (600 MHz, $CDCl_3$) δ 7.60–7.58 (m, ~0.2H), 7.53–7.46 (m, ~1.4H), 7.42–7.39 (m, ~0.4H), 7.32–7.22 (m, ~3.4H), 7.21–7.09 (m, ~3.6H), 7.04 (m, ~0.2H), 7.02–6.94 (m, 2H), 6.88 (m, 1H), 6.83–6.77 (m, 1H), 5.29–5.23 (m, ~0.6H), 5.14–5.01 (m, ~0.5H), 4.96 (s, ~0.2H), 4.91 (s, ~0.3H), 4.84 (s, ~0.5H), 4.50–4.43 (m,

~0.5H), 4.40–4.30 (m, 1H), 4.22–4.06 (m, ~2.5H), 3.79 (s, 2H), 3.78 (s, ~1H), 3.61–3.46 (m, 1H), 3.29–3.19 (m, ~0.5H), 3.10–2.98 (m, ~0.5H), 2.21–2.12 (m, ~0.4H), 2.05–1.92 (m, ~1H), 1.92–1.76 (m, ~2H), 1.68 (s, ~0.6H), 1.63 (s, ~0.9H). **$^{13}$C NMR** (126 MHz, $CDCl_3$) δ 170.6, 170.3, 170.2, 159.2, 159.1, 159.0, 142.9, 142.7, 142.0, 141.0, 139.0, 138.9, 138.8, 138.2, 137.6, 137.5, 137.2, 133.4, 133.1, 132.9, 132.6, 130.6, 130.5, 130.5, 130.2, 130.1, 130.0, 129.9, 129.5, 129.5, 129.1, 128.9, 128.6, 128.5, 128.3, 128.3, 128.3, 128.2, 128.2, 128.1, 128.1, 128.0, 127.9, 127.8, 127.4, 127.3, 127.2, 127.0, 127.0, 127.0, 119.8, 119.5, 119.4, 119.0, 114.9, 114.7, 114.7, 114.4, 113.8, 113.6, 113.5, 72.7, 72.4, 72.1, 67.5, 67.4, 67.0, 66.6, 61.1, 60.5, 56.4, 55.6, 55.3, 55.3, 55.3, 49.9, 49.3, 45.7, 45.0, 32.5, 31.6, 31.3, 30.7, 22.6, 22.4, 21.7. **IR** (film, $cm^{-1}$) 3063, 3030, 3003, 2936, 2861, 1636, 1516, 1410, 1302, 1248, 1173, 1144, 1101, 1080, 1030. **HRMS** (DART) calculated 508.14873 (found 508.15041 for $C_{28}H_{31}BrNO_3$). Enantiomeric purity was determined by HPLC analysis in comparison with racemic material $T_R = 20.3$ (major) $T_R = 22.8$ (minor) (> 99:1 er shown; Chiracel AD-H column, 10% $^i$PrOH in hexanes 1.0 mL/min, 210 nm). $[\alpha]_D^{20}$ −86.475 (c = 0.555, $CHCl_3$).

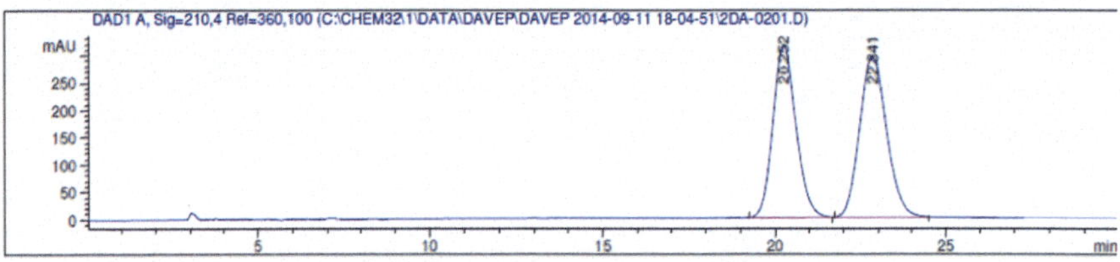


Signal 1: DAD1 A, Sig=210,4 Ref=360,100

| Peak # | RetTime [min] | Type | Width [min] | Area [mAU*s] | Height [mAU] | Area % |
|---|---|---|---|---|---|---|
| 1 | 20.252 | BB | 0.7263 | 1.47514e4 | 315.02893 | 48.2165 |
| 2 | 22.841 | BB | 0.8343 | 1.58427e4 | 294.18256 | 51.7835 |

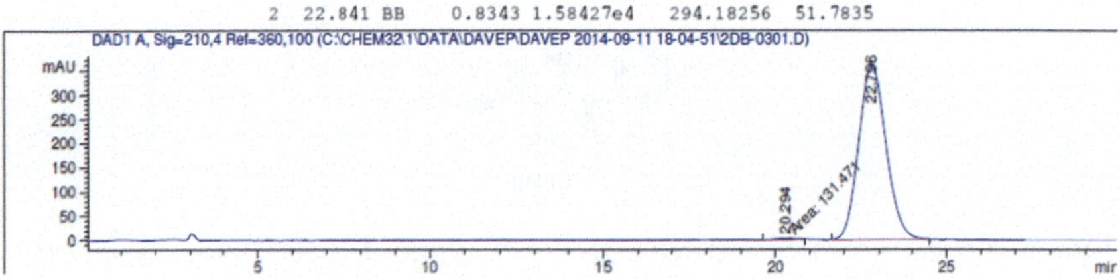


Signal 1: DAD1 A, Sig=210,4 Ref=360,100

| Peak # | RetTime [min] | Type | Width [min] | Area [mAU*s] | Height [mAU] | Area % |
|---|---|---|---|---|---|---|
| 1 | 20.294 | MM | 0.5883 | 131.47110 | 2.79527 | 0.6669 |
| 2 | 22.796 | BB | 0.8414 | 1.95831e4 | 359.54208 | 99.3331 |

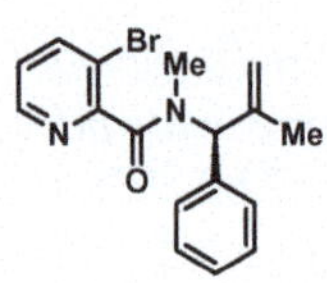


(*S*)-3-bromo-*N*-methyl-*N*-(2-methyl-1-phenylallyl)picolinamide (2.85o)

Prepared according to **GP1** using 3-bromopicolinic acid (202 mg, 1.0 mmol) and (*S*)-*N*,2-dimethyl-1-phenylprop-2-en-1-amine (177 mg, 1.1 mmol). The carboxamide was purified by flash column chromatography using hexanes: EtOAc (3:2 v:v) and was obtained as a clear orange oil (183 mg, 0.53 mmol, 53%).

*$^1$H NMR analysis showed the desired compound to be present as a mixture of 2 rotamers, and only approximate integration values are displayed.*

**$^1$H NMR** (500 MHz, $CDCl_3$) δ 8.60–8.48 (m, 1H), 7.96–7.86 (m, 1H), 7.40–7.22 (m, 5H), 7.22–7.14 (m, 1H), 6.39 (s, ~0.7H), 5.23–5.14 (m, 1H), 5.09 (s, ~0.3H), 5.07–4.81 (m, 1H), 3.11–2.47 (m, 3H), 1.91–1.54 (m, 3H). **$^{13}$C NMR** (126 MHz, $CDCl_3$) δ 168.2, 167.6, 155.1, 154.6, 148.1, 147.5, 141.9, 141.5, 140.8, 140.5, 136.7, 136.6, 129.4, 128.9, 128.5, 128.4, 127.8, 127.7, 124.7, 117.9, 117.0, 115.9, 114.2, 67.0, 61.8, 32.5, 30.2, 21.7, 21.5. **IR** (film, $cm^{-1}$) 3086, 3028, 2974, 2918, 1645, 1456, 1335, 1128, 1088, 1053. **HRMS** (DART) calculated 345.06025 m/z (found 345.05981 for $C_{17}H_{18}BrN_2O$). Enantiomeric purity was determined by HPLC analysis in comparison with racemic material $T_R = 28.3$ (minor) $T_R = 32.1$ (major) (99:1 er shown; Chiracel AD-H column, 10% $^i$PrOH in hexanes 0.35 mL/min, 210 nm). $[\alpha]_D^{20}$ −76.21 (c = 0.64, $CHCl_3$).

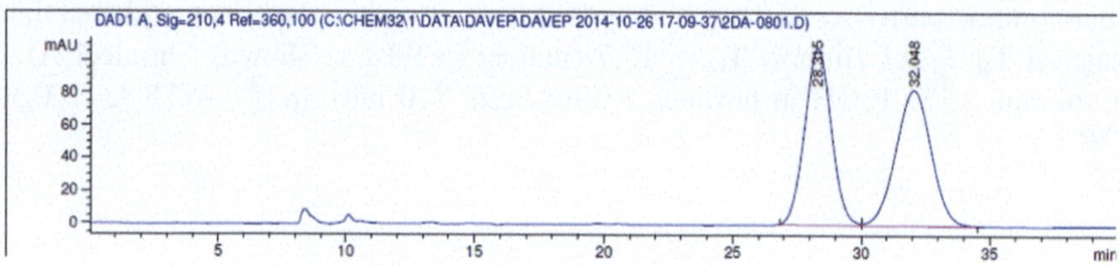


Signal 1: DAD1 A, Sig=210,4 Ref=360,100

| Peak # | RetTime [min] | Type | Width [min] | Area [mAU*s] | Height [mAU] | Area % |
|---|---|---|---|---|---|---|
| 1 | 28.315 | BV | 1.0980 | 7795.91602 | 108.01302 | 49.3578 |
| 2 | 32.048 | VB | 1.4193 | 7998.79736 | 83.07344 | 50.6422 |

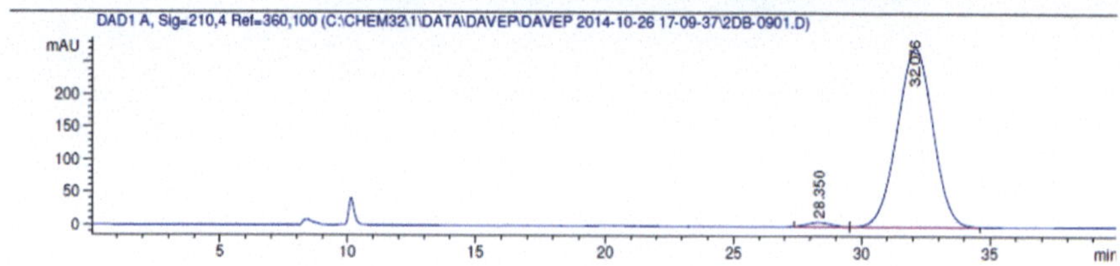


Signal 1: DAD1 A, Sig=210,4 Ref=360,100

| Peak # | RetTime [min] | Type | Width [min] | Area [mAU*s] | Height [mAU] | Area % |
|---|---|---|---|---|---|---|
| 1 | 28.350 | BV | 0.7607 | 409.10474 | 6.42606 | 1.5192 |
| 2 | 32.076 | VB | 1.4712 | 2.65197e4 | 273.50394 | 98.4808 |

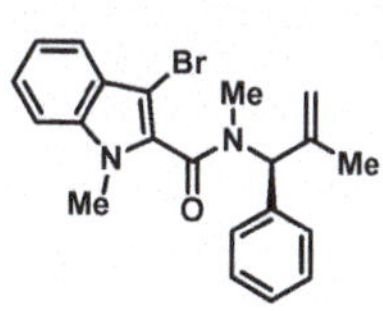


(*S*)-3-bromo-*N*,1-dimethyl-*N*-(2-methyl-1-phenylallyl)-1*H*-indole-2-carboxamide (2.85q)

Prepared according to **GP1** using 3-bromo-1-methyl-1*H*-indole-2-carboxylic acid (254 mg, 1 mmol) and (*S*)-*N*,2-dimethyl-1-phenylprop-2-en-1-amine (181 mg, 1.1 mmol). The carboxamide was purified by flash column chromatography using hexanes:EtOAc (3:1 v:v) and was obtained as a pale yellow oil (181 mg, 0.456 mmol, 45%).

*$^1$H NMR analysis showed the desired compound to be present as a complex mixture of 4 rotamers, and only approximate integration values are displayed.*

**$^1$H NMR** (500 MHz, $CDCl_3$) δ 7.62–7.51 (m, 1H), 7.47–7.12 (m, 7H), 6.90–6.80 (m, ~0.4H), 6.43 (s, ~0.4H), 6.31 (s, ~0.4H), 5.52–5.04 (m, 2H), 4.96–4.87 (m, ~0.5H), 4.59 (s, ~0.2H), 3.86–3.62 (m, ~2.4H), 3.26 (s, ~0.5H), 3.10–2.71 (m, 3H), 2.02–1.80 (m, ~2.4H), 1.47 (s, ~0.4H). **$^{13}$C NMR** (126 MHz, $CDCl_3$) δ 165.0, 164.5, 163.5, 163.2, 144.1, 142.6, 141.7, 141.6, 139.5, 137.2, 136.6, 136.4, 136.3, 136.2, 135.7, 132.0, 131.7, 129.8, 129.5, 128.9, 128.7, 128.6, 128.5, 128.2, 128.0, 127.8, 127.6, 126.4, 126.3, 123.9, 123.8, 123.7, 123.6, 120.9, 120.8, 119.8, 119.8, 119.7, 118.2, 115.1, 113.9, 112.6, 109.9, 109.8, 89.7, 89.2, 88.6, 88.2, 68.2, 67.5, 62.9, 62.4, 33.3, 32.9, 31.9, 31.4, 31.3, 31.1, 30.9, 22.3, 21.9, 21.5, 21.3. **IR** (neat, $cm^{-1}$) 3086, 3059, 3028, 2970, 2940, 2922, 2245, 1636, 1530,1464, 1447, 1399, 1319, 1065. **HRMS** (DART) calculated 397.09155 m/z (found 397.09117 m/z for $C_{21}H_{22}BrN_2O$). Enantiomeric purity was determined by HPLC analysis in comparison with racemic material $T_R$ = 9.1 (minor) $T_R$ = 15.7 (major) (> 99:1 er shown; Chiralcel AD-H column, 15% $^i$PrOH in hexanes 1.0 mL/min, 210 nm). $[\alpha]_D^{20}$ −62.8 (c = 0.5, $CHCl_3$).

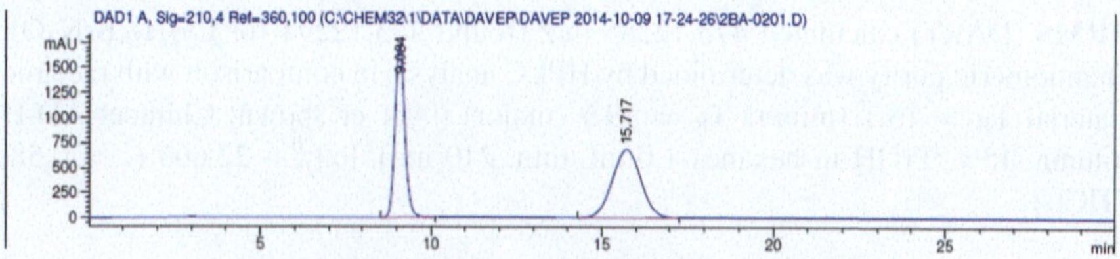


Signal 1: DAD1 A, Sig=210,4 Ref=360,100

| Peak # | RetTime [min] | Type | Width [min] | Area [mAU*s] | Height [mAU] | Area % |
|---|---|---|---|---|---|---|
| 1 | 9.084 | BB | 0.3074 | 3.41675e4 | 1722.37573 | 46.4159 |
| 2 | 15.717 | BB | 0.9009 | 3.94441e4 | 681.87775 | 53.5841 |

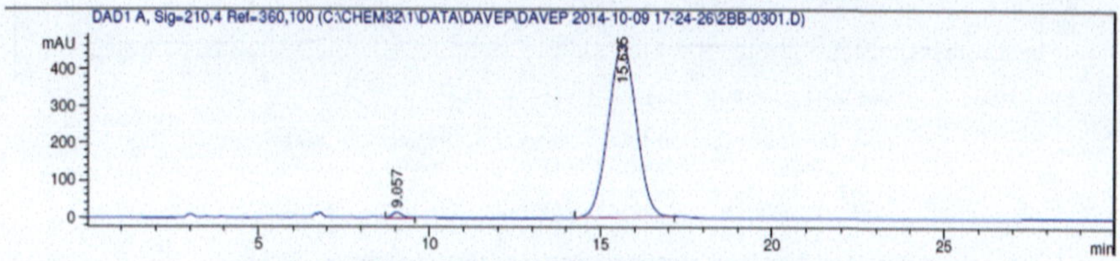


Signal 1: DAD1 A, Sig=210,4 Ref=360,100

| Peak # | RetTime [min] | Type | Width [min] | Area [mAU*s] | Height [mAU] | Area % |
|---|---|---|---|---|---|---|
| 1 | 9.057 | BB | 0.2901 | 258.10187 | 13.80371 | 0.9683 |
| 2 | 15.635 | BB | 0.8919 | 2.63982e4 | 463.89978 | 99.0317 |

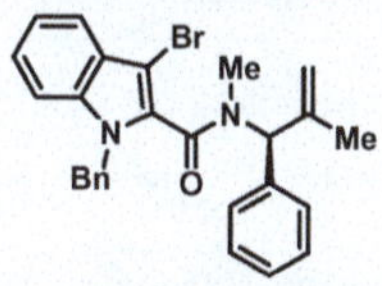


(*S*)-1-benzyl-3-bromo-*N*-methyl-*N*-(2-methyl-1-phenylallyl)-1*H*-indole-2-carboxamide (2.85r)

Prepared according to **GP1** using 1-benzyl-3-bromo-1*H*-indole-2-carboxylic acid (330 mg, 1 mmol) and (*S*)-*N*,2-dimethyl-1-phenylprop-2-en-1-amine (177 mg, 1.1 mmol). The carboxamide was purified by flash column chromatography using hexanes:DCM:EtOAc (20:10:1 v:v:v) and was obtained as a hygroscopic white foam (410 mg, 0.87 mmol, 87%).

*$^1$H NMR analysis showed the desired compound to be present as a mixture of 2 rotamers, and only approximate integration values are displayed.*

**$^1$H NMR** (500 MHz, $CDCl_3$) δ 7.66–7.58 (m, 1H), 7.42–7.13 (m, ~10H), 7.06–6.92 (m, 3H), 6.33 (s, ~0.5H), 6.18 (s, ~0.5H), 5.58 (d, $J = 17.0$ Hz, ~0.5H), 5.53–5.41 (m, $J = 2$H), 5.21 (m, ~0.5H), 5.05 (s, ~0.5H), 4.92–4.90 (m, ~0.5H), 4.22 (s, ~0.5), 2.66 (s, ~1.5), 2.63 (s, ~1.3H), 1.91 (s, ~1.5H), 1.68–1.66 (m, ~1.3H). **$^{13}$C NMR** (126 MHz, $CDCl_3$) δ 163.4, 163.1, 141.9, 140.8, 137.5, 137.3, 136.8, 136.8, 136.7, 136.3, 131.3, 131.0, 129.8, 128.9, 128.7, 128.7, 128.5, 128.4, 127.9, 127.6, 127.6, 127.5, 126.4, 126.3, 126.3, 126.2, 124.4, 124.3, 121.1, 120.0, 120.0, 115.4, 112.8, 110.2, 110.2, 91.6, 91.1, 63.0, 62.3, 47.6, 47.6, 33.2, 32.8, 22.1, 21.3. **IR** (film, $cm^{-1}$) 3086, 3061, 3030, 2922, 1634, 1526, 1452, 1402, 1320, 1250, 1161, 1117, 1071.

**HRMS** (DART) calculated 473.12285 m/z (found 473.12294 for $C_{27}H_{26}BrN_2O$). Enantiomeric purity was determined by HPLC analysis in comparison with racemic material $T_R$ = 13.1 (minor) $T_R$ = 19.9 (major) (99:1 er shown; Chiracel AD-H column, 15% $^i$PrOH in hexanes 1.0 mL/min, 210 nm). $[\alpha]_D^{20}$ −22.605 (c = 0.58, $CHCl_3$).

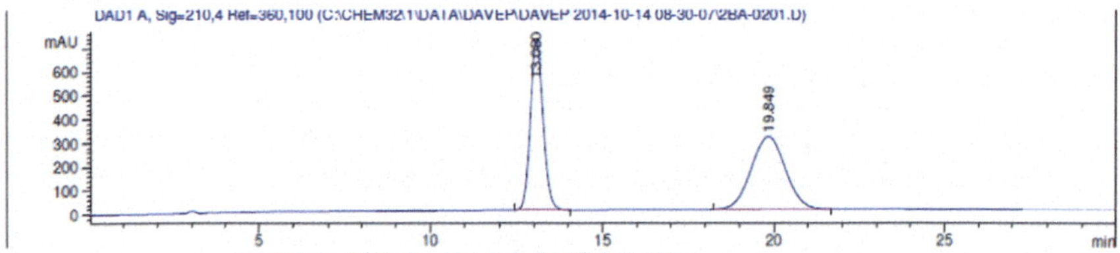


Signal 1: DAD1 A, Sig=210,4 Ref=360,100

| Peak # | RetTime [min] | Type | Width [min] | Area [mAU*s] | Height [mAU] | Area % |
|---|---|---|---|---|---|---|
| 1 | 13.080 | BB | 0.4034 | 1.85438e4 | 713.05786 | 46.7444 |
| 2 | 19.849 | BB | 1.0756 | 2.11268e4 | 305.18262 | 53.2556 |

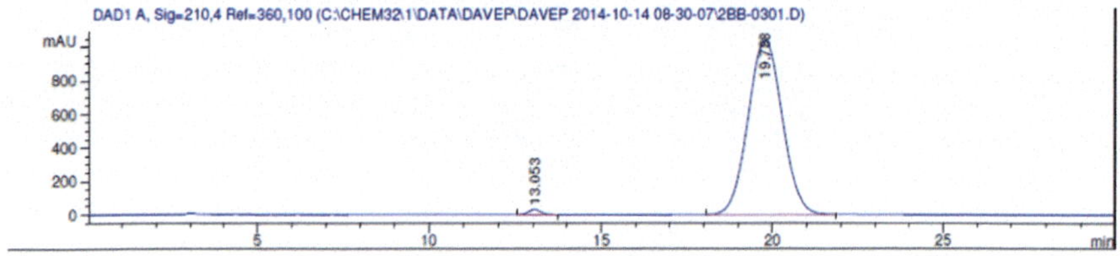


Signal 1: DAD1 A, Sig=210,4 Ref=360,100

| Peak # | RetTime [min] | Type | Width [min] | Area [mAU*s] | Height [mAU] | Area % |
|---|---|---|---|---|---|---|
| 1 | 13.053 | BB | 0.3940 | 829.68463 | 32.26740 | 1.1392 |
| 2 | 19.788 | BB | 1.0842 | 7.19987e4 | 1036.74780 | 98.8608 |

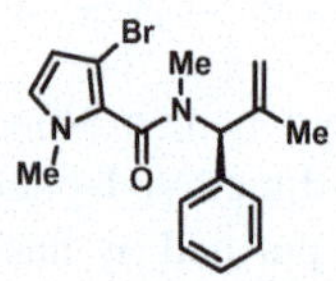


(*S*)-3-bromo-*N*,1-dimethyl-*N*-(2-methyl-1-phenylallyl)-1*H*-pyrrole-2-carboxamide (2.85s)

Prepared according to **GP1** using 3-bromo-1-methyl-1*H*-pyrrole-2-carboxylic acid (245 mg, 1.2 mmol) and (*S*)-*N*,2-dimethyl-1-phenylprop-2-en-1-amine (213 mg, 1.32 mmol). The carboxamide was purified by flash column chromatography using hexanes: EtOAc (3:1 v:v) and was obtained as an off-white solid (375 mg, 1.08 mmol, 90% (MP = 78–80 °C).

*$^1$H NMR Analysis of the Pure Title Compound in CDCl3 at 300 K Revealed a Broad Series of Peaks Which Precluded Assignment, for This Reason Only a $^1$H NMR Spectra Has Been Provided.*

**IR** (film, $cm^{-1}$) 33109, 3061, 3028, 1628, 1526, 1485, 1441, 1304, 1067, 1017. **HRMS** (DART) calculated 347.07590 m/z (found 347.07593 m/z for $C_{17}H_{20}BrN_2O$. Enantiomeric purity was determined by HPLC analysis in comparison with racemic material $T_R = 5.3$ (minor) $T_R = 7.2$ (major) (99:1 er shown; Chiracel AD-H column, 15% $^i$PrOH in hexanes 1.0 mL/min, 210 nm). $[\alpha]_D^{20}$ −69.8 (c = 1.0, $CHCl_3$).

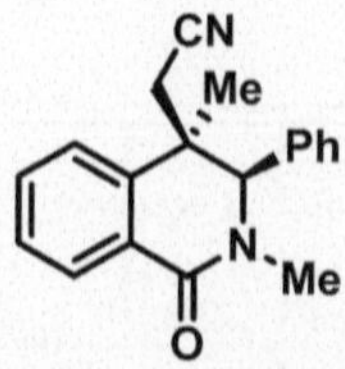

2-((3*R*,4*R*)-2,4-dimethyl-1-oxo-3-phenyl-1,2,3,4-tetrahydroisoquinolin-4-yl)acetonitrile (2.95a)

Prepared according to **GP2** using (*S*)-2-bromo-*N*-methyl-*N*-(2-methyl-1-phenylallyl)benzamide (103 mg, 0.3 mmol). The alkyl nitrile was purified by flash column chromatography using hexanes:EtOAc (3:1 v:v) and was obtained as a white solid (78 mg, 0.27 mmol, 90%, 20:1 dr (cis:trans)). **Major Diastereomer:** Isolated as a white solid (MP = 129–130 °C).

**$^1$H NMR** (500 MHz, $CDCl_3$) δ 8.31–8.25 (m, 1H), 7.54–7.45 (m, 2H), 7.31–7.25 (m, 1H), 7.24–7.17 (m, 2H), 7.08–6.97 (m, 3H), 4.43 (s, 1H), 3.09 (s, 3H), 2.89 (d, $J$ = 16.5 Hz, 1H), 2.23 (d, $J$ = 16.5 Hz, 1H), 1.72 (s, 3H). **$^{13}$C NMR** (126 MHz, $CDCl_3$) δ 162.9, 140.4, 135.7, 132.6, 129.0, 128.9, 128.8, 128.4, 128.0, 127.9, 122.5, 117.6, 71.5, 40.5, 34.4, 29.7, 25.8. **IR** ($cm^{-1}$, film). **HRMS** (ESI +) calculated 392.0506 m/z (found, 392.0511 m/z for $C_{18}H_{19}INO$). Enantiomeric purity was determined by HPLC analysis in comparison with racemic material $T_R = 8.1$ (minor) $T_R = 8.8$ (major) (> 99:1 er shown; Chiracel AD-H column, 20% $^i$PrOH in hexanes 1.0 mL/min, 210 nm). $[\alpha]_D^{20}$ +312.55 (c = 0.5, $CHCl_3$).

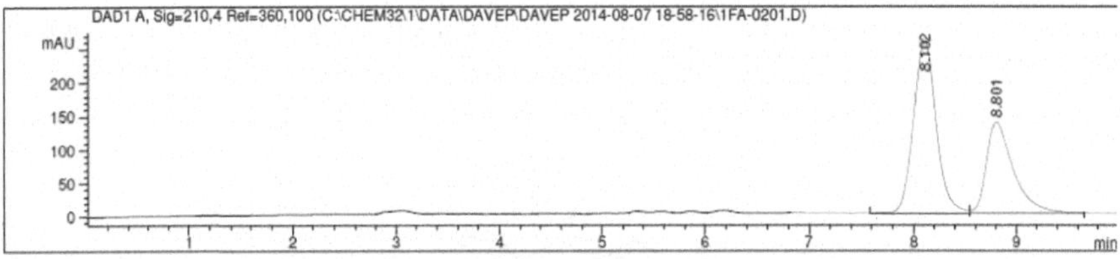


Signal 1: DAD1 A, Sig=210,4 Ref=360,100

| Peak # | RetTime [min] | Type | Width [min] | Area [mAU*s] | Height [mAU] | Area % |
|---|---|---|---|---|---|---|
| 1 | 8.102 | BV | 0.2359 | 3913.54883 | 256.71878 | 61.2241 |
| 2 | 8.801 | VB | 0.2675 | 2478.61719 | 136.71233 | 38.7759 |

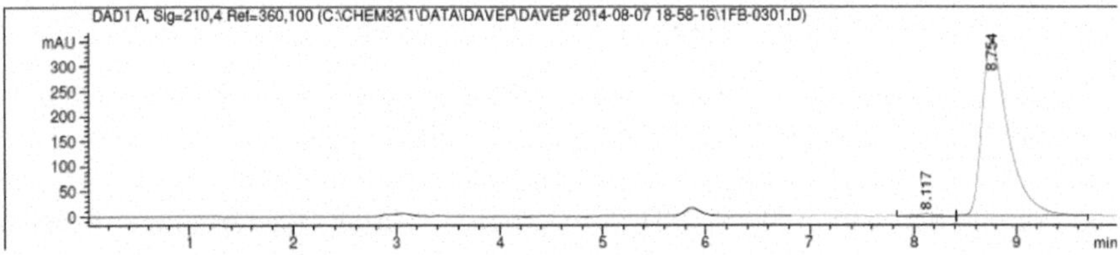


Signal 1: DAD1 A, Sig=210,4 Ref=360,100

| Peak # | RetTime [min] | Type | Width [min] | Area [mAU*s] | Height [mAU] | Area % |
|---|---|---|---|---|---|---|
| 1 | 8.117 | BV | 0.2249 | 45.47831 | 3.17915 | 0.7328 |
| 2 | 8.754 | VB | 0.2620 | 6160.58301 | 348.95831 | 99.2672 |

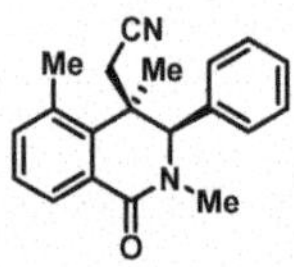

2-((3*R*,4*R*)-2,4,5-trimethyl-1-oxo-3-phenyl-1,2,3,4-tetrahydroisoquinolin-4-yl)acetonitrile (2.95b)

Prepared according to **GP2** using (*S*)-2-bromo-*N*,3-dimethyl-*N*-(2-methyl-1-phenylallyl)benzamide (107 mg, 0.3 mmol). The alkyl nitrile was purified by flash column chromatography using hexanes:EtOAc (3:2 v:v) and was obtained as a white solid (81 mg, 0.266 mmol, 89%, > 20:1 dr (*cis:trans*)). An analytically pure sample of the major diastereomer was obtained by recrystallization in EtOAc/hexanes at − 20 °C. **Major Diastereomer:** Isolated as a white solid (MP = 174–175 °C).

**$^1$H NMR** (500 MHz, $CDCl_3$) δ 8.20 (dd, $J = 7.5$, 1.5 Hz, 1H), 7.37–7.32 (m, 1H), 7.31–7.26 (m, 2H), 7.24–7.18 (m, 2H), 7.12–7.06 (m, 2H), 4.33 (s, 1H), 3.37 (d, $J =$ 16.5 Hz, 1H), 3.06 (s, 3H), 2.52 (d, $J = 16.5$ Hz, 1H), 2.43 (s, 3H), 1.76 (s, 3H). **$^{13}$C NMR** (126 MHz, $CDCl_3$) δ 163.3, 137.9, 137.4, 136.3, 133.7, 130.1, 129.0, 128.9, 128.7, 127.9, 127.8, 117.5, 72.6, 42.6, 34.3, 27.6, 27.3, 23.7. **IR** ($cm^{-1}$, film) 3063, 3007, 2974, 2930, 2874, 2247, 1636, 1589, 1454, 1427, 1402, 1329, 1271, 1072. **HRMS** (DART) calculated 406.06678 m/z (found, 406.06645 m/z for $C_{19}H_{21}INO$). Enantiomeric purity was determined by HPLC analysis in comparison with racemic material $T_R = 8.6$ (minor) $T_R = 7.8$ (major) (>99:1 er shown; Chiracel AD-H column, 20% $^i$PrOH in hexanes 1.0 mL/min, 225 nm). $[\alpha]_D^{20}$ +356.53 (c = 0.58, $CHCl_3$).

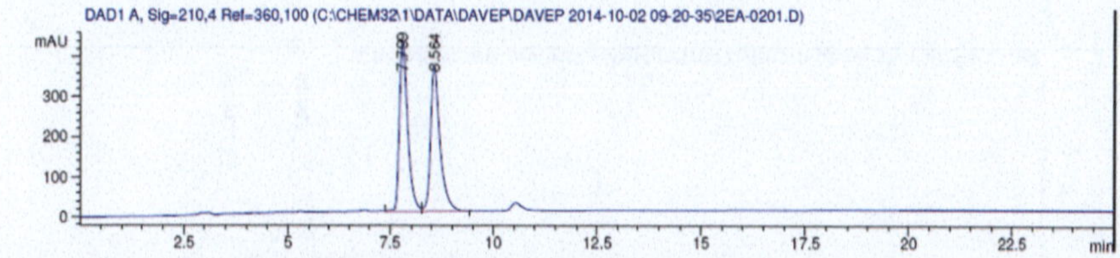


Signal 1: DAD1 A, Sig=210,4 Ref=360,100

| Peak # | RetTime [min] | Type | Width [min] | Area [mAU*s] | Height [mAU] | Area % |
|---|---|---|---|---|---|---|
| 1 | 7.799 | VV | 0.2031 | 5710.20459 | 423.98380 | 49.6348 |
| 2 | 8.564 | VB | 0.2343 | 5794.24414 | 362.84360 | 50.3652 |

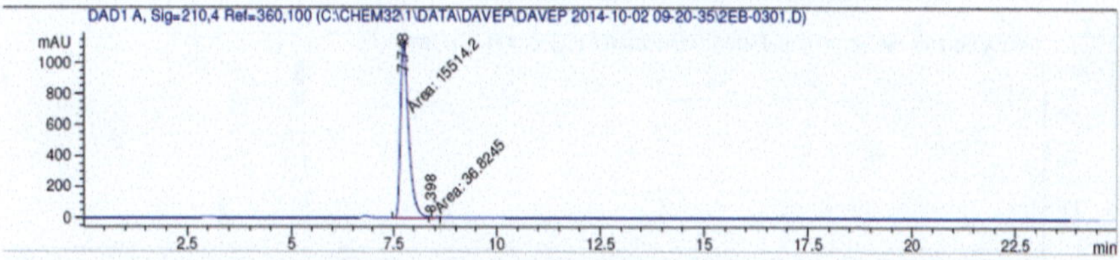


Signal 1: DAD1 A, Sig=210,4 Ref=360,100

| Peak # | RetTime [min] | Type | Width [min] | Area [mAU*s] | Height [mAU] | Area % |
|---|---|---|---|---|---|---|
| 1 | 7.748 | MM | 0.2286 | 1.55142e4 | 1130.99084 | 99.7632 |
| 2 | 8.398 | MM | 0.0573 | 36.82447 | 10.70500 | 0.2368 |

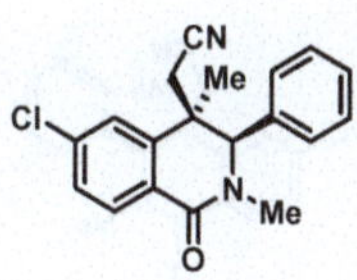


2-((3*R*,4*R*)-6-chloro-2,4-dimethyl-1-oxo-3-phenyl-1,2,3,4-tetrahydroisoquinolin-4-yl)acetonitrile (2.95c)

Prepared according to **GP2** using (*S*)-2-bromo-4-chloro-*N*-methyl-*N*-(2-methyl-1-phenylallyl)benzamide (114 mg, 0.3 mmol). The alkyl nitrile was purified by flash column chromatography using hexanes:EtOAc (3:2 v:v) and was obtained as a white solid (88 mg, 0.27 mmol, 90%, 20:1 dr (*cis:trans*)). **Major Diastereomer:** Isolated as a white solid (MP = 178–180 °C).

**$^1$H NMR** (500 MHz, $CDCl_3$) δ 8.22 (d, $J$ = 8.5 Hz, 1H), 7.45 (dd, $J$ = 8.5, 2.0 Hz, 1H), 7.33–7.19 (m, 3H), 7.07–6.98 (m, 3H), 4.44 (s, 1H), 3.07 (s, 3H), 2.87 (d, $J$ = 16.5 Hz, 1H), 2.22 (d, $J$ = 16.5 Hz, 1H), 1.72 (s, 3H). **$^{13}$C NMR** (126 MHz, $CDCl_3$) δ 162.2, 142.3, 138.9, 135.4, 130.6, 129.2, 129.0, 128.4, 127.9, 126.9, 123.2, 117.2, 71.4, 40.6, 34.4, 29.6, 25.8. **IR** ($cm^{-1}$, film) 3067, 3032, 2967, 2930, 2247, 1651, 1595, 1568, 1493, 1452, 1281, 1262, 1165. **HRMS** (DART) calculated 325.11077 m/z (found, 325.10950 m/z for $C_{19}H_{18}ClN_2O$). Enantiomeric purity was determined by HPLC analysis in comparison with racemic material $T_R$ = 14.3 (major) $T_R$ = 15.7 (minor) (99:1 er shown; Chiracel AD-H column, 20% $^i$PrOH in hexanes 1.0 mL/min, 210 nm). $[\alpha]_D^{20}$ +335.2 (c = 0.59, $CHCl_3$).

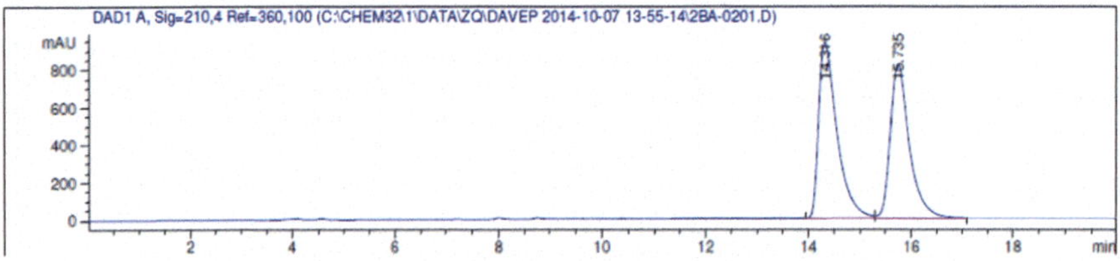


Signal 1: DAD1 A, Sig=210,4 Ref=360,100

| Peak # | RetTime [min] | Type | Width [min] | Area [mAU*s] | Height [mAU] | Area % |
|---|---|---|---|---|---|---|
| 1 | 14.316 | BV | 0.3729 | 2.33690e4 | 937.10870 | 52.8134 |
| 2 | 15.735 | VB | 0.3817 | 2.08792e4 | 818.18298 | 47.1866 |

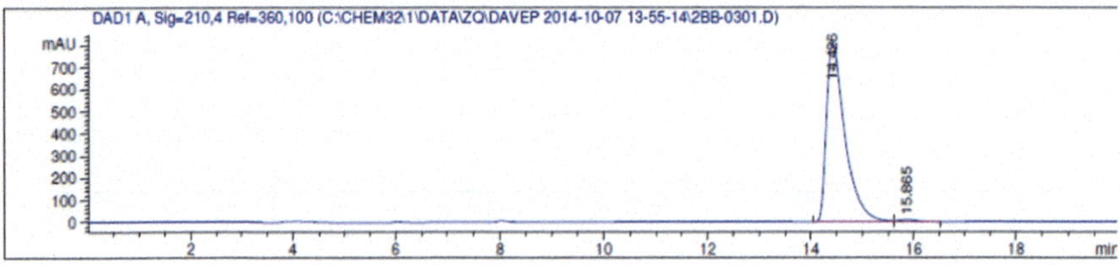


Signal 1: DAD1 A, Sig=210,4 Ref=360,100

| Peak # | RetTime [min] | Type | Width [min] | Area [mAU*s] | Height [mAU] | Area % |
|---|---|---|---|---|---|---|
| 1 | 14.426 | BB | 0.3763 | 2.03747e4 | 807.81110 | 98.7269 |
| 2 | 15.865 | BB | 0.3807 | 262.74625 | 10.26140 | 1.2731 |

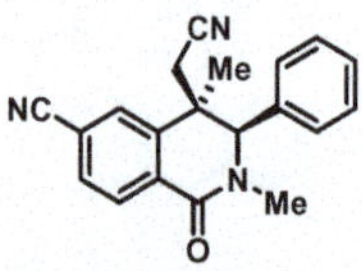


(3*R*,4*R*)-4-(cyanomethyl)-2,4-dimethyl-1-oxo-3-phenyl-1,2,3,4-tetrahydroisoquinoline-6-carbonitrile (2.95c′)

Prepared according to **GP2** using (*S*)-2-bromo-4-chloro-*N*-methyl-*N*-(2-methyl-1-phenylallyl)benzamide (114 mg, 0.3 mmol) and employing 105 mol% of $Zn(CN)_2$. The alkyl nitrile was purified by flash column chromatography using hexanes: EtOAc (1:1 v:v) and was obtained as a white solid (64 mg, 0.20 mmol, 67%, 19:1 dr (*cis:trans*)). An analytically pure sample of the major diastereomer was obtained by recrystallization in EtOAc/hexanes at −20 °C. **Major Diastereomer**: Isolated as a white solid (MP = 153–154 °C).

**$^1$H NMR** (500 MHz, $CDCl_3$) δ 8.37 (dd, *J* = 8.0, 0.5 Hz, 1H), 7.76 (dd, *J* = 8.0, 1.5 Hz, 1H), 7.37 (d, *J* = 1.5 Hz, 1H), 7.30–7.26 (m, 1H), 7.21 (ddd, *J* = 8.0, 7.5, 1.0 Hz, 2H), 6.95 (d, *J* = 7.0 Hz, 2H), 4.48 (s, 1H), 3.07 (s, 3H), 2.93 (d, *J* = 16.5 Hz, 1H), 2.24 (dd, *J* = 16.5, 1.0 Hz, 1H), 1.71 (d, *J* = 1.0 Hz, 3H). **$^{13}$C NMR** (126 MHz, $CDCl_3$) δ 161.1, 141.6, 134.7, 132.0, 131.8, 129.5, 129.3, 129.0, 127.6,

126.7, 117.7, 116.8, 116.0, 71.1, 40.6, 34.5, 29.4, 25.6. **IR** ($cm^{-1}$, film) 3067, 3034, 2969, 2932, 2247, 2232, 1651, 1609, 1456, 1427, 1399, 1331, 1289, 1261, 1082. **HRMS** (DART) calculated 316.14499 m/z (found 316.14426 m/z for $C_{20}H_{18}N_3O$). Enantiomeric purity was determined by HPLC analysis in comparison with racemic material $T_R = 10.1$ (minor) $T_R = 9.2$ (major) (99:1 er shown; Chiracel AD-H column, 25% $^i$PrOH in hexanes 1.0 mL/min, 210 nm). $[\alpha]_D^{20}$ +362.32 (c = 0.59, $CHCl_3$).

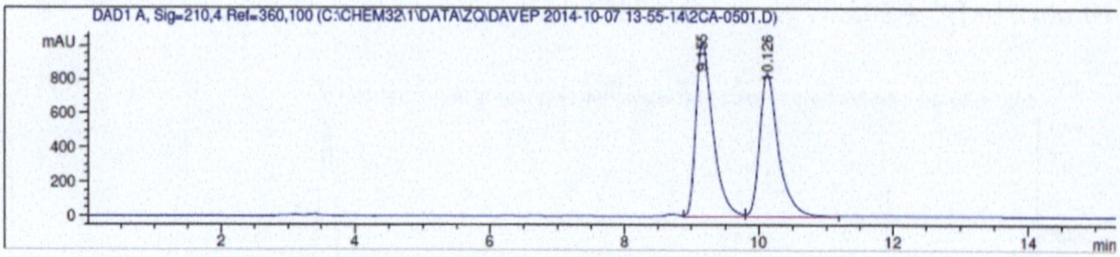


Signal 1: DAD1 A, Sig=210,4 Ref=360,100

| Peak # | RetTime [min] | Type | Width [min] | Area [mAU*s] | Height [mAU] | Area % |
|---|---|---|---|---|---|---|
| 1 | 9.155 | VV | 0.2683 | 1.81606e4 | 1017.26532 | 52.9405 |
| 2 | 10.126 | VB | 0.2887 | 1.61432e4 | 838.06323 | 47.0595 |

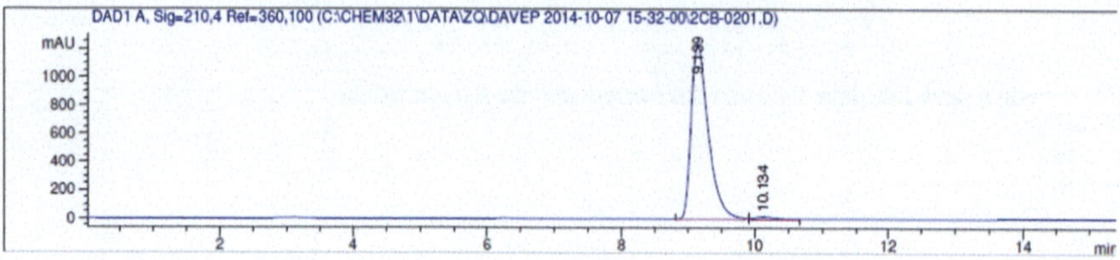


Signal 1: DAD1 A, Sig=210,4 Ref=360,100

| Peak # | RetTime [min] | Type | Width [min] | Area [mAU*s] | Height [mAU] | Area % |
|---|---|---|---|---|---|---|
| 1 | 9.139 | BV | 0.2690 | 2.19027e4 | 1234.48669 | 98.6458 |
| 2 | 10.134 | VB | 0.2877 | 300.67569 | 15.26951 | 1.3542 |

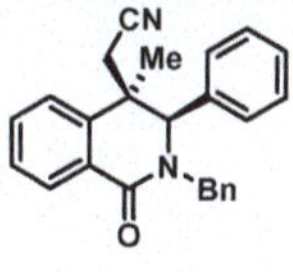


2-((3*R*,4*R*)-2-benzyl-4-methyl-1-oxo-3-phenyl-1,2,3,4-tetrahydroisoquinolin-4-yl)acetonitrile (2.95d)

Prepared according to **GP2** using (*S*)-*N*-benzyl-2-bromo-*N*-(2-methyl-1-phenylallyl)benzamide (126 mg, 0.3 mmol). The alkyl nitrile was purified by flash column chromatography using hexanes:DCM:EtOAc (20:20:1 v:v:v) and was obtained as a white solid (94 mg, 0.26 mmol, 85%, 10:1 dr (*cis:trans*)). **Major Diastereomer:** Isolated as a white solid (MP = 185–186 °C).

**$^1$H NMR** (500 MHz, $CDCl_3$) δ 8.41–8.29 (m, 1H), 7.54–7.46 (m, 2H), 7.40–7.25 (m, 6H), 7.24–7.17 (m, 2H), 7.08–7.03 (m, 1H), 7.03–6.95 (m, 2H), 5.67 (d, $J$ = 14.5 Hz, 1H), 4.34 (s, 1H), 3.52 (d, $J$ = 14.5 Hz, 1H), 2.75 (d, $J$ = 16.5 Hz, 1H),

2.16 (d, $J$ = 16.5 Hz, 1H), 1.34 (s, 3H). **$^{13}$C NMR** (126 MHz, $CDCl_3$) δ 162.8, 140.5, 136.3, 135.9, 132.8, 129.3, 129.2, 129.0, 128.9, 128.8, 128.5, 128.2, 128.1, 127.9, 122.8, 117.3, 67.2, 48.4, 40.0, 29.7, 25.9. **IR** ($cm^{-1}$, film) 3030, 2968, 2920, 2243, 1647, 1600, 1471, 1446, 1263. **HRMS** (DART) calculated 367.18104 m/z (found, 367.18043 m/z for $C_{25}H_{23}N_2O$). Enantiomeric purity was determined by HPLC analysis in comparison with racemic material $T_R$ = 17.3 (major) $T_R$ = 20.8 (minor) (>98:2 er shown; Chiralpack AS column, 10% $^i$PrOH in hexanes 1.0 mL/min, 210 nm). $[\alpha]_D^{20}$ +205.37 (c = 1.0, $CHCl_3$).

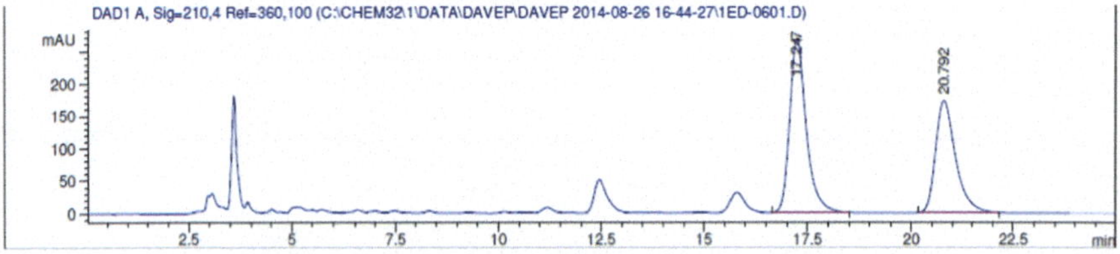


Signal 1: DAD1 A, Sig=210,4 Ref=360,100

| Peak # | RetTime [min] | Type | Width [min] | Area [mAU*s] | Height [mAU] | Area % |
|---|---|---|---|---|---|---|
| 1 | 17.247 | BB | 0.4243 | 7532.91797 | 267.68378 | 55.7761 |
| 2 | 20.792 | BB | 0.5219 | 5972.71777 | 173.36302 | 44.2239 |

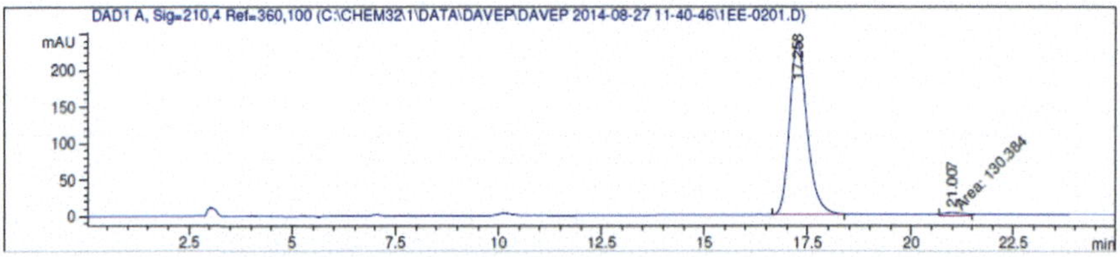


Signal 1: DAD1 A, Sig=210,4 Ref=360,100

| Peak # | RetTime [min] | Type | Width [min] | Area [mAU*s] | Height [mAU] | Area % |
|---|---|---|---|---|---|---|
| 1 | 17.258 | BB | 0.4526 | 7006.93115 | 237.23668 | 98.1732 |
| 2 | 21.007 | MM | 0.6115 | 130.38400 | 3.55348 | 1.8268 |

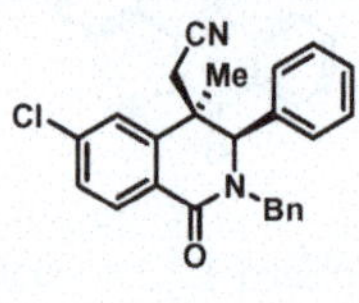


2-((3*R*,4*R*)-2-benzyl-6-chloro-4-methyl-1-oxo-3-phenyl-1,2,3,4-tetrahydroisoquinolin-4-yl)acetonitrile (2.95e)

Prepared according to **GP2** using (*S*)-*N*-benzyl-2-bromo-*N*-(2-methyl-1-phenylallyl)benzamide (136 mg, 0.3 mmol). The alkyl nitrile was purified by flash column chromatography using hexanes:EtOAc (5:1 v:v) and was obtained as a white solid (85 mg, 0.21 mmol, 70%, 10:1 dr (*cis:trans*)). **Major Diastereomer**: Isolated as a white solid (MP = 223–223 °C).

**$^1$H NMR** (500 MHz, $CDCl_3$) δ 8.29 (d, $J$ = 8.5 Hz, 1H), 7.47 (dd, $J$ = 8.5, 2.0 Hz, 1H), 7.42–7.28 (m, 6H), 7.26–7.21 (m, 2H), 7.06–6.94 (m, 3H), 5.65 (d, $J$ = 14.5 Hz, 1H), 4.35 (s, 1H), 3.52 (d, $J$ = 14.5 Hz, 1H), 2.71 (d, $J$ = 16.5 Hz, 1H), 2.15 (d, $J$ = 16.5 Hz, 1H), 1.34 (s, 3H). **$^{13}$C NMR** (126 MHz, $CDCl_3$) δ 162.0, 142.3, 139.1, 136.0, 135.4, 130.9, 129.2, 129.2, 129.1, 128.8, 128.4, 128.1, 128.1, 127.0, 123.3, 116.9, 67.1, 48.4, 40.1, 29.5, 25.8. **IR** ($cm^{-1}$, film) 30885, 3065, 3028, 3011, 2968, 2930, 2249, 1636, 1593, 1568, 1456, 1418, 1262, 1155. **HRMS** (DART) calculated 401.14207 m/z (found, 401.14256 m/z for $C_{25}H_{22}ClN_2O$). Enantiomeric purity was determined by HPLC analysis in comparison with racemic material $T_R$ = 20.4(major) $T_R$ = 23.6 (minor) (> 99:1 er shown; Chiracel AD-H column, 10% $^i$PrOH in hexanes 1.0 mL/min, 210 nm). $[\alpha]_D^{20}$ +200.58 (c = 1.0, $CHCl_3$).

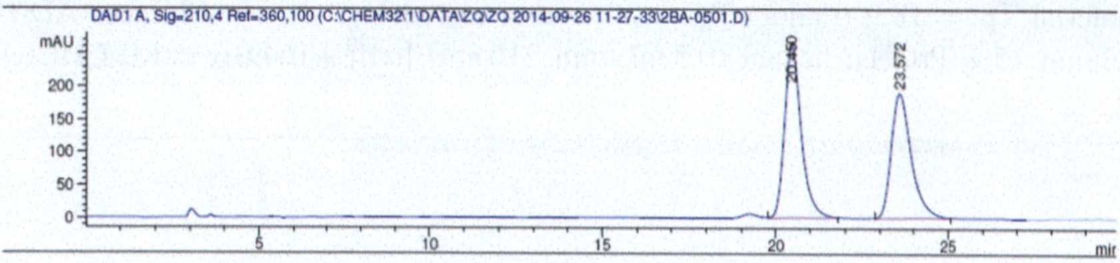


Signal 1: DAD1 A, Sig=210,4 Ref=360,100

| Peak # | RetTime [min] | Type | Width [min] | Area [mAU*s] | Height [mAU] | Area % |
|---|---|---|---|---|---|---|
| 1 | 20.450 | VB | 0.5129 | 8923.60352 | 263.60165 | 53.8723 |
| 2 | 23.572 | BB | 0.6106 | 7640.76514 | 189.42638 | 46.1277 |

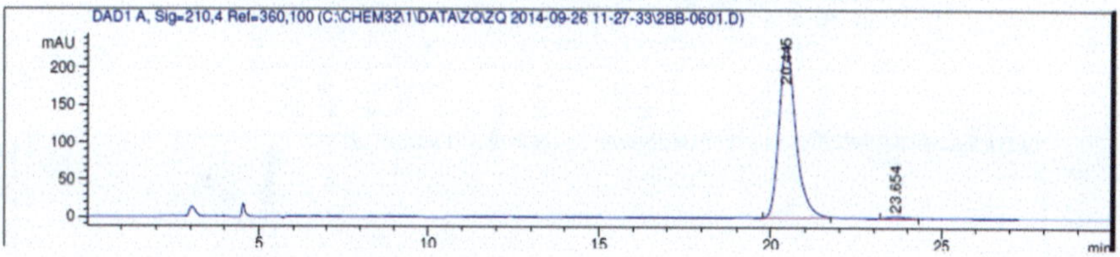


Signal 1: DAD1 A, Sig=210,4 Ref=360,100

| Peak # | RetTime [min] | Type | Width [min] | Area [mAU*s] | Height [mAU] | Area % |
|---|---|---|---|---|---|---|
| 1 | 20.445 | BB | 0.5075 | 7792.95508 | 231.02809 | 99.1329 |
| 2 | 23.654 | BB | 0.4084 | 68.16359 | 2.02645 | 0.8671 |

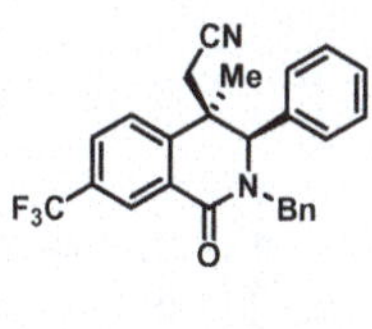


2-((3*R*,4*R*)-2-benzyl-4-methyl-1-oxo-3-phenyl-7-(trifluoromethyl)-1,2,3,4-tetrahydroisoquinolin-4-yl)acetonitrile (2.95f)

Prepared according to **GP2** using (*S*)-*N*-benzyl-2-bromo-*N*-(2-methyl-1-phenylallyl)-5-(trifluoromethyl)benzamide (136 mg, 0.3 mmol). The alkyl nitrile was purified by flash column chromatography using hexanes:DCM:EtOAc (20:10:1

v:v:v) and was obtained as a white solid (102 mg, 0.234 mmol, 78%, 9:1 dr (*cis:trans*)). **Major Diastereomer**: Isolated as a white solid (MP = 239–241 °C).

**$^{1}$H NMR** (500 MHz, $CDCl_3$) δ 8.67–8.61 (m, 1H), 7.80–7.73 (m, 1H), 7.41–7.29 (m, 6H), 7.27–7.19 (m, 3H), 7.02–6.92 (m, 2H), 5.67 (d, $J$ = 14.5 Hz, 1H), 4.37 (s, 1H), 3.55 (d, $J$ = 14.5 Hz, 1H), 2.75 (d, $J$ = 16.5 Hz, 1H), 2.22 (d, $J$ = 16.5 Hz, 1H), 1.35 (s, 3H). **$^{13}$C NMR** (125 MHz, $CDCl_3$) δ 161.4, 144.2, 135.8, 135.2, 130.7 (q, $J$ =33.5 Hz), 129.4 (q, 3.5 Hz), 129.4 (overlapping quartet), 129.3, 129.2, 129.1, 128.8, 128.1, 128.1, 126.4 (q, $J$ = 4.0), 124.7 (q, $J$ = 272 Hz) 123.7, 116.8, 67.1, 48.5, 40.3, 29.6, 25.9. **$^{19}$F NMR** (565 MHz, $CDCl_3$) δ −62.81. **IR** ($cm^{-1}$, film) 3090, 3069, 3026, 2967, 2243, 1657, 1651, 1618, 1452, 1333, 1296, 1252, 1227, 1175, 1128. **HRMS** (DART) calculated 435.16842 m/z (found 435.16976 m/z for $C_{26}H_{22}F_3N_2O$). Enantiomeric purity was determined by HPLC analysis in comparison with racemic material $T_R$ = 18.9 (major) $T_R$ = 20.1 (minor) (99:1 er shown; Chiracel AD-H column, 15% $^{i}$PrOH in hexanes 0.5 mL/min, 210 nm). $[\alpha]_D^{20}$ +164.0 (c = 1.0, $CHCl_3$).

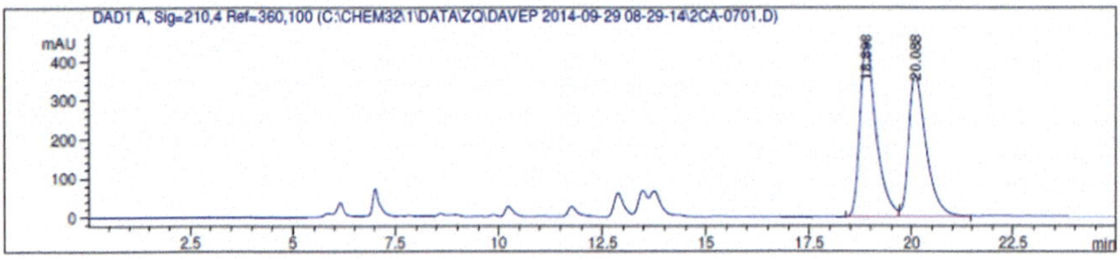


Signal 1: DAD1 A, Sig=210,4 Ref=360,100

| Peak # | RetTime [min] | Type | Width [min] | Area [mAU*s] | Height [mAU] | Area % |
|---|---|---|---|---|---|---|
| 1 | 18.898 | BV | 0.4033 | 1.20467e4 | 448.64743 | 52.7272 |
| 2 | 20.088 | VB | 0.4500 | 1.08005e4 | 362.10080 | 47.2728 |

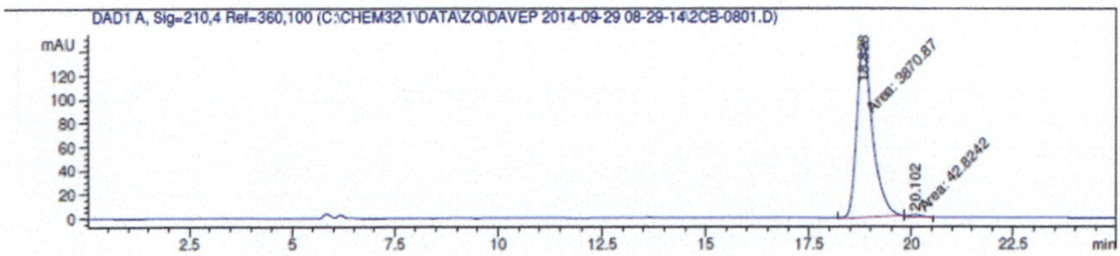


Signal 1: DAD1 A, Sig=210,4 Ref=360,100

| Peak # | RetTime [min] | Type | Width [min] | Area [mAU*s] | Height [mAU] | Area % |
|---|---|---|---|---|---|---|
| 1 | 18.828 | MM | 0.4404 | 3870.87109 | 146.50534 | 98.9058 |
| 2 | 20.102 | MM | 0.4149 | 42.82417 | 1.72019 | 1.0942 |

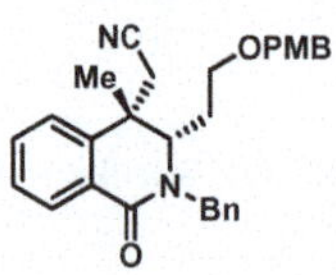


2-((3*S*,4*S*)-2-benzyl-3-(2-((4-methoxybenzyl)oxy)ethyl)-4-methyl-1-oxo-1,2,3,4-tetrahydroisoquinolin-4-yl)acetonitrile (2.95k)

Prepared according to **GP2** using (*S*)-*N*-benzyl-2-bromo-*N*-(5-((4-methoxybenzyl)oxy)-2-methylpent-1-en-3-yl)benzamide (153 mg, 0.3 mmol). The alkyl nitrile was purified by flash column chromatography using hexanes:EtOAc (3:2 v:v) and was obtained as a white semisolid (76 mg, 0.2 mmol, 66%, 1:1 dr (*cis:trans*)).

**Major Diastereomer**: Isolated as a white semisolid. **$^1$H NMR** (600 MHz, $CDCl_3$) δ 8.12 (dd, *J* = 8.0, 1.5 Hz, 1H), 7.50 (td, *J* = 8.0, 1.5 Hz, 1H), 7.41 (dd, *J* = 8.0, 1.0 Hz, 1H), 7.40–7.36 (m, 2H), 7.34–7.25 (m, 6H), 6.93–6.86 (m, 2H), 5.76 (d, *J* = 14.0 Hz, 1H), 4.48–4.40 (m, 2H), 3.80 (s, 3H), 3.77 (d, *J* = 14.0 Hz, 1H), 3.60 (dd, *J* = 9.5, 4.0 Hz, 1H), 3.52–3.47 (m, 1H), 3.43–3.40 (m, 1H), 2.10 (d, *J* = 17.0 Hz, 1H), 2.03–1.95 (m, 1H), 1.83 (d, *J* = 17.0 Hz, 1H), 1.55–1.45 (m, 4H). **$^{13}$C NMR** (151 MHz, $CDCl_3$) δ 162.6, 159.4, 140.0, 136.8, 132.5, 129.7, 129.6, 129.6, 128.8, 128.7, 128.1, 128.1, 128.1, 125.1, 116.7, 113.8, 72.9, 65.7, 59.4, 55.2, 49.0, 39.2, 31.0, 30.2, 20.7. **IR** ($cm^{-1}$, film) 3032, 2952, 2934, 2862, 2245, 1647, 1514, 1447, 1250, 1173, 1105, 1034. **HRMS** (DART): calculated 455.23347 m/z (found 455.23473 for $C_{29}H_{31}N_2O_3$). Enantiomeric purity was determined by HPLC analysis in comparison with racemic material $T_R$ = 12.7 (major) $T_R$ = 17 (minor) (>99:1 er shown; Chiracel OD-H column, 20% $^i$PrOH in hexanes 1.0 mL/min, 210 nm). $[\alpha]_D^{20}$ −185.6 (c = 0.5, $CHCl_3$).

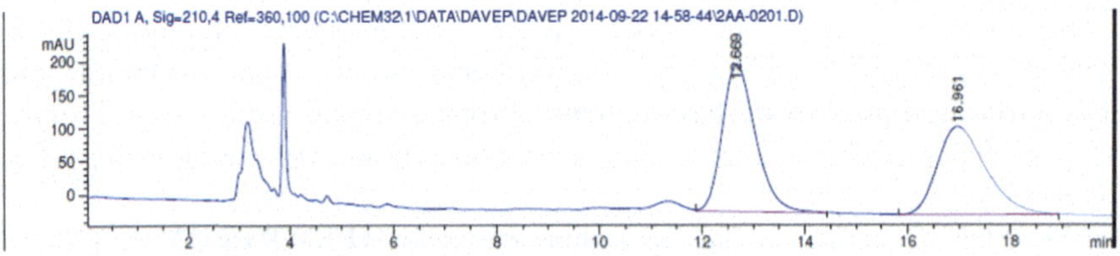


Signal 1: DAD1 A, Sig=210,4 Ref=360,100

| Peak # | RetTime [min] | Type | Width [min] | Area [mAU*s] | Height [mAU] | Area % |
|---|---|---|---|---|---|---|
| 1 | 12.669 | VB | 0.6962 | 1.00332e4 | 221.61809 | 53.3873 |
| 2 | 16.961 | BB | 1.0115 | 8760.01172 | 132.81671 | 46.6127 |

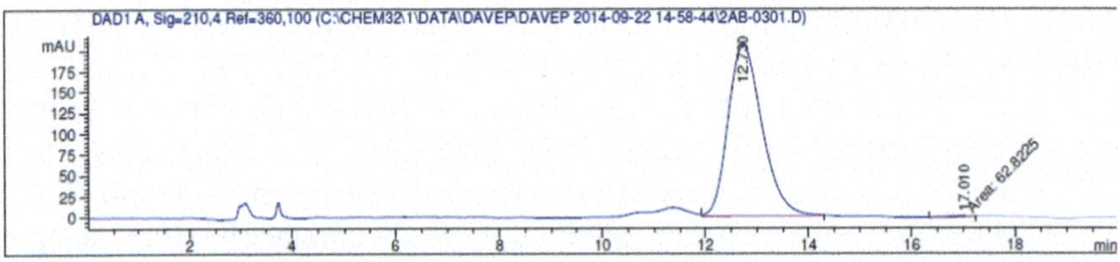


Signal 1: DAD1 A, Sig=210,4 Ref=360,100

| Peak # | RetTime [min] | Type | Width [min] | Area [mAU*s] | Height [mAU] | Area % |
|---|---|---|---|---|---|---|
| 1 | 12.730 | VB | 0.7035 | 9415.21973 | 206.67462 | 99.3372 |
| 2 | 17.010 | MM | 0.8026 | 62.82248 | 1.30461 | 0.6628 |

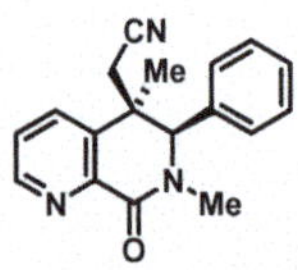


2-((5*R*,6*R*)-5,7-dimethyl-8-oxo-6-phenyl-5,6,7,8-tetrahydro-1,7-naphthyridin-5-yl)acetonitrile (2.95o)

Prepared according to **GP2** using (*S*)-3-bromo-*N*-methyl-*N*-(2-methyl-1-phenylallyl)picolinamide (104 mg, 0.3 mmol). The alkyl nitrile was purified by flash column chromatography using DCM:MeOH (20:1 v:v) and was obtained as a white solid (61 mg, 0.21 mmol, 70%, 13:1 dr (*cis:trans*)). **Major Diastereomer**: Isolated as an off white solid.

**$^{1}$H NMR** (500 MHz, $CDCl_3$) δ 8.82 (dd, $J = 4.5$, 1.5 Hz, 1H), 7.48 (dd, $J = 8.0$, 1.5 Hz, 1H), 7.42 (dd, $J = 8.0$, 4.5 Hz, 1H), 7.30–7.26 (m, 1H), 7.23–7.19 (m, 2H), 7.02–6.97 (m, 2H), 4.46 (s, 1H), 3.12 (s, 3H), 2.82 (d, $J = 16.5$ Hz, 1H), 2.26 (dd, $J =$ 16.5, 1.0 Hz, 1H), 1.76 (d, $J = 1.0$ Hz, 3H). **$^{13}$C NMR** (126 MHz, $CDCl_3$) δ 161.5, 149.6, 145.5, 136.7, 134.9, 131.7, 129.3, 129.0, 127.8, 126.5, 117.1, 71.1, 40.5, 34.9, 29.3, 25.6. **IR** ($cm^{-1}$, film) 3057, 3034, 2978, 2965, 2247, 1661, 1572, 1481, 1450, 1424, 1398, 1268, 1229. **HRMS** (DART): calculated 292.14499 m/z (found 292.14457 m/z for $C_{18}H_{18}N_3O$). Enantiomeric purity was determined by HPLC analysis in comparison with racemic material $T_R = 81.9$ (minor) $T_R = 89.6$ (major) (> 99:1 er shown; Chiracel AD-H column, 60% $^{i}$PrOH in hexanes 0.125 mL/min, 210 nm). $[\alpha]_D^{20}$ +280 (c = 0.525, $CHCl_3$).

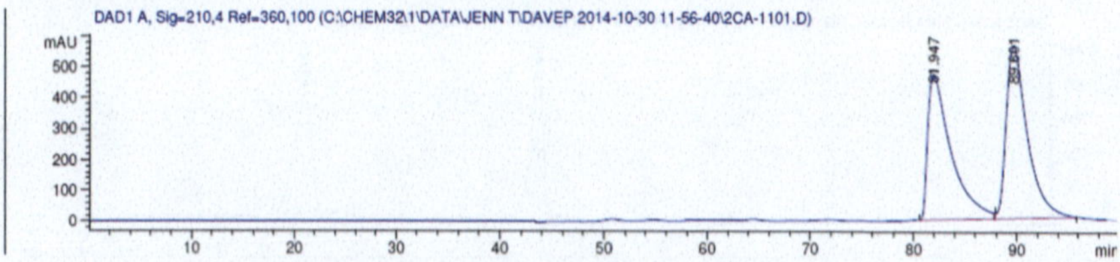


Signal 1: DAD1 A, Sig=210,4 Ref=360,100

| Peak # | RetTime [min] | Type | Width [min] | Area [mAU*s] | Height [mAU] | Area % |
|---|---|---|---|---|---|---|
| 1 | 81.947 | BB | 2.1662 | 7.44843e4 | 469.68030 | 48.3000 |
| 2 | 89.601 | BB | 2.0690 | 7.97273e4 | 559.99365 | 51.7000 |

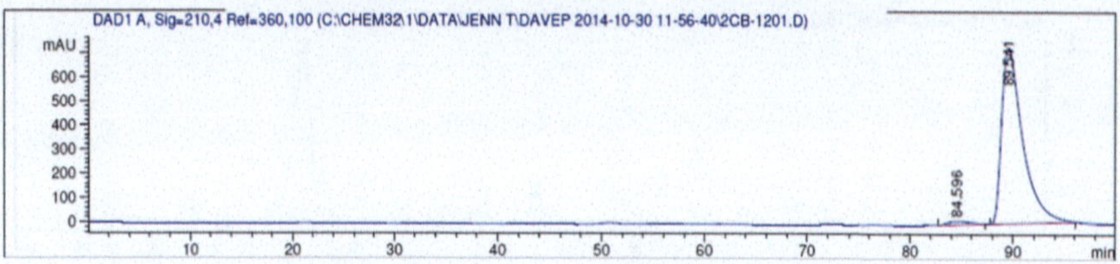


Signal 1: DAD1 A, Sig=210,4 Ref=360,100

| Peak # | RetTime [min] | Type | Width [min] | Area [mAU*s] | Height [mAU] | Area % |
|---|---|---|---|---|---|---|
| 1 | 84.596 | BB | 1.7082 | 1868.93787 | 12.87029 | 1.7973 |
| 2 | 89.541 | BB | 2.0409 | 1.02119e5 | 726.04938 | 98.2027 |

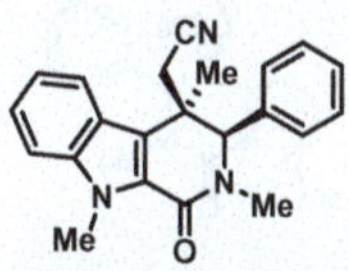


2-((3*R*,4*R*)-2,4,9-trimethyl-1-oxo-3-phenyl-2,3,4,9-tetrahydro-1*H*-pyrido[3,4-*b*]indol-4-yl)acetonitrile (2.95q)

Prepared according to **GP2** using (*S*)-3-bromo-*N*,1-dimethyl-*N*-(2-methyl-1-phenylallyl)-1*H*-indole-2-carboxamide (119 mg, 0.3 mmol). The alkyl nitrile was purified by flash column chromatography using hexanes: EtOAc (2:1 v:v) and was obtained as a white foam (79 mg, 0.23 mmol, 77%, 3:2 dr (*cis:trans*)).

***cis* Diastereomer**: Isolated as a white foam. **$^1$H NMR** (500 MHz, $CDCl_3$) δ 7.55 (dt, $J$ = 8.5, 1.0 Hz, 1H), 7.47 (m, 1H), 7.39–7.35 (m, 1H), 7.33–7.27 (m, 1H), 7.26–7.22 (m, 4H), 7.14 (m, 1H), 4.45 (s, 1H), 4.25 (s, 3H), 3.36 (d, $J$ = 16.5 Hz, 1H), 3.04 (s, 3H), 2.46 (dd, $J$ = 16.5, 0.5 Hz, 1H), 1.80 (d, $J$ = 1.0 Hz, 3H). **$^{13}$C NMR** (126 MHz, $CDCl_3$) δ 160.1, 139.3, 136.0, 129.0, 128.9, 128.5, 125.4, 124.5, 122.5, 120.7, 120.3, 119.5, 117.6, 111.0, 74.0, 39.9, 33.5, 31.5, 28.5, 26.3. **IR** ($cm^{-1}$, film) 2960, 2924, 2247, 1647, 1533, 1489, 1452, 1204. **HRMS** (ESI +) calculated 344.1757 m/z (found 344.1755 m/z for $C_{22}H_{22}N_3O$). Enantiomeric purity was determined by HPLC analysis in comparison with racemic material $T_R$ = 7.4 (minor) $T_R$ = 10.8 (major) (> 99:1 er shown; Chiracel AD-H column, 20% $^i$PrOH in hexanes 1.0 mL/min, 210 nm). $[\alpha]_D^{20}$ +320.6 (c = 0.315, $CHCl_3$).

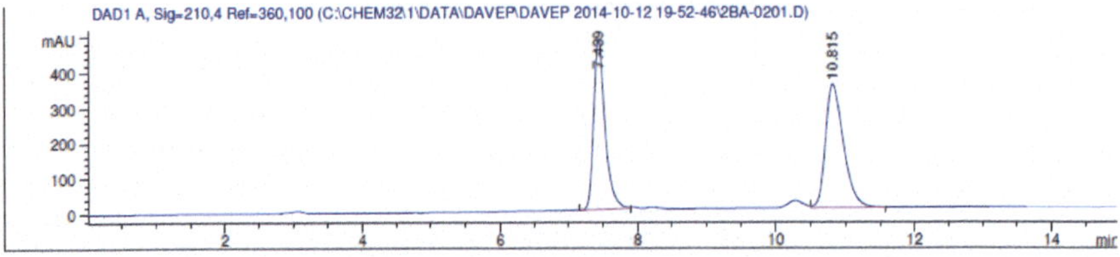


Signal 1: DAD1 A, Sig=210,4 Ref=360,100

| Peak # | RetTime [min] | Type | Width [min] | Area [mAU*s] | Height [mAU] | Area % |
|---|---|---|---|---|---|---|
| 1 | 7.439 | BB | 0.1715 | 5497.58789 | 480.38940 | 46.8473 |
| 2 | 10.815 | VB | 0.2726 | 6237.53076 | 348.83115 | 53.1527 |

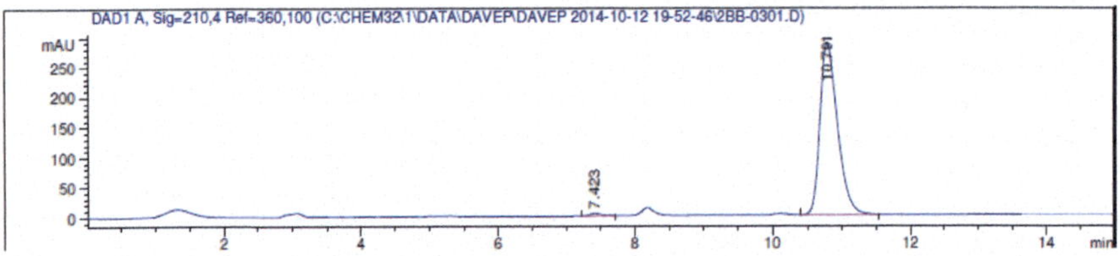


Signal 1: DAD1 A, Sig=210,4 Ref=360,100

| Peak # | RetTime [min] | Type | Width [min] | Area [mAU*s] | Height [mAU] | Area % |
|---|---|---|---|---|---|---|
| 1 | 7.423 | BB | 0.1742 | 45.96484 | 3.93812 | 0.9003 |
| 2 | 10.791 | VB | 0.2693 | 5059.33447 | 287.52872 | 99.0997 |

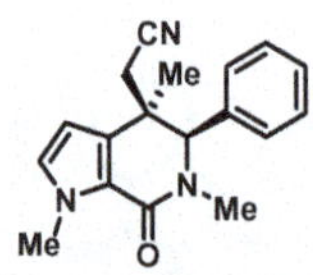


2-((4R,5R)-1,4,6-trimethyl-7-oxo-5-phenyl-4,5,6,7-tetrahydro-1Hpyrrolo[2,3-c]pyridin-4-yl)acetonitrile (2.95s)

Prepared according to **GP2** using (*S*)-3-bromo-*N*,1-dimethyl-*N*-(2-methyl-1-phenylallyl)-1*H*-pyrrole-2-carboxamide (104 mg, 0.3 mmol). The alkyl nitrile was purified by flash column chromatography using hexanes: EtOAc (1:1 v:v) and was obtained as an off white semisolid (81 mg, 0.276 mmol, 92%, 1:1 dr (*cis:trans*)).

***cis* Diastereomer**: Isolated as an off-white semi-solid. $^1$**H NMR** (500 MHz, $CDCl_3$) δ 7.30–7.21 (m, 3H), 7.13–7.08 (m, 2H), 6.68 (d, $J = 2.5$ Hz, 1H), 5.90 (d, $J = 2.5$ Hz, 1H), 4.27 (s, 1H), 4.02 (s, 3H), 2.93 (s, 3H), 2.53 (d, $J = 16.5$ Hz, 1H), 2.11 (d, $J = 16.5$ Hz, 1H), 1.65 (s, 3H). $^{13}$**C NMR** (126 MHz, $CDCl_3$) δ 159.7, 136.5, 130.5, 128.8, 128.8, 128.1, 119.8, 117.8, 102.7, 74.3, 38.2, 36.2, 33.1, 29.3, 26.2. **IR** ($cm^{-1}$, film) 3106, 3030, 2926, 2243, 1636, 1539, 1512, 1435, 1398, 1344, 1134, 1028. **HRMS** (DART): calculated 294.16064 m/z (found 294.16148 m/z for $C_{18}H_{20}N_3O$). Enantiomeric purity was determined by HPLC analysis in comparison with racemic material $T_R = 5.9$ (minor) $T_R = 7.0$ (major) (99:1 er shown; Chiracel AD-H column, 20% $^i$PrOH in hexanes 1.0 mL/min, 210 nm). $[\alpha]_D^{20}$ +193.8 (c = 0.52, $CHCl_3$).

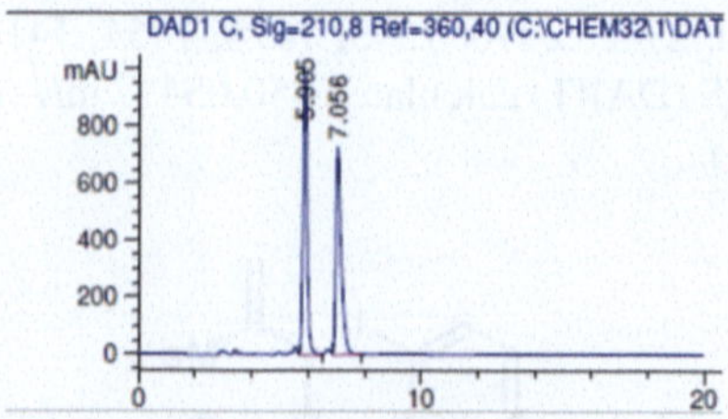


Signal 3: DAD1 C, Sig=210,8 Ref=360,40

| Peak # | RetTime [min] | Type | Width [min] | Area [mAU*s] | Height [mAU] | Area % |
|---|---|---|---|---|---|---|
| 1 | 5.905 | VV | 0.1469 | 9563.98145 | 996.88293 | 49.9601 |
| 2 | 7.056 | VB | 0.1951 | 9579.25781 | 729.97864 | 50.0399 |

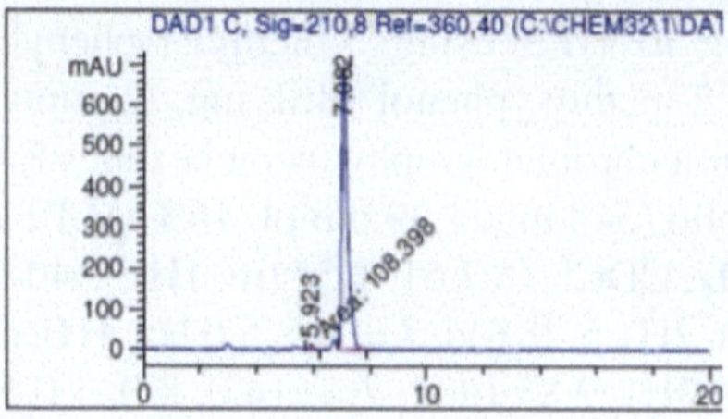


Signal 3: DAD1 C, Sig=210,8 Ref=360,40

| Peak # | RetTime [min] | Type | Width [min] | Area [mAU*s] | Height [mAU] | Area % |
|---|---|---|---|---|---|---|
| 1 | 5.923 | MM | 0.1530 | 108.39830 | 11.80430 | 1.1236 |
| 2 | 7.082 | VB | 0.2032 | 9538.72949 | 694.73193 | 98.8764 |

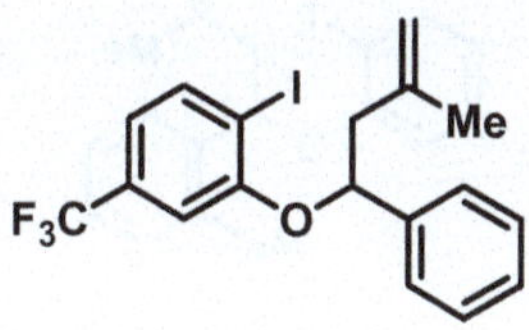


1-iodo-2-((3-methyl-1-phenylbut-3-en-1-yl)oxy)-4-(trifluoromethyl)benzene (2.101b)

Prepared according to **GP3** using 3-methyl-1-phenylbut-3-en-1-ol (487 mg, 3 mmol) and 2-iodo-5-(trifluoromethyl)phenol (864 mg, 3 mmol). The aryl iodide was purified by flash column chromatography using hexanes:DCM (30:1 v:v) and was obtained as a clear colourless oil (801 mg, 1.85 mmol, 62%).

**$^1$H NMR** (500 MHz, $CDCl_3$) δ 7.88–7.85 (m, 1H), 7.43–7.33 (m, 4H), 7.31–7.26 (m, 1H), 6.90–6.86 (m, 1H), 6.86–6.84 (m, 1H), 5.37 (dd, $J$ = 8.2, 5.0 Hz, 1H), 4.90–4.88 (m, 1H), 4.87–4.85 (m, 1H), 2.90 (ddd, $J$ = 14.2, 8.2, 1.0 Hz, 1H), 2.60 (ddd, $J$ = 14.3, 5.0, 1.1 Hz, 1H), 1.83 (dd, $J$ = 1.5, 0.9 Hz, 3H). **$^{13}$C NMR** (126 MHz, $CDCl_3$) δ 156.7, 141.1, 140.0, 139.9, 131.5 (q, $J$ = 32.5 Hz), 128.8, 128.1, 126.1, 123.6 (q, $J$ = 272.5 Hz), 118.8 (q, $J$ = 4.0 Hz), 114.3, 110.0 (q, $J$ = 4.0 Hz), 91.7 (q, $J$ = 1.5 Hz), 80.8, 46.6, 23.2. **$^{19}$F NMR** (564 MHz, $CDCl_3$) δ −63.00. **IR** (neat film, cm$^{-1}$) 3074, 3032, 2970, 2942, 2916, 1595, 1478, 1454, 1416, 1325, 1236, 1171, 1132, 1082, 1020. **HRMS** (DART) calculated 450.05417 m/z (found 450.05359 m/z for $C_{18}H_{16}$-$F_3IO$ + $NH_4$).

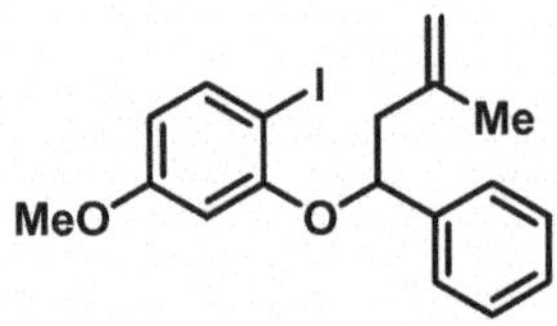


1-iodo-4-methoxy-2-((3-methyl-1-phenylbut-3-en-1-yl)oxy)benzene (2.101c)

Prepared according to **GP3** using 3-methyl-1-phenylbut-3-en-1-ol (487 mg, 3 mmol) and 2-iodo-5-methoxyphenol (861 mg, 3 mmol). The aryl iodide was purified by flash column chromatography using hexanes:EtOAc (10:1 v:v) and was obtained as a white solid (548 mg, 1.39 mmol, 46%, M.P.: 74–75 °C).

**$^1$H NMR** (500 MHz, $CDCl_3$) δ 7.61–7.57 (m, 1H), 7.40–7.31 (m, 4H), 7.28–7.23 (m, 1H), 6.26–6.20 (m, 2H), 5.28 (dd, $J = 8.5, 5.0$ Hz, 1H), 4.87–4.84 (m, 1H), 4.83–4.80 (m, 1H), 3.63 (s, 3H), 2.85 (ddd, $J = 14.0, 8.0, 1.0$ Hz, 1H), 2.55 (ddd, $J = 14.0, 5.0, 1.0$ Hz, 1H), 1.81 (dd, $J = 1.5, 1.0$ Hz, 3H). **$^{13}$C NMR** (126 MHz, $CDCl_3$) δ 160.8, 157.1, 141.4, 140.8, 139.0, 128.6, 127.7, 126.0, 113.9, 107.2, 101.7, 80.3, 76.0, 55.3, 46.7, 23.3. **IR** (neat film, cm$^{-1}$) 3077, 3009, 2938, 2833, 1573, 1477, 1451, 1418, 1300, 1252, 1197, 1152, 1141, 1051, 1010. **HRMS** (ESI +) calculated 417.0322 m/z (found 417.0312 m/z for $C_{18}H_{19}$-$IO_2$ + Na).

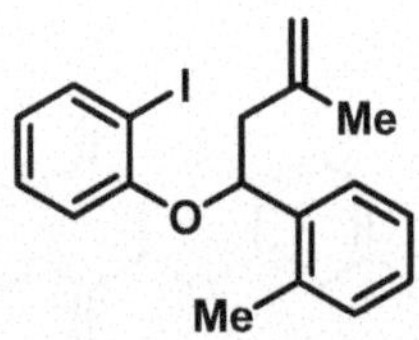


1-iodo-2-((3-methyl-1-(o-tolyl)but-3-en-1-yl)oxy)benzene (2.101 h)

Prepared according to **GP3** using 3-methyl-1-(*o*-tolyl)but-3-en-1-ol (530 mg, 3 mmol) and 2-iodophenol (660 mg, 3 mmol). The aryl iodide was purified by flash column chromatography using hexanes:DCM (10:1 v:v) and was obtained as a clear colourless oil (536 mg, 1.36 mmol, 45%).

**$^1$H NMR** (500 MHz, $CDCl_3$) δ 7.76 (dd, $J = 8.0, 1.5$ Hz, 1H), 7.43–7.37 (m, 1H), 7.17 (d, $J = 3.0$ Hz, 3H), 7.07 (ddd, $J = 8.5, 7.5, 1.5$ Hz, 1H), 6.60 (ddd, $J = 8.0, 7.5, 1.5$ Hz, 1H), 6.41 (dd, $J = 8.5, 1.5$ Hz, 1H), 5.48 (dd, $J = 9.0, 3.5$ Hz, 1H), 4.92–4.88 (m, 2H), 2.78 (ddd, $J = 14.5, 9.0, 1.0$ Hz, 1H), 2.52–2.45 (m, 4H). **$^{13}$C NMR** (126 MHz, $CDCl_3$) δ 156.3, 141.9, 139.5, 139.1, 133.7, 130.5, 129.1, 127.4, 126.6, 125.6, 122.1, 113.7, 113.0, 86.8, 77.3, 45.3, 23.3, 19.2. **IR** (neat film, cm$^{-1}$) 3069, 2969, 2943, 2918, 1582, 1568, 1489, 1467, 1439, 1289, 1242, 1121,

1047, 1017, 993. **HRMS** (DART) calculated 396.08243 m/z (found 396.08193 m/z for $C_{18}H_{19}$-IO + $NH_4$).

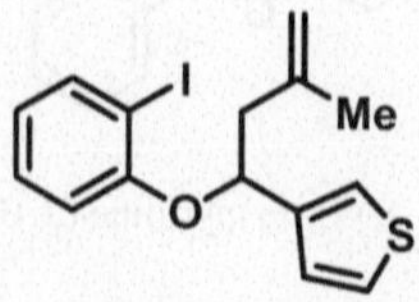

1-3-(1-(2-iodophenoxy)-3-methylbut-3-en-1-yl)thiophene (2.101i)

Prepared according to **GP3** using 3-methyl-1-(thiophen-3-yl)but-3-en-1-ol (505 mg, 3 mmol) and 2-iodophenol (660 mg, 3 mmol). The aryl iodide was purified by flash column chromatography using hexanes:EtOAc (10:1 v:v) and was obtained as a clear colourless oil (370 mg, 1.00 mmol, 33%).

**$^1$H NMR** (500 MHz, $CDCl_3$) δ 7.76 (dd, $J$ = 8.0, 1.5 Hz, 1H), 7.24 (ddd, $J$ = 5.0, 1.0, 0.5 Hz, 1H), 7.17 (ddd, $J$ = 8.0, 7.5, 1.5 Hz, 1H), 7.01 (ddd, $J$ = 3.5, 1.0, 0.5 Hz, 1H), 6.93 (dd, $J$ = 5.0, 3.5 Hz, 1H), 6.82–6.77 (m, 3H), 6.67 (ddd, $J$ = 8.0, 7.5, 1.5 Hz, 1H), 5.59 (dd, $J$ = 7.5, 6.0 Hz, 1H), 4.88–4.85 (m, 1H), 4.85–4.82 (m, 1H), 2.98 (ddd, $J$ = 14.0, 7.5, 1.0 Hz, 1H), 2.69 (ddd, $J$ = 14.0, 6.0, 1.0 Hz, 1H), 1.81 (dd, $J$ = 1.5, 1.0 Hz, 3H). **$^{13}$C NMR** (126 MHz, $CDCl_3$) δ 156.3, 144.1, 141.0, 139.5, 129.0, 126.4, 125.1, 125.0, 122.9, 114.3, 114.1, 87.8, 76.4, 46.8, 23.1. **IR** (neat film, cm$^{-1}$) 3073, 2969, 2932, 2915, 2361, 2324, 1582, 1468, 1438, 1371, 1271, 1236, 1119, 1047, 1018, 746, 700. **HRMS** (EI +) calculated 369.9888 m/z (found 369.9884 m/z for $C_{15}H_{15}$-IOS).

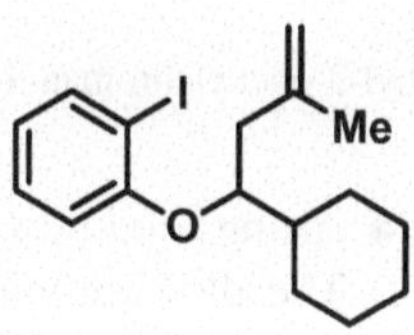

1-((1-cyclohexyl-3-methylbut-3-en-1-yl)peroxy)-2-iodobenzene (2.101j)

Prepared according to **GP3** using 1-cyclohexyl-3-methylbut-3-en-1-ol (505 mg, 3 mmol) and 2-iodophenol (660 mg, 3 mmol). The aryl iodide was purified by flash column chromatography using hexanes:DCM (8:1 v:v) and was obtained as a clear colourless oil (120 mg, 0.31 mmol, 10%).

**$^1$H NMR** (500 MHz, $CDCl_3$) δ 7.76 (dd, $J$ = 8.0, 1.5 Hz, 1H), 7.25 (ddd, $J$ = 8.5, 7.5, 1.5 Hz, 1H), 6.78 (ddd, $J$ = 8.5, 1.5, 0.5 Hz, 1H), 6.65 (ddd, $J$ = 8.0, 7.5, 1.5 Hz, 1H), 4.82–4.80 (m, 2H), 4.30 (td, $J$ = 6.5, 3.5 Hz, 1H), 2.47–2.34 (m, 2H), 1.94–1.60 (m, 9H), 1.42–1.12 (m, 3H). **$^{13}$C NMR** (126 MHz, $CDCl_3$) δ 157.2, 142.4, 139.6, 129.2, 121.9, 113.2, 112.7, 87.8, 82.0, 40.9, 38.8, 29.3, 27.4, 26.5, 26.4, 26.4, 23.2. **IR** (neat film, cm$^{-1}$) 3073, 2928, 2853, 2795, 1582, 1468, 1440, 1375, 1273, 1244, 1119, 1045, 1017. **HRMS** (DART) calculated 371.08718 m/z (found 371.08692 m/z for $C_{17}H_{24}$-IO).

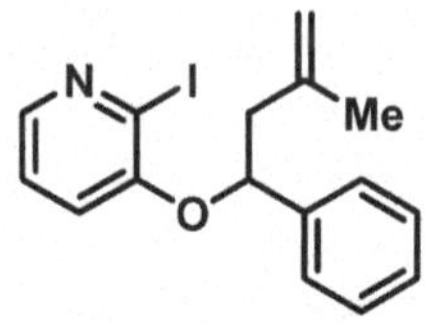

2-iodo-3-((3-methyl-1-phenylbut-3-en-1-yl)oxy)pyridine (2.101l)

Prepared according to **GP2** using 3-methyl-1-phenylbut-3-en-1-ol (243 mg, 1.5 mmol) and 2-iodopyridin-3-ol (332 mg, 1.5 mmol). The aryl iodide was purified by flash column chromatography using hexanes:EtOAc (8:1 v:v) and was obtained as a clear colourless oil (163 mg, 0.447 mmol, 30%).

**$^1$H NMR** (500 MHz, $CDCl_3$) δ 7.89 (dd, $J$ = 4.5, 1.5 Hz, 1H), 7.36–7.30 (m, 4H), 7.30–7.23 (m, 1H), 6.96 (dd, $J$ = 8.0, 4.5 Hz, 1H), 6.75 (ddd, $J$ = 8.0, 1.5, 0.5 Hz, 1H), 5.28 (dd, $J$ = 8.0, 5.0 Hz, 1H), 4.89–4.85 (m, 1H), 4.84–4.81 (m, 1H), 2.86 (ddd, $J$ = 14.5, 8.0, 1.0 Hz, 1H), 2.57 (ddd, $J$ = 14.5, 5.0, 1.0 Hz, 1H), 1.80 (dd, $J$ = 1.5, 1.0 Hz, 3H). **$^{13}$C NMR** (126 MHz, $CDCl_3$) δ 153.6, 142.5, 141.0, 139.8, 128.8, 128.1, 126.0, 123.1, 119.5, 114.3, 112.7, 80.8, 46.7, 23.3. **IR** (neat film, $cm^{-1}$) 3061, 2969, 2940, 1557, 1495, 1441, 1404, 1279, 1196, 1125, 1067, 1042, 988. **HRMS** (DART) calculated 366.03548 m/z (found 366.03476 m/z for $C_{16}H_{17}$-INO).

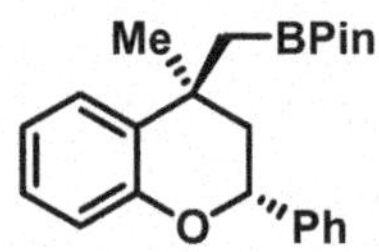

4,4,5,5-tetramethyl-2-(((2*R*,4*S*)-4-methyl-2-phenylchroman-4-yl)methyl)-1,3,2-dioxaborolane (2.105a)

Prepared according to **GP4** using1-iodo-2-((3-methyl-1-phenylbut-3-en-1-yl)oxy)benzene (73 mg, 0.2 mmol). The alkyl boronate was purified by flash column chromatography using hexanes:EtOAc (10:1 v:v) and was obtained as a white solid (60 mg, 0.164 mmol, 82%, M.P.: 74–75 °C).

**$^1$H NMR** (500 MHz, $CDCl_3$) δ 7.51–7.47 (m, 2H), 7.43–7.39 (m, 2H), 7.39–7.35 (m, 1H), 7.35–7.31 (m, 1H), 7.11 (ddd, $J$ = 8.0, 7.0, 1.5 Hz, 1H), 6.93–6.89 (m, 2H), 5.28 (dd, $J$ = 12.0, 2.0 Hz, 1H), 2.27 (dd, $J$ = 14.0, 2.0 Hz, 1H), 1.93 (ddd, $J$ = 14.0, 12.0, 1.5 Hz, 1H), 1.47 (s, 3H), 1.44–1.40 (m, 2H), 1.27 (d, $J$ = 1.5 Hz, 12H). **$^{13}$C NMR** (126 MHz, $CDCl_3$) δ 153.5, 142.4, 132.8, 128.4, 127.6, 127.1, 126.6, 125.9, 120.4, 117.1, 83.2, 74.4, 45.6, 33.5, 29.3, 25.0, 24.9. **$^{11}$B NMR** (128 MHz, $CDCl_3$) δ 33.00. **IR** (neat film, $cm^{-1}$) 3065, 3032, 2976, 2930, 1607, 1578, 1485, 1447, 1356, 1327, 1220, 1145, 1066, 1011, 970. **HRMS** (DART) calculated 382.25535 m/z (found 382.25588 m/z for $C_{23}H_{29}$-$BO_3$ + $NH_4$).

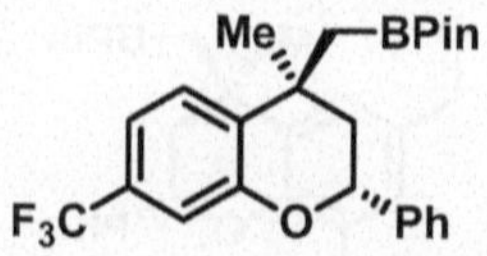

4,4,5,5-tetramethyl-2-(((2*R*,4*S*)-4-methyl-2-phenyl-7-(trifluoromethyl)chroman-4-yl)methyl)-1,3,2-dioxaborolane (2.105b)

Prepared according to **GP4** using 3-iodo-5-methoxy-4-((3-methyl-1-phenylbut-3-en-1-yl)oxy)benzaldehyde (85 mg, 0.2 mmol). The alkyl boronate was purified by flash column chromatography using hexanes:EtOAc (5:1 v:v) and was obtained as a clear, colourless oil (76 mg, 0.176 mmol, 88%).

**$^{1}$H NMR** (500 MHz, $CDCl_3$) δ 7.52–7.47 (m, 3H), 7.46–7.41 (m, 2H), 7.38–7.33 (m, 1H), 7.22–7.19 (m, 1H), 7.19–7.15 (m, 1H), 5.34 (dd, $J$ = 12.0, 2.0 Hz, 1H), 2.33 (dd, $J$ = 14.0, 2.0 Hz, 1H), 1.95 (ddd, $J$ = 14.0, 12.0, 1.0 Hz, 1H), 1.50 (s, 3H), 1.47–1.38 (m, 2H), 1.29 (s, 12H). **$^{13}$C NMR** (126 MHz, $CDCl_3$) δ 153.6, 141.6, 136.5 (q, $J$ = 1.0 Hz), 129.4 (q, $J$ = 32.5 Hz), 128.5, 127.8, 127.2, 125.8, 124.0 (q, $J$ = 272.1 Hz), 116.8 (q, $J$ = 4.0 Hz), 114.3 (q, $J$ = 4.0 Hz), 83.3, 74.8, 45.1, 33.6, 29.0, 24.9, 24.8. **$^{11}$B NMR** (128 MHz, $CDCl_3$) δ 32.79. **$^{19}$F NMR** (377 MHz, $CDCl_3$) δ −62.56. **IR** (neat film, $cm^{-1}$) 3032, 2978, 2932, 1578, 1506, 1456, 1427, 1325, 1275, 1215, 1124, 1089, 1067, 1017, 970. **HRMS** (DART) calculated 450.24273 m/z (found 450.24332 m/z for $C_{24}H_{31}$-$BF_3O_3$ + $NH_4$).

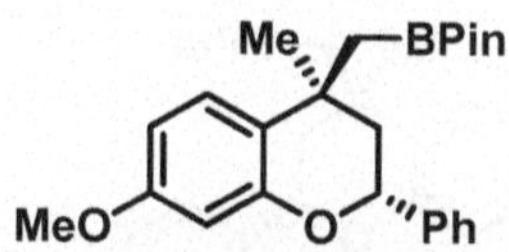

2-(((2*R*,4*S*)-7-methoxy-4-methyl-2-phenylchroman-4-yl)methyl)-4,4,5,5-tetramethyl-1,3,2-dioxaborolane (2.105c)

Prepared according to **GP4** using 1-iodo-2-((3-methyl-1-phenylbut-3-en-1-yl)oxy)-4-(trifluoromethyl)benzene (86 mg, 0.176 mmol). The alkyl boronate was purified by flash column chromatography using hexanes:EtOAc (10:1 v:v) and was obtained as a clear, colourless oil (49 mg, 0.124 mmol, 62%).

**$^{1}$H NMR** (500 MHz, $CDCl_3$) δ 7.54–7.50 (m, 2H), 7.46–7.41 (m, 2H), 7.38–7.32 (m, 1H), 7.30 (d, $J$ = 8.5 Hz, 1H), 6.55 (dd, $J$ = 8.5, 2.5 Hz, 1H), 6.50 (d, $J$ = 2.5 Hz, 1H), 5.30 (dd, $J$ = 12.0, 2.0 Hz, 1H), 3.79 (s, 3H), 2.26 (dd, $J$ = 14.0, 2.0 Hz, 1H), 1.94 (dd, $J$ = 12.0, 1.5 Hz, 1H), 1.47 (s, 3H), 1.41 (s, 2H), 1.29 (s, 12H). **$^{13}$C NMR** (126 MHz, $CDCl_3$) δ 158.7, 154.2, 142.2, 128.4, 127.6, 127.3, 125.9, 125.0, 107.5, 101.3, 83.0, 74.7, 55.1, 45.8, 33.0, 29.4, 24.9, 24.8. **$^{11}$B NMR** (128 MHz, $CDCl_3$) δ 33.11. **IR** (neat film, $cm^{-1}$) 3030, 2976, 2932, 1620, 1504, 1418, 1354, 1321, 1248, 1198, 1161, 1144, 1119, 1082, 1041. **HRMS** (DART) calculated 395.23936 m/z (found 395.23874 m/z for $C_{24}H_{32}BO_4$).

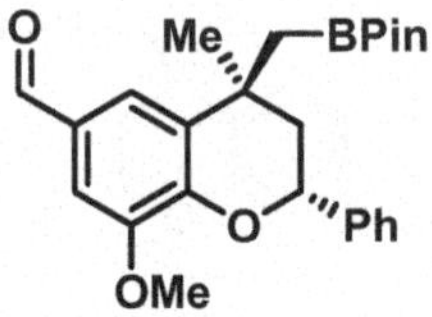

(2*R*,4*S*)-8-methoxy-4-methyl-2-phenyl-4-((4,4,5,5-tetramethyl-1,3,2-dioxaborolan-2-yl)methyl)chroman-6-carbaldehyde (2.105e)

Prepared according to **GP4** using 3-iodo-5-methoxy-4-((3-methyl-1-phenylbut-3-en-1-yl)oxy)benzaldehyde (85 mg, 0.2 mmol). The alkyl boronate was purified by flash column chromatography using hexanes:EtOAc (5:1 v:v) and was obtained as a white solid (61 mg, 0.144 mmol, 72%, M.P.: 103–104 °C).

**$^1$H NMR** (500 MHz, $CDCl_3$) δ 9.83 (s, 1H), 7.55 (d, *J* = 2.0 Hz, 1H), 7.48–7.44 (m, 2H), 7.42–7.36 (m, 2H), 7.34–7.29 (m, 1H), 7.27–7.25 (m, 1H), 5.36 (dd, *J* = 12.0, 2.0 Hz, 1H), 3.90 (s, 3H), 2.30 (dd, *J* = 14.0, 2.0 Hz, 1H), 1.96–1.87 (m, 1H), 1.49 (s, 3H), 1.40 (q, *J* = 15.0 Hz, 2H), 1.25 (d, *J* = 5.5 Hz, 12H). **$^{13}$C NMR** (126 MHz, $CDCl_3$) δ 191.1, 149.4, 149.1, 141.2, 133.1, 128.9, 128.5, 127.8, 125.9, 124.5, 107.1, 83.3, 75.5, 56.1, 45.2, 33.5, 29.0, 24.9, 24.9. **$^{11}$B NMR** (128 MHz, $CDCl_3$) δ 33.51. **IR** (neat film, $cm^{-1}$) 2976, 2933, 2837, 1684, 1582, 1480, 1452, 1391, 1354, 1278, 1261, 1236, 1215, 1144, 1078, 1009. **HRMS** (DART) calculated 423.23428 m/z (found 423.23405 m/z for $C_{25}H_{32}$-$BO_5$).

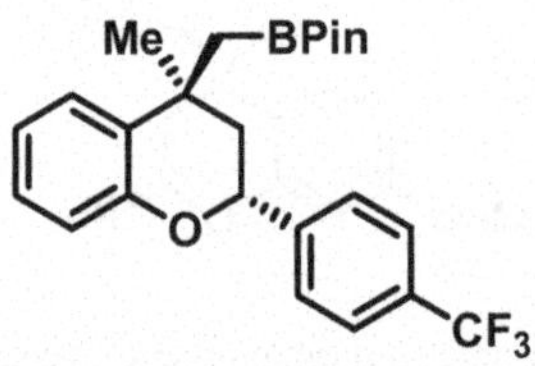

4,4,5,5-tetramethyl-2-(((2*R*,4*S*)-4-methyl-2-(4-(trifluoromethyl)phenyl)chroman-4-yl)methyl)-1,3,2-dioxaborolane (2.105f)

Prepared according to **GP4** using 1-iodo-2-((3-methyl-1-(4-(trifluoromethyl)phenyl)but-3-en-1-yl)oxy)benzene (85 mg, 0.2 mmol). The alkyl boronate was purified by flash column chromatography using hexanes:EtOAc (10:1 v:v) and was obtained as a clear, colourless oil (76 mg, 0.176 mmol, 88%).

**$^1$H NMR** (500 MHz, $CDCl_3$) δ 7.75–7.61 (m, 4H), 7.44–7.38 (m, 1H), 7.18–7.12 (m, 1H), 7.00–6.93 (m, 2H), 5.39 (dd, *J* = 12.0, 1.0 Hz, 1H), 2.35 (dd, *J* = 14.0, 2.0 Hz, 1H), 1.92 (ddd, *J* = 14.0, 12.0, 1.0 Hz, 1H), 1.51 (s, 3H), 1.49–1.44 (m, 2H), 1.32 (s, 12H). **$^{13}$C NMR** (126 MHz, $CDCl_3$) δ 153.0, 146.5 (q, *J* = 1.0 Hz), 132.7, 129.6 (q, *J* = 32.5 Hz), 127.2, 126.5, 126.0, 125.3 (q, *J* = 3.5 Hz), 123.8 (q, *J* = 272.0 Hz), 120.7, 117.0, 83.2, 73.8, 45.5, 33.4, 29.2, 24.9, 24.9. **$^{11}$B NMR** (128 MHz, $CDCl_3$) δ 32.78. **$^{19}$F NMR** (377 MHz, $CDCl_3$) δ −62.39. **IR** (neat film, $cm^{-1}$) 3030, 2978, 2930, 2911, 1622, 1578, 1485, 1449, 1356, 1325, 1221, 1165,

1125, 1066, 1017. **HRMS** (DART) calculated 450.24273 m/z (found 450.24317 m/z for $C_{24}H_{28}BF_3O_3 + NH_4$).

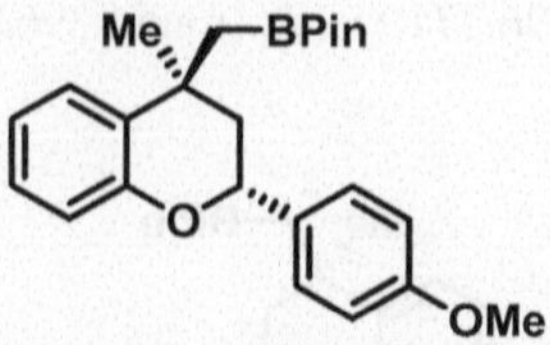


(2-(((2*R*,4*S*)-2-(4-methoxyphenyl)-4-methylchroman-4-yl)methyl)-4,4,5,5-tetramethyl-1,3,2-dioxaborolane (2.105g)

Prepared according to **GP4** using 1-iodo-2-((1-(4-methoxyphenyl)-3-methylbut-3-en-1-yl)oxy)benzene (79 mg, 0.2 mmol). The alkyl boronate was purified by flash column chromatography using hexanes:EtOAc (6:1 v:v) and was obtained as a white solid (58 mg, 0.148 mmol, 74%, M.P.: 82–84 °C).

**$^1$H NMR** (500 MHz, $CDCl_3$) δ 7.43–7.39 (m, 2H), 7.36 (dd, *J* = 8.0, 1.5 Hz, 1H), 7.10 (ddd, *J* = 8.0, 7.0, 1.5 Hz, 1H), 6.98–6.92 (m, 2H), 6.93–6.86 (m, 2H), 5.22 (dd, *J* = 12.0, 2.0 Hz, 1H), 3.84 (s, 3H), 2.22 (dd, *J* = 14.0, 2.0 Hz, 1H), 1.93 (ddd, *J* = 14.0, 12.0, 1.0 Hz, 1H), 1.47 (s, 3H), 1.40 (d, *J* = 5.5 Hz, 2H), 1.26 (s, 12H). **$^{13}$C NMR** (126 MHz, $CDCl_3$) δ 159.2, 153.6, 134.5, 132.7, 127.3, 127.0, 126.6, 120.3, 117.1, 113.9, 83.1, 74.1, 55.3, 45.4, 33.5, 29.3, 25.0, 24.9. **$^{11}$B NMR** (128 MHz, $CDCl_3$) δ 33.16. **IR** (neat film, $cm^{-1}$) 2976, 2930, 1578, 1516, 1485, 1447, 1327, 1302, 1248, 1221, 1175, 1144, 1119, 1076, 1037. **HRMS** (DART) calculated 412.26591 m/z (found 412.26627 m/z for $C_{24}H_{31}$-$BNO_4 + NH_4$).

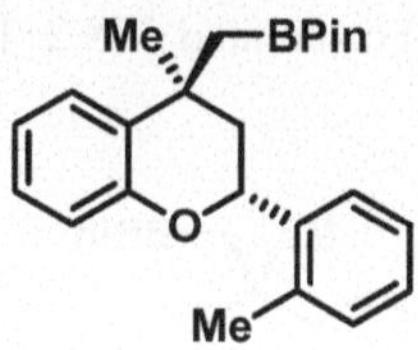


4,4,5,5-tetramethyl-2-(((2*R*,4*S*)-4-methyl-2-(*o*-tolyl)chroman-4-yl)methyl)-1,3,2-dioxaborolane (2.105 h)

Prepared according to **GP4** using 1-iodo-2-((3-methyl-1-(*o*-tolyl)but-3-en-1-yl)oxy)benzene (76 mg, 0.2 mmol). The alkyl boronate was purified by flash column chromatography using hexanes:EtOAc (10:1 v:v) and was obtained as a white solid (57 mg, 0.15 mmol, 75%, M.P.: 75–77 °C).

**$^1$H NMR** (500 MHz, $CDCl_3$) δ 7.61 (dd, *J* = 7.5, 1.0 Hz, 1H), 7.40 (dd, *J* = 7.5, 1.5 Hz, 1H), 7.33–7.27 (m, 1H), 7.27–7.22 (m, 1H), 7.22–7.18 (m, 1H), 7.13 (ddd, *J* = 8.0, 7.5, 1.5 Hz, 1H), 6.97–6.89 (m, 2H), 5.48 (dd, *J* = 12.0, 2.0 Hz, 1H), 2.45 (s, 3H), 2.33 (dd, *J* = 14.0, 2.0 Hz, 1H), 1.86 (ddd, *J* = 14.0, 12.0, 1.0 Hz, 1H), 1.49–1.43 (m, 4H), 1.26 (s, 12H). **$^{13}$C NMR** (126 MHz, $CDCl_3$) δ 153.7, 140.4, 134.5,

132.9, 130.3, 127.3, 127.1, 126.8, 126.3, 125.6, 120.4, 117.1, 83.2, 71.4, 44.2, 33.7, 29.4, 25.0, 24.9, 18.9. **$^{11}$B NMR** (128 MHz, $CDCl_3$) δ 33.17. **IR** (neat film, $cm^{-1}$) 2976, 29, 2928, 1598, 1485, 1449, 1356, 1325, 1227, 1165, 1144, 1119, 1011, 970. **HRMS** (DART) calculated 396.27100 m/z (found 396.27101 m/z for $C_{24}H_{31}$-$BO_3$ + $NH_4$).

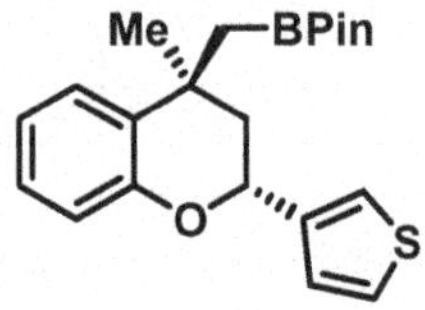


4,4,5,5-tetramethyl-2-(((2*R*,4*S*)-4-methyl-2-(thiophen-3-yl)chroman-4-yl)methyl)-1,3,2-dioxaborolane (2.105i)

Prepared according to **GP3** using 3-(1-(2-iodophenoxy)-3-methylbut-3-en-1-yl)thiophene (74 mg, 0.2 mmol). The alkyl boronate was purified by flash column chromatography using hexanes:EtOAc (10:1 v:v) and was obtained as a white solid (54 mg, 0.146 mmol, 73%).

**$^{1}$H NMR** (500 MHz, $CDCl_3$) δ 7.36 (dd, *J* = 8.0, 1.5 Hz, 1H), 7.31 (dd, *J* = 5.0, 1.0 Hz, 1H), 7.13 (ddd, *J* = 3.5, 1.0, 1.0 Hz, 1H), 7.09 (ddd, *J* = 8.0, 7.0, 1.5 Hz, 1H), 7.03 (dd, *J* = 5.0, 3.5 Hz, 1H), 6.91 (ddd, *J* = 8.0, 7.0, 1.5 Hz, 1H), 6.88 (dd, *J* = 8.0, 1.5 Hz, 1H), 5.54–5.50 (m, 1H), 2.39 (dd, *J* = 14.0, 2.0 Hz, 1H), 2.08 (dd, *J* = 14.0, 12.0 Hz, 1H), 1.48 (s, 4H), 1.38 (s, 2H), 1.25 (d, *J* = 1.5 Hz, 12H). **$^{13}$C NMR** (126 MHz, $CDCl_3$) δ 153.0, 145.4, 132.5, 127.1, 126.6, 126.6, 124.8, 124.2, 120.6, 117.1, 83.2, 70.6, 45.3, 33.4, 29.4, 25.0, 24.9. **$^{11}$B NMR** (128 MHz, $CDCl_3$) δ 32.96. **IR** (neat film, $cm^{-1}$) 2976, 2930, 1578, 1485, 1449. 1356, 1327, 1219, 1144, 1115, 1005, 847, 700. **HRMS** (DART) calculated 388.21177 m/z (found 388.21152 m/z for $C_{21}H_{27}$-$BO_3S$ + $NH_4$).

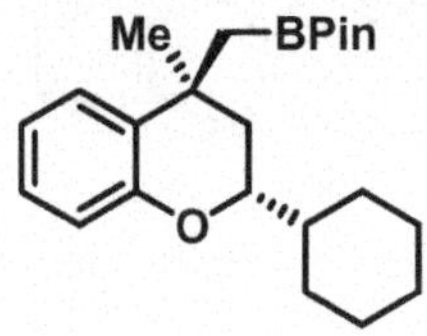


2-(((2*R*,4*S*)-2-cyclohexyl-4-methylchroman-4-yl)methyl)-4,4,5,5-tetramethyl-1,3,2-dioxaborolane (2.105j)

Prepared according to **GP3** using 1-((1-cyclohexyl-3-methylbut-3-en-1-yl)oxy)-2-iodobenzene (74 mg, 0.2 mmol). The alkyl boronate was purified by flash column chromatography using hexanes:EtOAc (10:1 v:v) and was obtained as a clear, colourless oil (42 mg, 0.112 mmol, 56%).

**$^{1}$H NMR** (500 MHz, $CDCl_3$) δ 7.31 (dd, *J* = 8.0, 1.5 Hz, 1H), 7.05 (ddd, *J* = 8.0, 7.0, 1.5 Hz, 1H), 6.88–6.81 (m, 1H), 6.79 (dd, *J* = 8.0, 1.0 Hz, 1H), 3.96 (ddd, *J* = 12.0, 6.0, 1.5 Hz, 1H), 2.11–2.00 (m, 2H), 1.88–1.77 (m, 3H), 1.78–1.70 (m,

1H), 1.68–1.56 (m, 2H), 1.45 (s, 3H), 1.41–1.12 (m, 19H). **$^{13}$C NMR** (126 MHz, $CDCl_3$) δ 153.9, 133.2, 126.8, 126.6, 119.8, 116.8, 83.0, 76.3, 42.6, 39.3, 32.9, 29.5, 28.6, 28.4, 26.7, 26.3, 26.2, 25.0, 24.9. **$^{11}$B NMR** (128 MHz, $CDCl_3$) δ 33.10. **IR** (neat film, $cm^{-1}$) 2976, 2926, 2853, 1578, 1485, 1449, 1356, 1325, 1269, 1233, 1163, 1144, 1111, 1069, 1045. **HRMS** (DART) calculated 371.27525 m/z (found 371.27648 m/z for $C_{23}H_{36}$-$BO_3$).

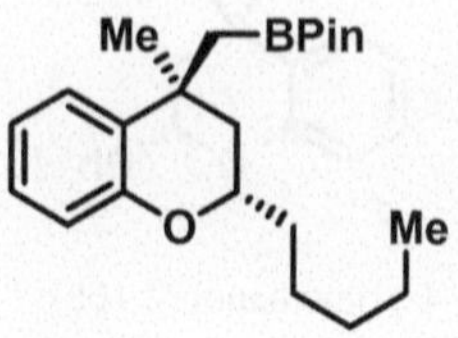


4,4,5,5-tetramethyl-2-(((2*S*,4*S*)-4-methyl-2-pentylchroman-4-yl)methyl)-1,3,2-dioxaborolane (2.105k)

Prepared according to **GP4** using 1-iodo-2-((2-methylnon-1-en-4-yl)oxy)benzene (72 mg, 0.2 mmol). The alkyl boronate was purified by flash column chromatography using hexanes:EtOAc (9:1 v:v) and was obtained as a clear, colourless oil (50 mg, 0.14 mmol, 70%).

**$^{1}$H NMR** (500 MHz, $CDCl_3$) δ 7.32 (dd, $J$ = 8.0, 1.5 Hz, 1H), 7.05 (ddd, $J$ = 8.0, 7.0, 1.5 Hz, 1H), 6.89–6.81 (m, 1H), 6.79 (dd, $J$ = 8.0, 1.5 Hz, 1H), 4.20–4.13 (m, 1H), 2.09 (dd, $J$ = 13.5, 2.0 Hz, 1H), 1.84–1.73 (m, 1H), 1.68–1.52 (m, 3H), 1.52–1.29 (m, 10H), 1.27 (d, $J$ = 2.5 Hz, 12H), 0.99–0.91 (m, 3H). **$^{13}$C NMR** (126 MHz, $CDCl_3$) δ 153.7, 133.1, 126.8, 126.7, 120.0, 116.8, 83.0, 72.3, 42.6, 36.0, 32.9, 31.9, 29.6, 25.0, 24.9, 24.9, 22.7, 14.1. **$^{11}$B NMR** (128 MHz, $CDCl_3$) δ 33.10. **IR** (neat film, $cm^{-1}$) 2976, 2457, 2932, 2861, 1578, 1485, 1447, 1387, 1356. 1327, 1231, 1165, 1144, 1121. **HRMS** (DART) calculated 359.27575 m/z (found 359.27579 m/z for $C_{22}H_{36}$-$BO_3$).

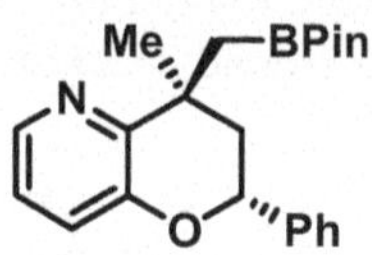


(2*R*,4*S*)-4-methyl-2-phenyl-4-((4,4,5,5-tetramethyl-1,3,2-dioxaborolan-2-yl)methyl)-3,4-dihydro-2H-pyrano[3,2-*b*]pyridine (2.105l)

Prepared according to **GP4** using 2-iodo-3-((3-methyl-1-phenylbut-3-en-1-yl)oxy)pyridine (73 mg, 0.2 mmol). The alkyl boronate was purified by flash column chromatography using hexanes:EtOAc (6:1 v:v) and was obtained as an off white solid (32 mg, 0.088 mmol, 44%, M.P.: 91–93 °C).

**$^{1}$H NMR** (500 MHz, $CDCl_3$) δ 8.17 (dd, $J$ = 4.5, 1.5 Hz, 1H), 7.48–7.44 (m, 2H), 7.43–7.38 (m, 2H), 7.36–7.31 (m, 1H), 7.15 (dd, $J$ = 8.0, 1.5 Hz, 1H), 7.03 (dd, $J$ = 8.0, 4.5 Hz, 1H), 5.25 (dd, $J$ = 12.0, 2.0 Hz, 1H), 2.34 (dd, $J$ = 14.0, 2.0 Hz, 1H), 2.06 (ddd, $J$ = 14.0, 12.0, 1.0 Hz, 1H), 1.54–1.49 (m, 3H), 1.26 (d, $J$ = 10.5 Hz, 12H). **$^{13}$C**

**NMR** (126 MHz, $CDCl_3$) δ 151.8, 149.7, 141.4, 141.3, 128.5, 127.8, 125.9, 124.0, 122.2, 83.1, 74.9, 45.8, 36.3, 27.9, 25.0, 24.9. **$^{11}$B NMR** (128 MHz, $CDCl_3$) δ 33.07. **IR** (neat film, $cm^{-1}$) 3063, 2978, 2930, 1570, 1441, 1356, 1329, 1285, 1267, 1231, 1144, 1113, 1094, 1067, 1007. **HRMS** (DART) calculated 366.22405 m/z (found 366.22475 m/z for $C_{22}H_{28}$-$BO_3$ + $NH_4$).

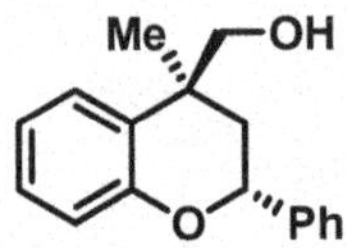


((2*R*,4*S*)-4-methyl-2-phenylchroman-4-yl)methanol (2.108)

To a 2 dram vial, 4,4,5,5-tetramethyl-2-(((2*R*,4*S*)-4-methyl-2-phenylchroman-4-yl)methyl)-1,3,2-dioxaborolane (73 mg, 0.2 mmol, 1 equiv) was added. THF (0.133 M) was added and the vial was cooled to 0 °C. Aqueous NaOH (0.8 mL, 3 M) and 30% $H_2O_2$ (0.4 mL) was added dropwise. A Teflon line screw cap was fitted on the 2 dram vial, sealed with Teflon tape and stirred at room temperature for 30 min. Once TLC analysis confirmed full conversion, the reaction was extracted with EtOAc/$H_2O$ (3x), dried over $Na_2SO_4$ and passed through a 2 cm plug of silica gel in a pipette using EtOAc. The chroman was purified by flash column chromatography using hexanes:EtOAc (3:1 v:v) and was obtained as an off-white solid (50 mg, 0.198 mmol, 99%, M.P.: 82–84 °C).

**$^{1}$H NMR** (500 MHz, $CDCl_3$) δ 7.51–7.45 (m, 2H), 7.45–7.38 (m, 2H), 7.36–7.32 (m, 1H), 7.32–7.28 (m, 1H), 7.19 (ddd, *J* = 8.2, 7.3, 1.7 Hz, 1H), 6.99–6.94 (m, 2H), 5.25 (dd, *J* = 12.0, 2.0 Hz, 1H), 3.85 (d, *J* = 11.0 Hz, 1H), 3.71 (d, *J* = 11.0 Hz, 1H), 2.27 (dd, *J* = 14.0, 2.5 Hz, 1H), 1.91 (dd, *J* = 14.0, 12.0 Hz, 1H), 1.66 (s, 1H), 1.35 (s, 3H). **$^{13}$C NMR** (126 MHz, $CDCl_3$) δ 155.1, 141.7, 128.5, 128.0, 127.9, 127.2, 126.6, 126.1, 120.7, 117.5, 74.7, 71.8, 41.0, 36.9, 25.1. **IR** (neat film, $cm^{-1}$) 3521, 2966, 2929, 2874, 1578, 1487, 1455, 1372, 1291, 1214, 1120, 1088, 1001. **HRMS** (DART) calculated 255.13850 m/z (found 255.13796 m/z for $C_{17}H_{19}O_2$).

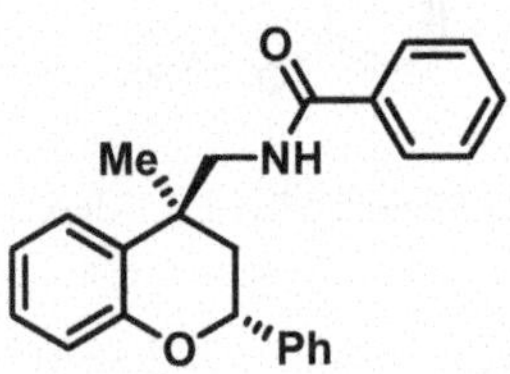


*N*-(((2*R*,4*S*)-4-methyl-2-phenylchroman-4-yl)methyl)benzamide (2.109)

To a flame dried or oven dried 1 dram vial cooled under argon, benzamide (24 mg, 0.2 mmol, 1 equiv), 4,4,5,5-tetramethyl-2-(((2*R*,4*S*)-4-methyl-2-phenylchroman-4-yl)methyl)-1,3,2-dioxaborolane (109 mg, 0.3 mmol, 1.5 equiv), CuBr (6 mg, 0.04 mmol, 20 mol%), NaOTMS (45 mg, 0.4 mmol, 2 equiv), and di-tert-butyl peroxide (110 μL, 0.6 mmol, 3 equiv) were added and allowed to purge for 10 min. *t*BuOH (0.4 M) was added. A Teflon line screw cap was fitted on the 1 dram vial, sealed with Teflon tape and place in a preheated oil bath at 75 °C for 48 h. The reaction mixture was cooled to room temperature. Once cooled, the reaction was passed through a 2 cm plug of Celite in a pipette using EtOAc. The chroman was purified by flash column chromatography using hexanes:EtOAc (3:1 v:v) and was obtained as an off-white solid (35 mg, 0.146 mmol, 49%, M.P.: 173–174 °C).

**$^1$H NMR** (500 MHz, $CDCl_3$) δ 7.80–7.72 (m, 2H), 7.55–7.48 (m, 3H), 7.48–7.42 (m, 2H), 7.42–7.35 (m, 2H), 7.35–7.28 (m, 2H), 7.24–7.18 (m, 1H), 7.03–6.95 (m, 2H), 5.39 (dd, $J$ = 12.0, 2.0 Hz, 1H), 4.14 (dd, $J$ = 14.0, 8.0 Hz, 1H), 3.47 (ddd, $J$ = 14.0, 5.5, 1.0 Hz, 1H), 2.17 (dd, $J$ = 14.0, 2.0 Hz, 1H), 1.86 (ddd, $J$ = 14.0, 12.0, 1.0 Hz, 1H), 1.40 (s, 3H). **$^{13}$C NMR** (126 MHz, $CDCl_3$) δ 167.9, 154.8, 141.7, 134.5, 131.6, 128.7, 128.5, 128.2, 127.7, 127.2, 126.9, 126.8, 126.0, 120.7, 117.7, 74.5, 49.7, 41.6, 36.5, 26.1. **IR** (neat film, cm$^{-1}$) 3414, 3322, 3063, 3032, 2965, 2926, 1640, 1578, 1534, 1487, 1447, 1295, 1229, 1119, 1063. **HRMS** (DART) calculated 358.18070 m/z (found 358.18011 m/z for $C_{24}H_{24}$-$NO_2$).

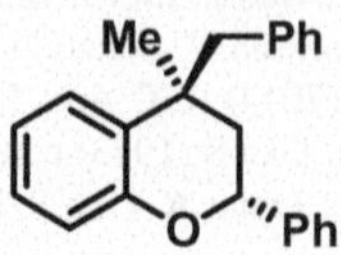


(2*R*,4*S*)-4-benzyl-4-methyl-2-phenylchroman (2.110)

To a flame dried or oven dried 2 dram vial cooled under argon, 4,4,5,5-tetramethyl-2-(((2*R*,4*S*)-4-methyl-2-phenylchroman-4-yl)methyl)-1,3,2-dioxaborolane (36 mg, 0.1 mmol, 1 equiv), $Pd(OAc)_2$ (2.26 mg, 0.01 mmol, 10 mol%), K*t*BuO (33.7 mg, 0.3 mmol, 3 equiv), and 1,3,5,7-tetramethyl-8-phenyl-2,4,6-trioxa-8-phosphaadamantane (4.7 mg, 0.01 mmol, 10 mol%) were added and allowed to purge for 10 min. PhMe (0.1 M) and iodobenzene (17 μL, 0.15 mmol, 1.5 equiv) was added. A Teflon line screw cap was fitted on the 2 dram vial, sealed with Teflon tape and place in a preheated oil bath at 100 °C for 12 h. The reaction mixture was cooled to room temperature. Once cooled, the reaction was passed through a 2 cm plug of silica gel in a pipette using EtOAc. The chroman was purified by flash column chromatography using hexanes:EtOAc (10:1 v:v) and was obtained as a clear colourless oil (13 mg, 0.042 mmol, 42%).

**$^1$H NMR** (500 MHz, $CDCl_3$) δ 7.45–7.39 (m, 4H), 7.38–7.22 (m, 4H), 7.21–7.16 (m, 1H), 7.12–7.07 (m, 2H), 7.00–6.93 (m, 2H), 4.94 (dd, $J$ = 12.0, 2.0 Hz, 1H), 3.20 (d, $J$ = 13.5 Hz, 1H), 2.99 (d, $J$ = 13.5 Hz, 1H), 2.12 (dd, $J$ = 14.0, 2.0 Hz, 1H), 1.88 (dd, $J$ = 14.0, 12.0 Hz, 1H), 1.37 (s, 3H). **$^{13}$C NMR** (126 MHz, $CDCl_3$) δ 154.27, 141.84, 138.09, 130.70, 129.86, 128.55, 128.06, 127.86, 127.49, 127.47, 126.56, 125.96, 120.52, 117.31, 74.26, 49.61, 42.81, 35.95, 28.65. **IR** (neat film, $cm^{-1}$) 3062, 3028, 2961, 2929, 1578, 1484, 1448, 1297, 1278, 1225, 1119, 1083, 1066, 1009. **HRMS** (DART) calculated 315.17489 m/z (found 315.17435 m/z for $C_{23}H_{23}$-O).

## References


1. (a) Yoon H, Petrone DA, Lautens M (2014) Org Lett 16:6420. (b) Yoon H, Jang YJ, Lautens M (2016) Synthesis 48:1483
2. Ellis GP, Romney-Alexander TM (1987) Chem Rev 87:779
3. Zanon J, Klapars A, Buchwald SL (2003) J Am Chem Soc 125:2890
4. For selected reviews, see: (a) Anbarasan P, Schareina T, Beller M (2011) Chem Soc Rev 40:5049. (b) Wen Q, Lu P, Wang Y (2014) RSC Adv 4:47806
5. Takagi K, Okamoto T, Sakakibara Y, Oka S (1973) Chem Lett 2:471
6. Takagi K, Okamoto T, Sakakibara Y, Ohno A, Oka S, Hayama N (1976) Bull Chem Soc Jpn 49:3177
7. Erhardt S, Grushin VV, Kilpatrick AH, Macgregor SA, Marshall WJ, Roe DC (2008) J Am Chem Soc 130:4828
8. Dobbs KD, Marshall WJ, Grushin VV (2007) J Am Chem Soc 129:30
9. Jin F, Confalone PN (2000) Tetrahedron Lett 41:3271
10. Schareina T, Zapf A, Beller M (2004) Chem Commun 2004:1388
11. Senecal TD, Shu W, Buchwald SL (2013) Angew Chem Int Ed 52:10035
12. For an overview of organoboron chemistry, see: (a) Boronic acids: preparation and applications in organic synthesis and medicine, 2nd ed. Wiley-VCH, Weinheim, 2011. (b) Stockland RA (2016) Practical functional group synthesis. Wiley, Waukegan, pp 515–555
13. Ishiyama T, Murata M, Miyaura N (1995) J Org Chem 60:7508
14. Sumimoto M, Iwane N, Takahama T, Sakaki S (2004) J Am Chem Soc 126:10457
15. Murata M, Watanabe S, Masuda Y (1997) J Org Chem 62:6458
16. Lam KC, Marder TB, Lin Z (1849) Organometallics 2010:29
17. (a) Billingsley KL, Barder TE, Buchwald SL (2007) Angew Chem Int Ed 119:5455. (b) Billingsley KL, Buchwald SL (2008) J Org Chem 73:5589
18. Molander GA, Trice SLJ, Kennedy SM, Dreher SD, Tudge MT (2012) J Am Chem Soc 134:11667
19. For selected reviews, see: (a) Vlaar T, Ruijter E, Orru RVA (2011) Adv Synth Catal 353:809. (b) Klein JEMN, Taylor RJK (2018) Eur J Org Chem 2011:6821. (c) Ohno H, Inuki S (2018) Synthesis 50:700. (d) Biemolt J, Ruijter E (2018) Adv Synth Catal 360:3821. https://doi.org/10.1002/adsc.201
20. For a detailed review on work by the Grigg group, see: Grigg R, Sridharan V (1995) Comprehensive organometallic chemistry II. In: Abel EW, Stone FGA, Wilkinson G (eds) Pergamon, Oxford, UK, vol 12, pp 299
21. Grigg R, Sansano JM, Santhakumar V, Sridharan V, Thangavelanthum R, Thornton-Pett M, Wilson D (1997) Tetrahedron 53:11803
22. Wilson JE (2012) Tetrahedron Lett 53:2308
23. Arcadi A, Blesi F, Cacchi S, Fabrizi G, Goggiamani A, Marinelli F (2013) J Org Chem 78:4490

24. René O, Lapointe D, Fagnou K (2009) Org Lett 11:4560
25. Kong W, Wang Q, Zhu J (2015) J Am Chem Soc 137:16028
26. Teplý F, Stará IG, Starý I, Kollárović A, Šaman D, Fiedler P (2002) Tetrahedron 58:9007
27. Wang D-C, Wang H-X, Hao E-J, Jiang X-H, Xie M-S, Qu G-R, Guo H-M (2016) Adv Synth Catal 358:494
28. Liu X, Ma X, Huang Y, Gu Z (2013) Org Lett 15:4814
29. Liu Z, Xia Y, Zhou S, Wang L, Zhang Y, Wang J (2013) Org Lett 15:5032
30. Liu X, Li B, Gu Z (2015) J Org Chem 80:7547
31. Shen C, Liu R-R, Fan R-J, Li Y-L, Xu T-F, Gao J-R, Jia Y-X (2015) J Am Chem Soc 137:4936
32. Petrone DA, Yen A, Zeidan N, Lautens M (2015) Org Lett 17:4838
33. Petrone DA, Kondo M, Zeidan N, Lautens M (2016) Chem Eur J 22:5684
34. Liu R-R, Wang Y-G, Li Y-L, Huang BB, Liang R-X, Jia Y-X (2017) Angew Chem Int Ed 56:7475
35. Grigg R, Santhakumar V, Sridharan V (1993) Tetrahedron Lett 34:3163
36. Pinto A, Jia Y, Neuville L, Zhu J (2007) Chem Eur J 13:961
37. Kobayashi Y, Kamisaki H, Yanada R, Takemoto Y (2006) Org Lett 8:2711
38. Yasui Y, Kamisaki H, Takemoto Y (2008) Org Lett 10:3303
39. Marco-Martinez J, Lopez-Carrillo V, Bunuel E, Simancas R, Cardenas DJJ (1874) Am Chem Soc 2007:129
40. (a) Persson AKÅ, Jiang T, Johnson MT, Bäckvall J-E (2011) Angew Chem Int Ed 50:6155. (b) Deng Y, Bartholomeyzik T, Bäckvall J-E (2013) Angew Chem Int Ed 52:6283
41. Jiang T, Bartholomeyzik T, Mazuela J, Willersinn J, Bäckvall J-E (2015) Angew Chem Int Ed 54:6024
42. Vachhani DD, Butani HH, Sharma N, Bhoya UC, Shah AK, Van der Eycken EV (2015) Chem Commun 51:14862
43. For a detailed review, see: Littke AF, Fu GC (2002) Angew Chem Int Ed 41:4176
44. Petrone DA, Malik HA, Clemenceau A, Lautens M (2012) Org Lett 14:4806
45. Morgan JB, Morken JP (2004) J Am Chem Soc 126:15338
46. Rossi SA, Shimkin KW, Xu Q, Mori-Quiroz LM, Watson DA (2013) Org Lett 15:2314
47. Dreher SD, Lim S-E, Sandrock DL, Molander GA (2009) J Org Chem 74:3626
48. Brenstrum T, Gerristma DA, Adjabeng GM, Frampton CS, Britten J, Robertson AJ, McNulty J, Capretta A (2004) J Org Chem 69:7635

# Chapter 3
# Pd-Catalyzed Spirocyclization via C–H Activation and Benzyne/Alkyne Insertion

## 3.1 C–H Functionalization

The selective functionalization of carbon–hydrogen bonds has been an ongoing challenge in the chemical community. Traditional methods to formally functionalize C–H bonds have included reactions such as Friedel–Crafts alkylation/acylation (C(sp$^2$)–H bond) and benzylic bromination (C(sp$^3$)–H bond) (Scheme 3.1). Unfortunately, these reactions generally require harsh conditions, stoichiometric reagents or are unselective.

Over the past several decades, investigation into transition metal catalyzed C–H functionalization has provided a new perspective in forging chemical bonds. The transformation is defined as "*the formation of an organometallic intermediate arising from the cleavage of a C–H bond*" [2]. A wide range of transition metals have been capable of activating C–H bonds (Pd, Rh, Ir, etc.); [3] with commonly accepted mechanisms that include concerted metalation-deprotonation (CMD) and σ-bond metathesis pathways.

The CMD mechanism was first proposed by the Fagnou group in 2006, [4] and the nature of biaryl cross coupling reactions were studied experimentally and computationally [5]. Key competition experiments between benzene vs. anisole or fluorobenzene were performed to provide evidence for the CMD pathway (Scheme 3.2). The fluorobenzene participated in the biaryl coupling forming **3.8a** faster than benzene with high regioselectivity (22:3:1 *o*:*m*:*p*) while the anisole reacted slower than benzene (2:1, **3.8**:**3.8b**). These findings suggest that the reaction does not proceed through an electrophilic aromatic substitution or radical process, but correlates to a proton-transfer pathway.

Parts of this chapter have been reproduced with permission from [1].



H. Yoon, *Palladium and Nickel Catalyzed Transformations Forming Functionalized Heterocycles*, Springer Theses, https://doi.org/10.1007/978-3-030-54077-7_3

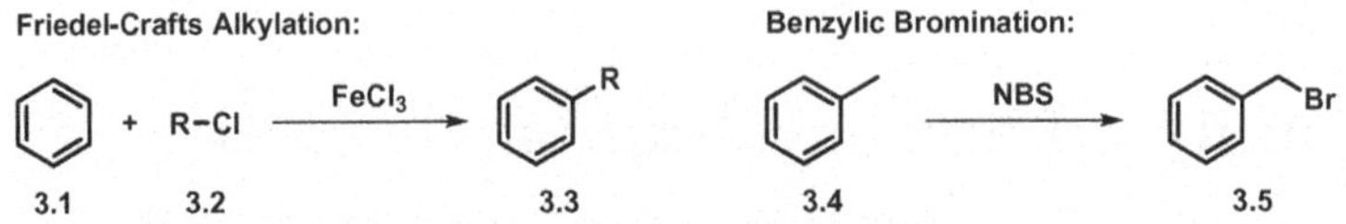


**Scheme 3.1** Traditonal methods to functionalize C–H bonds

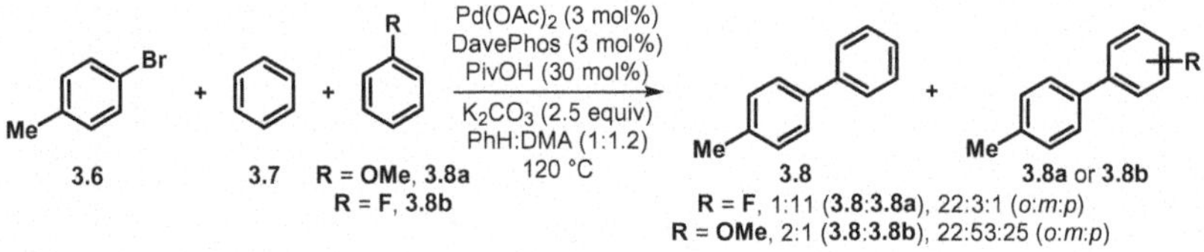


**Scheme 3.2** Competition experiment observing the rates of C–H activation of benzene, fluorobenzene and anisole

The proposed mechanism proceeds via a ligand exchange to give a pivalate bound Pd-complex **3.10**. A key six membered transition state is proposed between the desired C–H bond and metal (**3.11**) to give **3.12** and subsequent reductive elimination gives the biaryl product (**3.8**) (Scheme 3.3).

Alternatively, a σ-bond metathesis mechanism proceeds without the aid of an inorganic salt (Scheme 3.4). The metal center interacts with the desired C–H bond to form the σ-complex and afterwards, it is postulated that **3.16** is formed from the

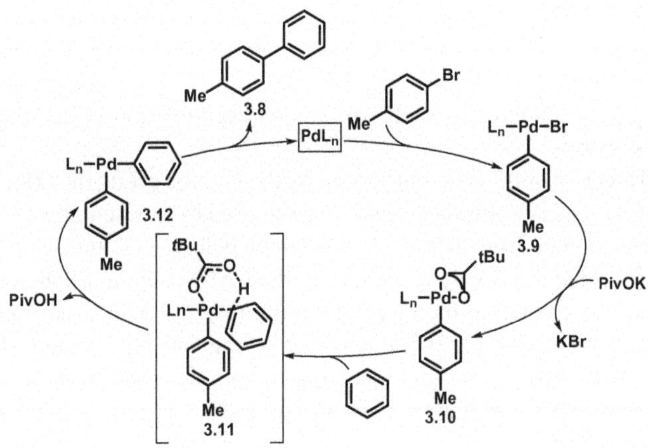


**Scheme 3.3** CMD mechanism for the biaryl coupling

**Scheme 3.4** σ-bond metathesis mechanism

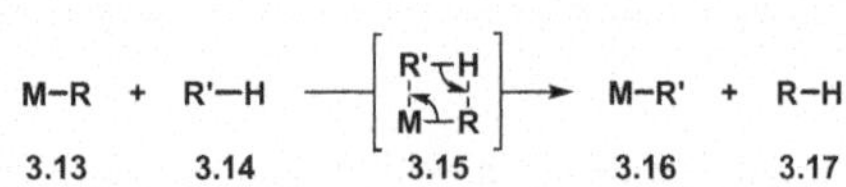

**Scheme 3.5** Directed *ortho* C–H functionalization

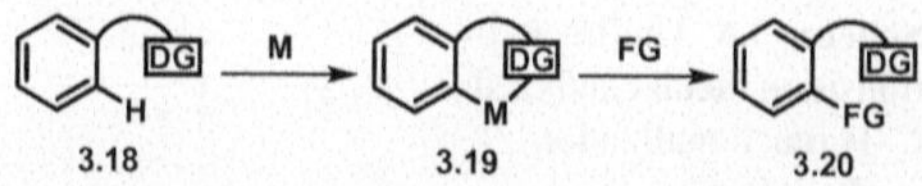

transition state **3.15** [6]. Typically, $d^0$ transition metals react in this manner; however, recent studies have found that late transition metals such as Ir and Ru can access this pathway [7].

## 3.2 Modes of Transition-Metal Catalyzed C–H Functionalization

### 3.2.1 Directed C–H Functionalization

There are several ways in which transition-metal catalyzed C–H functionalization is applied. Among them, the most intensely explored is the *directed ortho* C–H functionalization of arenes. This method exploits a directing group embedded within the substrate, which interacts and guides the metal center towards the desired C–H bond, generating metallacycle **3.19.** A subsequent series of transformations generates the *ortho*-functionalized arene **3.20**. *Meta* and *para* functionalizations have been reported wherein fine-tuning the directing group was essential for selectivity [8] (Scheme 3.5).

### 3.2.2 Undirected C–H Functionalization

Despite the significant advances made in *directed* C–H functionalization, a limitation revolves around the pre-installation of an appropriate directing group, prompting catalyst or ligand design in undirected C–H functionalization. Several methodologies have been reported including the Ir-catalyzed borylation (**3.22**), [9] Rh-catalyzed silylation (**3.23**) [10] and Pd-catalyzed arylation (**3.24**) [11, 12] or olefination (**3.25**) [13] (Scheme 3.6).

### 3.2.3 Catellani Reaction

The use of transient directing group processes, wherein the directing group is not incorporated in the final product, has become a powerful tool in C–H bond activation methodologies. Most notably, in the 1990s, the Catellani group reported a Pd-catalyzed polyfunctionalization of iodoarenes (Scheme 3.7) [14]. Using norbornene

**Scheme 3.6** Undirected transition metal catalyzed C–H functionalization methodologies

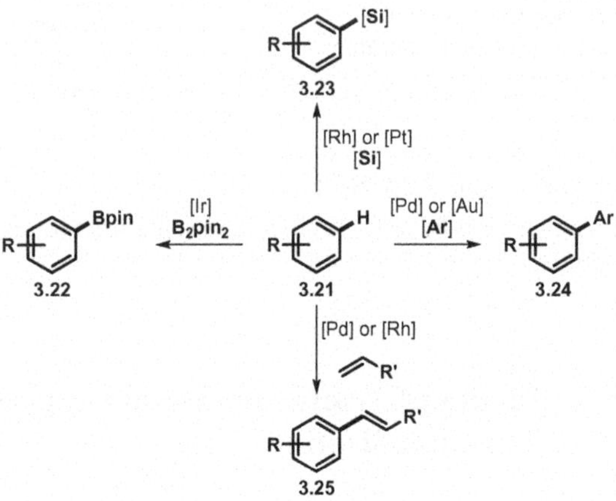


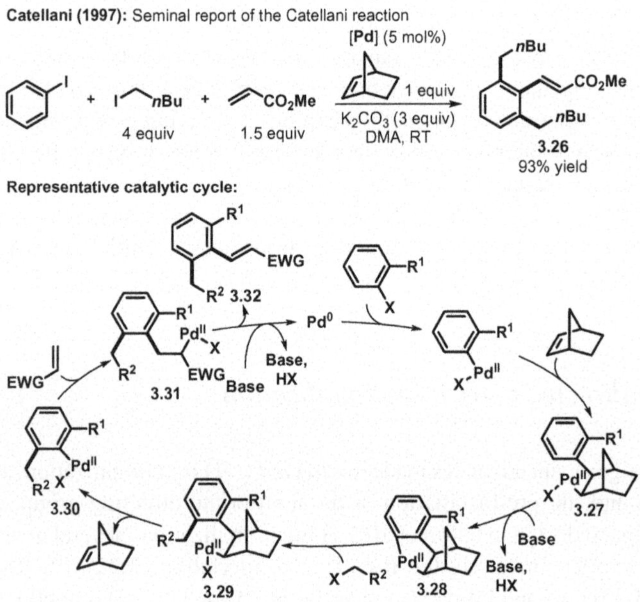


**Scheme 3.7** Seminal report of the Catellani reaction and representative catalytic cycle for the Catellani reaction

as a transient directing group, *ortho* and *ipso* functionalization of the iodoarene was realized. The catalytic cycle proceeds via an oxidative addition into the aryl halide followed by a carbopalladation across the exogenous norbornene and generating alkylpalladium intermediate **3.27**. A base mediated C–H activation and selective $Csp^2$–$sp^3$ bond formation via a $Pd^{IV}$ or dinuclear $Pd^{II}$ intermediate gives **3.29**. β-carbon elimination releases the norbornene, generating arylpalladium intermediate

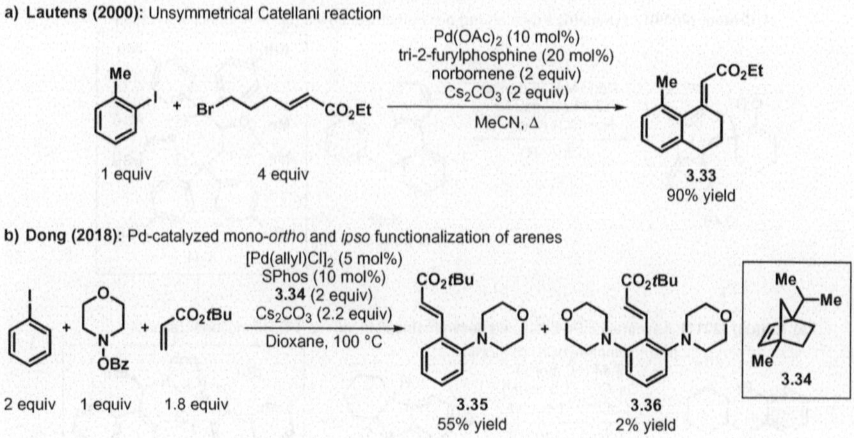


**Scheme 3.8** Advancements in the Catellani reaction

**3.30**. Subsequent reaction of **3.30** with a compatible coupling partner furnishes **3.32** with concomitant catalyst regeneration (Scheme 3.7) [15].

Following the initial reports, our group demonstrated the value of phosphine ligands applied in the unsymmetrical Catellani reaction by using a preinstalled *ortho* substituent on the aryl halide and a subsequent Heck reaction as the terminating step to give **3.33** in excellent yield (Scheme 3.8a) [16]. Unfortunately, the Catellani reaction suffers from a significant limitation wherein selective mono-*ortho* functionalization cannot be accessed unless an *ortho* substituent is preinstalled. To overcome this intrinsic issue, the Dong group reported the mono-*ortho* amination of arenes using an unique norbornene scaffold (**3.34**) (Scheme 3.8b). The installed substituents on the bridgehead of the norbornene increases the energy barrier (steric hindrance) for the second C–H activation and thus promotes the mono-functionalized arenes (**3.35**) [17].

## 3.2.4 *Intramolecular C–H Functionalization*

Intramolecular C–H functionalization places the metal center in close proximity to the desired C–H bond within the structure via a traditional organometallic process such as oxidative addition or transmetallation (Scheme 3.9).

**Scheme 3.9** Intramolecular C–H functionalization

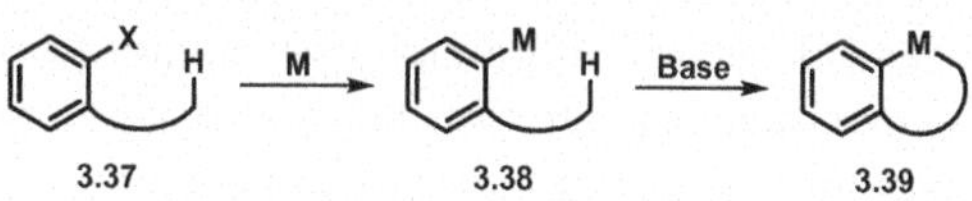

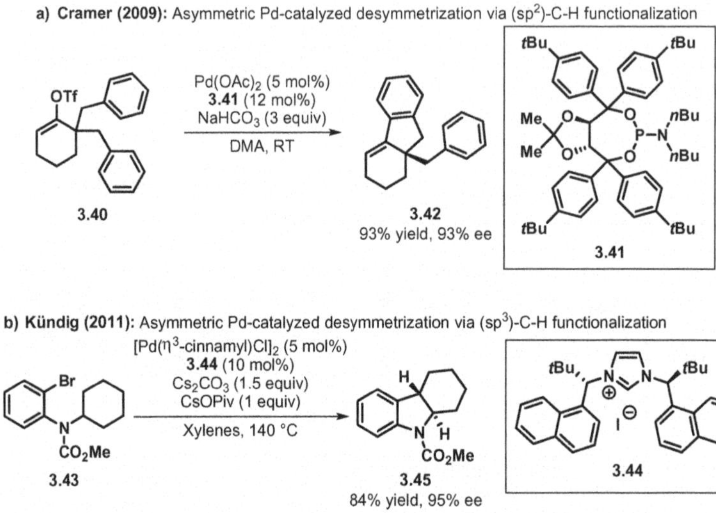


**Scheme 3.10** Pd-catalyzed intramolecular C($sp^2$) and ($sp^3$)–H functionalization

Initially reported by Ames in 1982, [18] the intramolecular Pd-catalyzed C–H functionalization reaction has evolved to activating $sp^3$ C–H bonds or enantioselectively generating C–C bond bonds. In 2009, the Cramer group showcased a Pd-catalyzed desymmetrization of cyclohexenyl triflates involving an intramolecular C($sp^2$)–H activation (Scheme 3.10a). Using a TADDOL-based monodentate phosphoramidite ligand **3.41**, vinyl triflate **3.40** was transformed into indane **3.42** in excellent yield and ee [19]. In 2011, Kündig disclosed the first enantioselective C($sp^2$)–C($sp^3$) coupling via C–H functionalization. The generation of the chiral indoline **3.45** required the use of a chiral NHC ligand (**3.44**), and elevated temperatures were required to activate the C($sp^3$)–H bond (140 °C or 160 °C, Scheme 3.10b) [20]. Since these initial reports by the Cramer and Kündig group, several ($sp^2$) and ($sp^3$) C–H functionalizations have been employed in the synthesis of various complex heterocycles and natural products [21].

### 3.2.5 *Pd-Catalyzed Cascade Reactions Involving C–H Functionalization of Alkyl-Pd Intermediates*

The combination of C–H functionalization in a domino Pd-catalyzed process has extended the versatility and applicability of the reaction as it enables the metal centre to interact with C–H bonds that are otherwise unreactive. The seminal report concerning the utilization of this method was by Grigg (Scheme 3.11), where the tetracyclic and spirocyclic scaffolds (**3.47** and **3.49** respectively) were generated from the C–H activation of the pendant aromatic group [22].

**Scheme 3.11** Grigg's seminal reports on the Pd-catalyzed domino C–H functionalization reaction

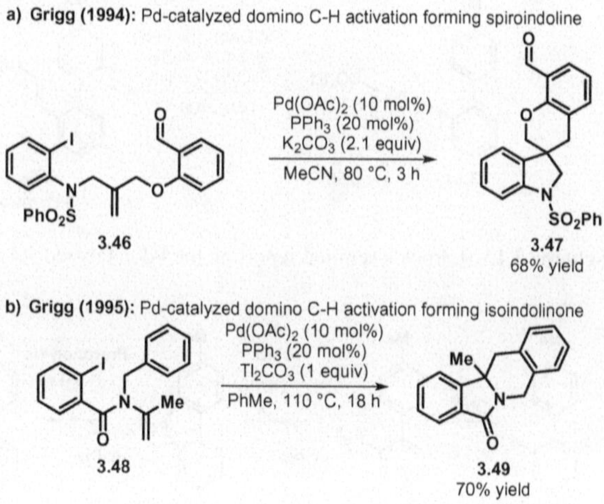


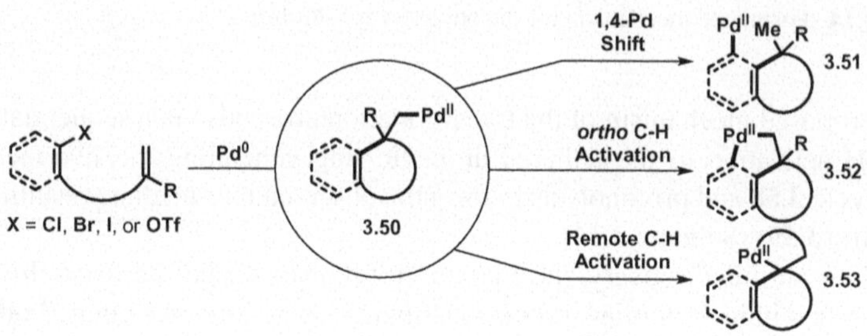


**Scheme 3.12** Potential routes of intramolecular C–H activation

The general mechanism for this class of transformation proceeds through the formation of the alkyl-Pd$^{II}$ species **3.50** mentioned in Chap. 2 (Scheme 3.12). In the presence of spatially available C(sp$^2$)–H or C(sp$^3$)–H bonds, C–H activation furnishes the respective palladium complex **3.51** or palladacycles **3.52** and **3.53**. Once generated, various pathways are possible to access functionalized heterocycles including direct reductive elimination, transmetallation followed by reductive elimination, or insertion of unsaturated π-systems.

### 3.2.5.1 1,4-Pd Shift

The 1,4-Pd shift is among the most commonly observed migration in Pd-catalysis. In this elementary step, the Pd shifts from one carbon to another via a five-membered palladacycle. In 2002, the Larock group disclosed the first report of a 1,4-Pd shift in *o*-iodobiaryls [23]. In the presence of CsOPiv, a 1:1 mixture of olefinated biaryls **3.56** and **3.57** arising from the direct Heck reaction or a sequential 1,4-shift and Heck reaction was found (Scheme 3.13).

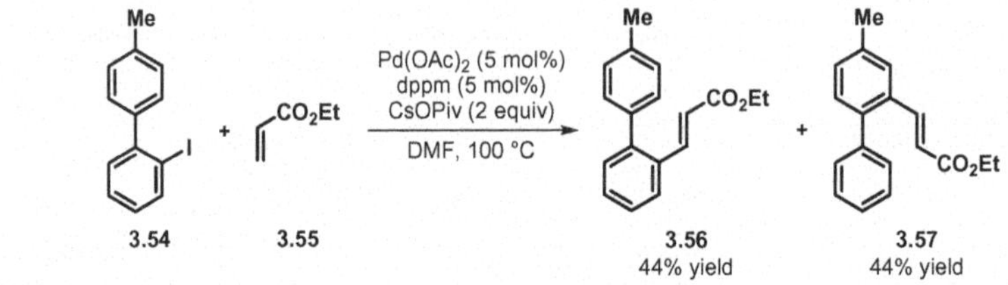


**Scheme 3.13** Larock's seminal report on the Pd-catalyzed 1,4-Pd shift utilizing *o*-iodobiaryls

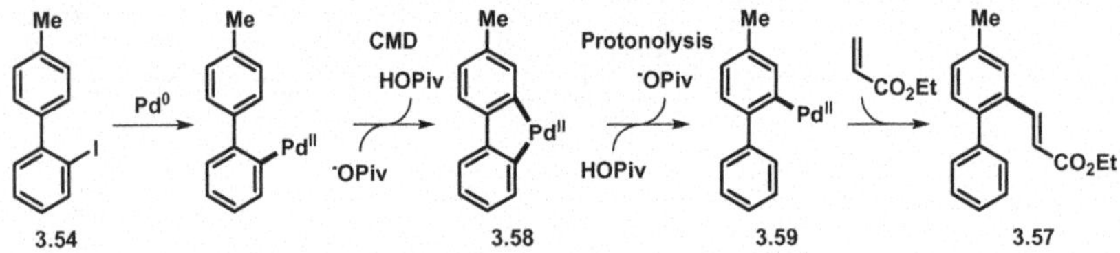


**Scheme 3.14** Postulated mechanism for the observed 1,4-Pd shift

The proposed mechanism of the transformation proceeds via a sequential CMD-type C–H activation to the tethered aromatic ring generating the five-membered palladacycle **3.58** and protonolysis to form the observed intermediate resulting from a 1,4-shift (**3.59**) (Scheme 3.14).

Building on this discovery, the Larock group investigated the feasibility of the 1,4-Pd shift within a domino process. Utilizing the *o*-iodoaryl ether **3.60** in the presence of Pd and 4 equivalents of CsOPiv, a sequential carbopalladation, 1,4-Pd shift and C–H activation occurs to generate **3.61** in good yield (Scheme 3.15a) [24]. Following this report, the Jia group reported a divergent, solvent dependent, method to obtain the 1,4-shift products (**3.63b**) or the anion-capture cascade products (**3.63a**) (Scheme 3.15b) [25]. In the absence of water, **3.63a** was formed while a 95:5 mixture exclusively gave **3.63b**. Performing the reaction in $D_2O$ generated **3.63b** with >95% deuterium incorporation *ortho* to the starting arylhalide (**3.62**), suggesting that the palladacycle formed in situ undergoes a reversible protodemetallation. Similarly to Larock, the Zhu group expanded this methodology by subjecting the acrylamides to specific reaction conditions, and generate polycyclic oxindoles (Scheme 3.15c) [26].

#### 3.2.5.2 *Ortho* and Remote C–H Activation in a Domino Process

*Ortho* and remote C–H functionalization are the other commonly employed methods in a domino process. As opposed to the aforementioned 1,4-shift, this class of functionalization utilizes the palladacycle formed in situ to generate novel bonds. Our group, along with several others, have made significant contributions to advancing this field [27] Of note, we demonstrated the synthesis of polyfused isochromans arising from two aryl iodides and two C–H activations (Scheme 3.16) [28]. QPhos

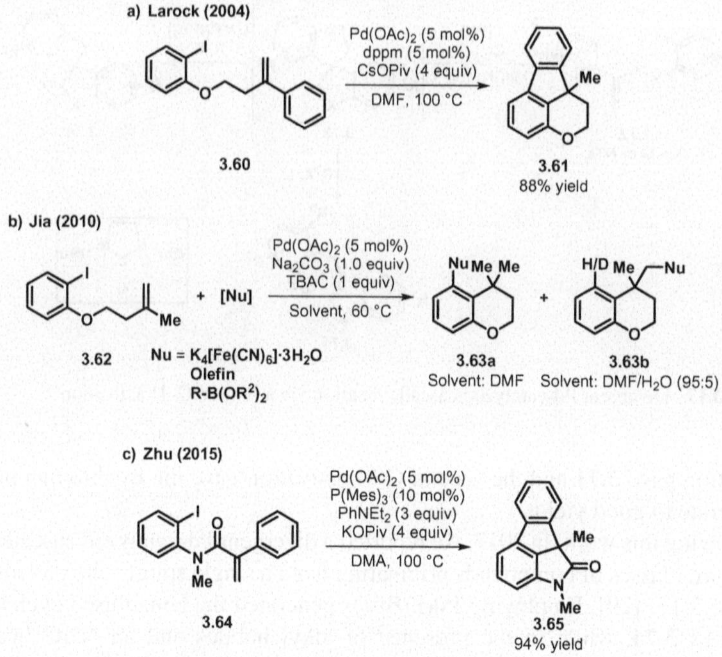


**Scheme 3.15** Pd-catalyzed domino reactions involving a 1,4-Pd shift

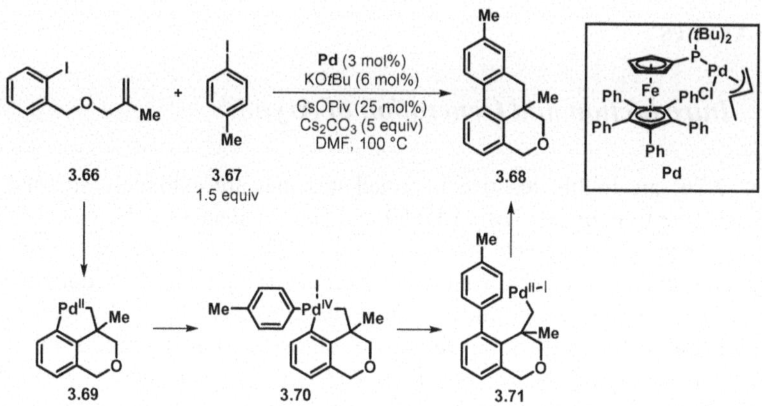


**Scheme 3.16** Pd-catalyzed synthesis of polycyclic isochromans via a double C–H activation

was the optimal ligand for this transformation to enable reversible oxidative addition between the two aryl iodides used in the reaction (**3.66** and **3.67**). Upon generation of intermediate **3.69** from an *ortho* C–H activation, an oxidative addition into the exogenous aryl iodide **3.67** generated the proposed $Pd^{IV}$ intermediate **3.70**. Reductive

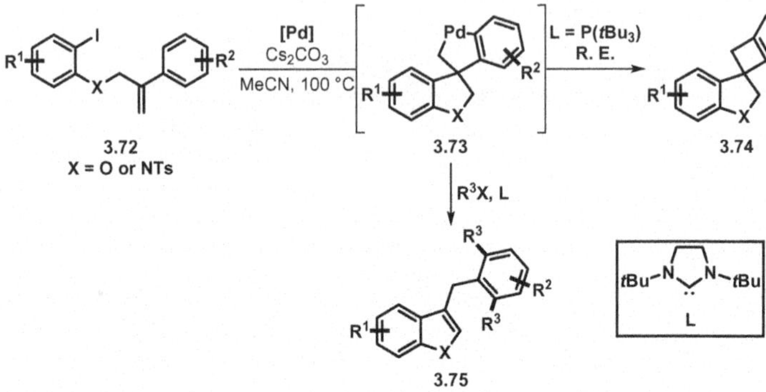


**Scheme 3.17** Divergent Pd-catalyzed cascade reaction via a remote C–H activation

elimination gave **3.71** and the second C–H activation gave the isochroman product in moderate to good yields.

Following this work, in 2017, we reported a divergent Pd-catalyzed cascade reaction of two classes of compounds originating from a single spirocyclic palladacycle (Scheme 3.17) [29]. Employing $Pd(PtBu_3)_2$ generated the benzofused cyclobutene derivatives **3.74**, while in the presence of alkyl halides and an NHC ligand, a Catellani-like reaction sequence occurred to form benzofuran **3.75**.

## 3.3 Arynes

### 3.3.1 *Introduction and Generation of Arynes*

Arynes (**3.76**) are highly reactive intermediates that are extremely strained and shortlived. Due to their low lying LUMO and strain induced by the sp-hybridized carbons, arynes exhibit high electrophilicity [30]. Originally, arynes were generated through an elimination of HX from halobenzenes [31]. Due to the requirement of a strong base, alternative and milder methods have been developed for their generation (Scheme 3.18). In 1983, the Kobayashi group developed a mild synthesis of arynes in situ by subjecting **3.77** to a fluoride source [32]. This method enabled the controlled generation of aryne by modifying the fluoride source and its solubility.

### 3.3.2 *Arynes in Palladium-Catalyzed Methodologies*

Low valent transition-metals such as $Pd^0$ have been shown to undergo oxidative cyclizations with arynes [33]. Similar to alkynes, cyclotrimerization and carbometallation across arynes have been reported. The Castedo group disclosed the generation

**Scheme 3.18** Methods to generate arynes

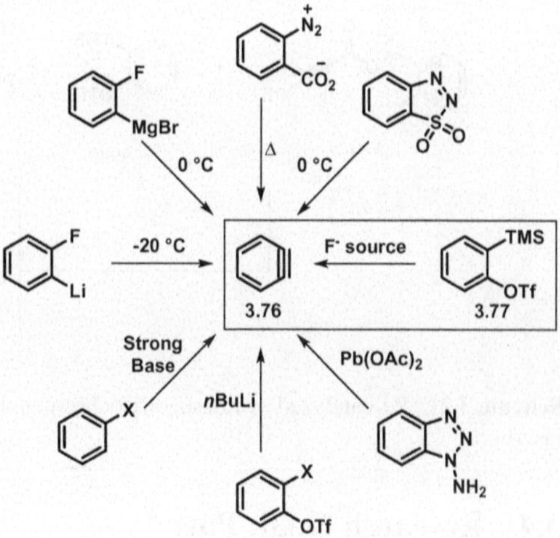


of triphenylene (**3.78**) originating from the Kobayashi precursor (Scheme 3.19) [34]. Of note, the authors did not detect triphenylene in the reaction mixture in the absence of $Pd(PPh_3)_4$.

To provide insight into the trimerization, the Wenger group isolated the first Pd-bound aryne complex (Scheme 3.20) [35]. Using the 2-bromophenylboronate **3.79** in the presence of stoichiometric amounts of Pd and *t*BuOK, palladacycle **3.80** was generated.

Understanding both C–H functionalization and benzyne reactivity in Pd-catalysis, the Cheng group successfully synthesized isochromenones (Scheme 3.21) [36]. The postulated mechanism proceeded through the generation of palladacycle **3.83** via sequential oxidative addition, carbopalladation across the alkyne, and intramolecular *ortho* C–H activation. Subsequent, benzyne insertion and reductive elimination yielded **3.82**.

**Scheme 3.19** Initial studies on the trimerization of arynes

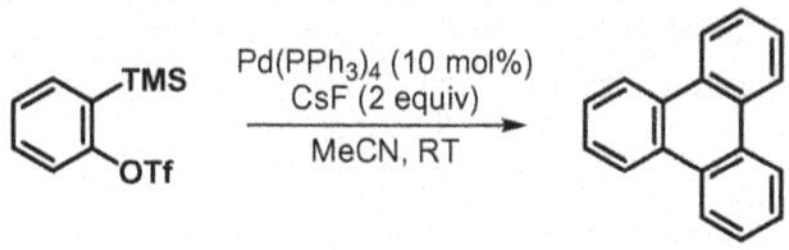


**3.77** **3.78**

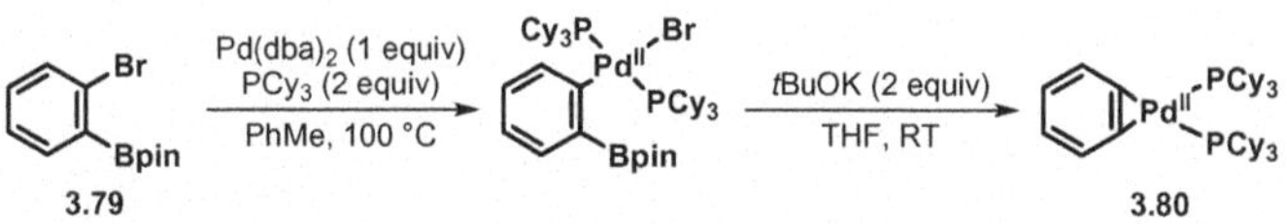


**3.79** **3.80**

**Scheme 3.20** Generation and isolation of the Pd-bound benzyne

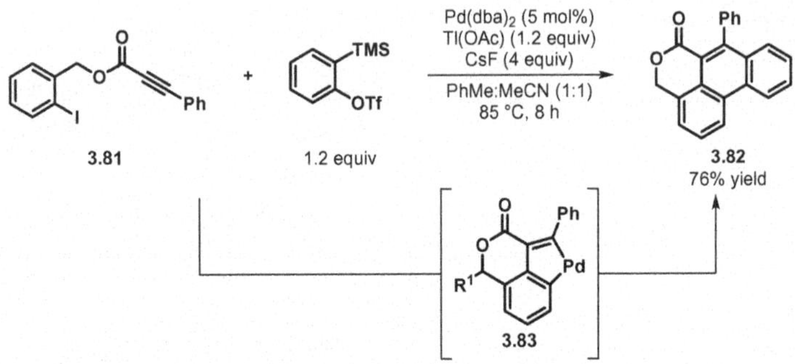


**Scheme 3.21** Pd-catalyzed synthesis of isochromenones via C–H functionalization and benzyne insertion

## 3.4 Research Goal: Part 1

Over the past decade, several groups have combined the intramolecular Pd-catalyzed Heck reaction with C–H activation to furnish complex heterocycles. Dr. Juntao Ye and co-workers from our group disclosed a catalyst-dependent divergent cascade reaction to furnish two classes of compounds which originate from a singular spirocyclic palladacycle (Scheme 3.17).

Intrigued by the reactivity of the palladacycle formed in the catalytic reaction and gaining inspiration from the work of Cheng (Scheme 3.21), we envisioned a strategy further functionalizing the intermediate, by inserting highly reactive benzynes, to provide access to the respective spirocycles (Scheme 3.22). We decided to explore the generation of dihydrobenzofuran and oxindole motifs. In addition to forming the novel biaryl spirocycles, we were interested in probing the mechanism of the reaction.

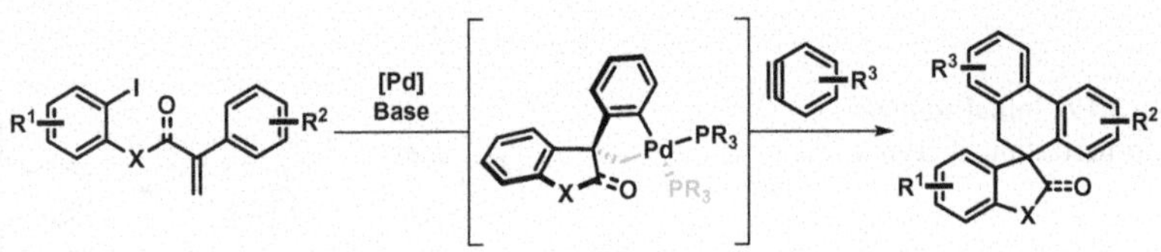


**Scheme 3.22** Proposed Pd-catalyzed spirocyclization via carbopalladation, C–H activation and benzyne insertion

## 3.5 Results and Discussion: Spirocyclization via Aryne Insertion

### 3.5.1 *Starting Material Preparation*

The synthesis of the dihydrobenzofurans and oxindoles was achieved by employing 2-iodoarylethers and 2-iodoacrylamides respectively. 2-Iodoarylether **3.85** was prepared by coupling allylic alcohols **3.84**, which were synthesized via a carbocupration of propargyl alcohol with aryl cuprates, with commercially available functionalized 2-iodophenols via a Mitsunobu reaction (Scheme 3.23).

The 2-iodoacrylamides were prepared in a similar method described in Chap. 1. A Schotten-Baumann type amide coupling of 2-phenylacryloyl chloride (**3.86**) with 2-iodoaniline generated the *N*-unsubstituted acrylamide **3.87**. Then, alkylation of these acrylamides were performed by using NaH in THF at 0 °C to generate the respective starting material (**3.88**) (Scheme 3.24).

The functionalized 2-phenylacrylic acids were synthesized from diethyl oxalate. A single acyl substitution with a preformed aryl Grignard reagent generated ethyl aryl-2-glyoxalate (**3.89**). A Wittig reaction with methyl triphenylphosphonium bromide generated the acrylate and hydrolysis with LiOH gave the desired arylacrylic acid **3.90** (Scheme 3.25).

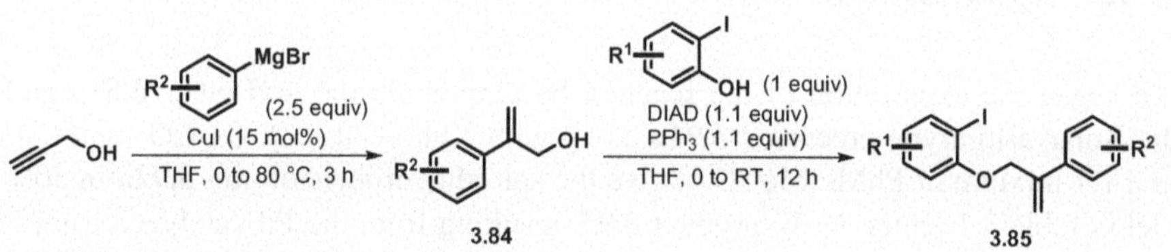


**Scheme 3.23** General route to 2-iodoarylethers

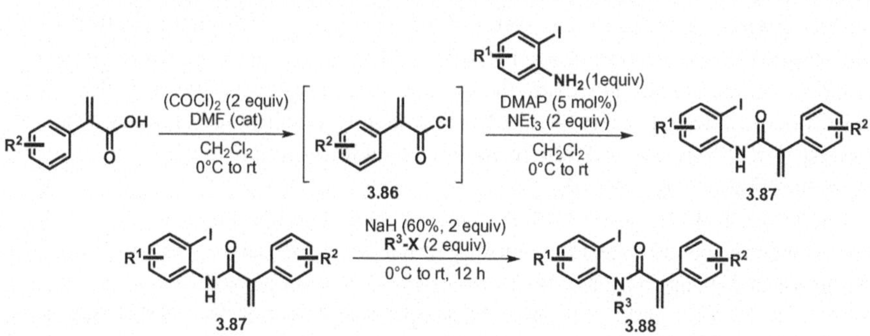


**Scheme 3.24** General route to 2-iodoacrylamides

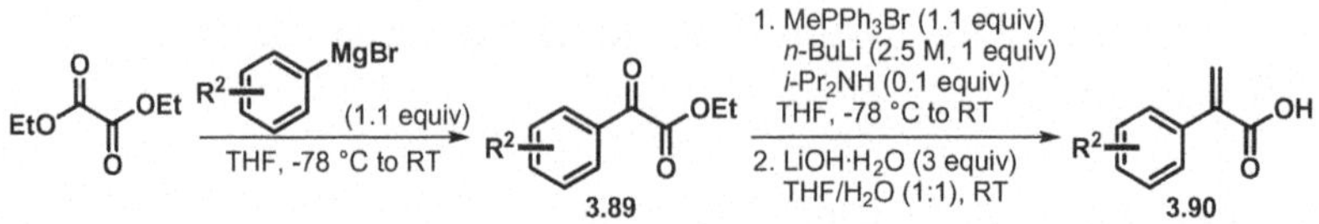


**Scheme 3.25** General route to substituted 2-phenylacrylic acids

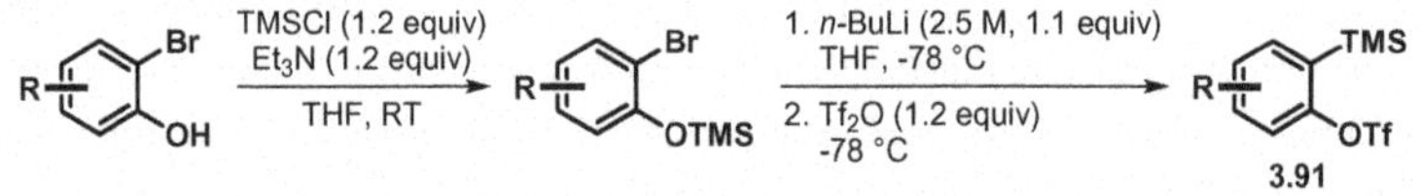


**Scheme 3.26** General route to 2-(trimethylsilyl)aryl trifluoromethanesulfonates

The Kobayashi aryne precursors (2-(trimethylsilyl)aryl trifluoromethanesulfonate, **3.91**) were synthesized according to a modified procedure outlined by the Guitián group [37]. A stepwise TMS protection of bromophenol, retro-Brook rearrangement, and quenching with trifluoroacetic anhydride gave the respective starting material (Scheme 3.26).

## 3.5.2 Optimization

We began the exploration of the reaction by employing the aryl ether **3.85a** and the Kobayashi aryne precursor **3.91a** along with XPhos-Pd -G2, $Cs_2CO_3$ and CsF in a 1:1 mixture of PhMe:MeCN to give the spirodihydrobenzofuran **3.92a** in 50% yield (Table 3.1, entry 1). Byproduct **3.93**, resulting from the Pd-catalyzed trimerization of the in situ generated benzynes, was obtained in 38% yield. Following this initial discovery, we began screening various Pd-catalysts and ligands. Screening another Buchwald precatalyst revealed increased selectivity to the unwanted side product. Switching to the precatalyst, $Pd(dppf)Cl_2$ gave the product in higher yields with slightly lower formation of the trimer (Table 3.1, entry 3). Other bidentate ligands were screened; however, the product was formed in lower yields (dppp and d*t*bpf) (Table 3.1, entry 4 and 5). Fortunately employing a readily available catalyst, $Pd(PPh_3)_4$, gave **3.92a** in 82% yield with minimal the formation of the byproduct (Table 3.1, entry 6).

Encouraged by the activity of the racemic bidentate ligands, a library of chiral ligands were screened to test the viability of the enantioselective variant (Table 3.2). The proof of concept was achieved when BINAP was employed to give the desired product in 5% ee (Table 3.2, entry 1). Employing Josiphos SL-J001-1 generated the product in low yields and 20% ee (Table 3.2, entry 2). Various commercially available chiral biaryl phosphine ligands were also used; however, the highest ee observed was 30% with difluorophos (Table 3.2, entries 3–6). It should be noted that the spirooxindoles failed to give product formation under these conditions.

**Table 3.1** Pd-catalyzed spirocyclization via C–H and benzyne insertion forming 3.92a: Pd catalysts

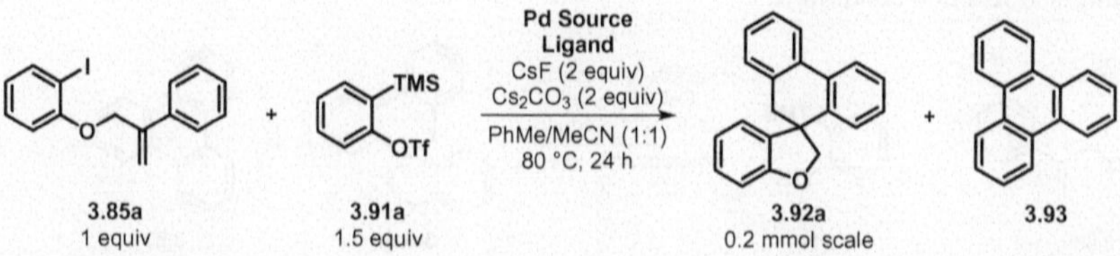


| Entry | Pd Source (10 mol%) | Ligand (20 mol%) | Conversion (%) | NMR yield of 3.92a (%)[a] | NMR yield of 3.93 (%)[a,b] |
|---|---|---|---|---|---|
| 1[c] | XPhos-Pd-G2 | – | >95 | (50) | 38 |
| 2 | RuPhos-Pd-G2 | – | >95 | 58 | 48 |
| 3[c] | Pd(dppf)$Cl_2$ | – | >95 | 75 (70) | 33 |
| 4 | $Pd(OAc)_2$ | d$^t$bpf | >95 | 53 | 42 |
| 5 | $Pd(OAc)_2$ | dppp | >95 | 55 | 34 |
| 6 | $Pd(PPh_3)_4$ | – | >95 | 82 | 5 |

Reactions were run on 0.2 mmol scale. [a]Determined by $^1$H NMR analysis of crude reaction mixtures using 1,3,5 trimethoxybenzene as internal standard. [b]Yield of 3.93 calculated with respect to quantity of aryne precursor. [c]Isolated yields in parenthesis

**Table 3.2** Pd-catalyzed spirocyclization via C–H and benzyne insertion forming 3.92a: Ligands

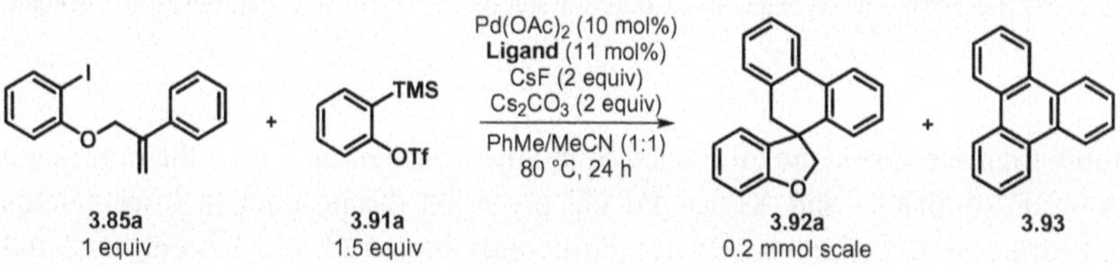


| Entry | Ligand | Conversion (%) | NMR yield of 3.92a (%)[a] | ee (%) |
|---|---|---|---|---|
| 1 | BINAP | >95 | 75 | 5 |
| 2 | Josiphos SL-J001-1 | >95 | 10 | 20 |
| 3 | Segphos | >95 | 36 | 27 |
| 4 | DTBM-Segphos | >95 | 20 | 15 |
| 5 | Ph-Garphos | >95 | 38 | 25 |
| 6 | Difluorophos | >95 | 49 | 30 |

Reactions were run on 0.2 mmol scale. [a]Determined by $^1$H NMR analysis of crude reaction mixtures using 1,3,5 trimethoxybenzene as internal standard

Upon screening other reaction parameters, it was found that the optimal conditions were 5 mol% $Pd(PPh_3)_4$, 2 equivalents of CsF, 1.5 equivalents of $Cs_2CO_3$, in a 2:1 mixture of PhMe:MeCN at 80 °C for 16 h (Table 3.3, entry 1). The spirocyclic structure of **3.92a** was unambiguously determined by spectroscopic analysis and single crystal X-ray crystallography. We next sought to alter the standard conditions

**Table 3.3** Pd-catalyzed spirocyclization via C–H and benzyne insertion forming 3.92a: Variation from optimized reaction conditions

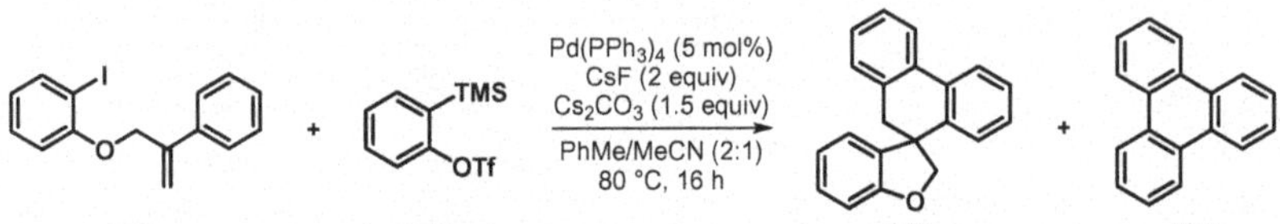


| Entry | Variation from "optimized conditions" | Yield 3.92a (%)[a, b] | Yield 3.93 (%)[a] |
|---|---|---|---|
| 1 | None | (84) | 5 |
| 2 | Pd(dppf)$Cl_2$ instead of $Pd(PPh_3)_4$ | 79 | 40 |
| 3 | XPhos Pd G2 instead of $Pd(PPh_3)_4$ | 57 | 48 |
| 4 | AgF instead of CsF | 34 | 17 |
| 5 | CsOPiv instead of $Cs_2CO_3$ | 80 | 11 |
| 6 | CsOAc instead of $Cs_2CO_3$ | 64 | 6 |
| 7 | -Br instead of -I | 10 | 5 |
| 8 | THF instead of PhMe/MeCN (2:1) | 54 | 3 |
| 9 | 1:1 instead of 2:1 PhMe/MeCN | 70 | 13 |
| 10 | 3:1 instead of 2:1 PhMe/MeCN | 30 | <1 |

Reactions were run on 0.2 mmol scale. [a]Determined by $^1$H NMR analysis of the crude reaction mixtures using 1,3,5-trimethoxybenzene as internal standard. [b]Yield in parentheses are isolated yields

to examine their effects on the efficiency of the transformation. Under the optimized conditions, Pd(dppf)$Cl_2$ and XPhos Pd G2 provided the product in lower yields and the formation of triphenylene was significantly higher (Table 3.3, entries 2 and 3). AgF was found to be inferior to CsF in the reaction (Table 3.3 entry 4). Other cesium bases such as CsOPiv and CsOAc gave the product in comparable yields; however, $Cs_2CO_3$ was found to be optimal as it was the easiest to handle in non-inert conditions (Table 3.3, entries 5 and 6). Aryl bromides yielded substantially less product and a possible explanation is that the temperature is too low for the oxidative addition to occur (Table 3.3, entry 7). Lastly, the solvent ratio of PhMe:MeCN was essential for suppressing the formation of **3.93**. This result can be rationalized by limiting the solubility of CsF in solution (Table 3.3 entries 8–10). By limiting the concentration of CsF, the concentration of available benzyne is also limited, thus lowering the trimerization.

Analogously, for the spirooxindole **3.94a**, the optimal catalyst was found to be 10 mol% $Pd(PPh_3)_4$, 2 equivalents of CsF, 1.5 equivalents of $Cs_2CO_3$, in a 1:1 mixture of PhMe:MeCN at 80 °C for 6 h. A variation of standard conditions was also performed to examine their effects. Similar trends were observed with respect to the spirooxindoles. However, decreasing the catalyst loading was detrimental to

**Table 3.4** Pd-catalyzed spirocyclization via C–H and benzyne insertion forming 3.94a: Variation from optimized reaction conditions

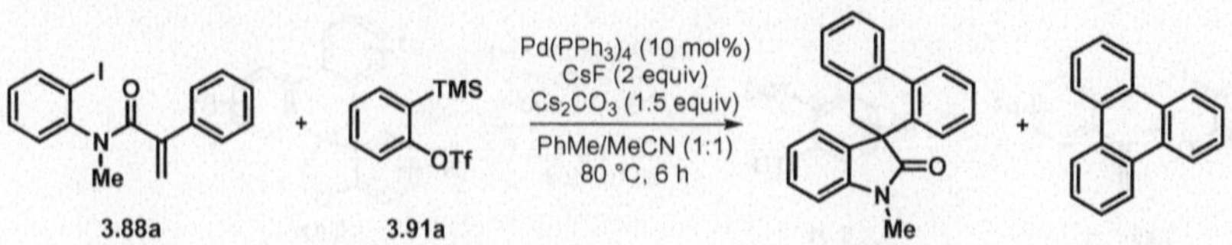


3.88a 1 equiv; 3.91a 1.5 equiv; 3.94a 0.2 mmol scale; 3.93

| Entry | Variation from "optimized conditions" | Yield 3.94a (%)[a, b] | Yield 3.93 (%)[a] |
|---|---|---|---|
| 1 | None | >95 (96) | 4 |
| 2 | $Pd(dppf)Cl_2$ instead of $Pd(PPh_3)_4$ | 48 | 23 |
| 3 | Pd XPhos G2 instead of $Pd(PPh_3)_4$ | 36 | 34 |
| 4 | AgF instead of CsF | 49 | 10 |
| 5 | CsOPiv instead of $Cs_2CO_3$ | 84 | 4 |
| 6 | CsOAc instead of $Cs_2CO_3$ | 81 | 5 |
| 7 | ArOTf instead of ArI | – | 4 |
| 8 | THF instead of PhMe/MeCN (1:1) | 51 | 3 |
| 9 | 5 mol% $Pd(PPh_3)_4$ | 53 | 11 |
| 10 | 5 mol% $Pd(PPh_3)_4$ and 2:1 PhMe:MeCN | 72 | 9 |

Reactions were run on 0.2 mmol scale. [a]Determined by $^1H$ NMR analysis of the crude reaction mixtures using 1,3,5-trimethoxybenzene as internal standard. [b]Yield in parentheses are isolated yields

the efficiency of the reaction. Modifying the solvent ratio along with lower catalyst loading also produced lower yields (Table 3.4).

### 3.5.3 *Substrate Scope*

With the optimized conditions in hand, we investigated a series of substituted 2-iodoarylethers. Chlorinated and phenyl substituted aryl iodides **3.85b** and **3.85c** reacted in excellent yields (94 and 93% respectively) (Table 3.5, entries 2 and 3). Naphthyl ether **3.85d** underwent the cyclization in 67% yield, suggesting that the reaction is sensitive to steric hindrance near the C–I bond (Table 3.5, entry 4). Electron poor pendant aromatic groups enhanced the cyclization to the spirodihydrobenzofurans (**3.92e**, 94% yield) (Table 3.5, entry 5). It was found that unsymmetrical tethered aryl groups undergo the C–H activation regioselectively to minimize steric interaction on the palladacycle (**3.92f** and **3.92 g**, 6.7:1 rr and 9:1 rr respectively) (Table 3.5, entry 6 and 7). In addition to exploring the scope of the 2-iodoarylethers, substituted aryne precursors were explored. Sterically encumbered and electron rich benzyne precursors **3.91b** and **3.91c** were incorporated to give **3.92 h** and **3.92i** in

**Table 3.5** Pd-catalyzed spirocyclization via C–H and benzyne insertion forming spirodihydrobenzofurans. Substrate scope

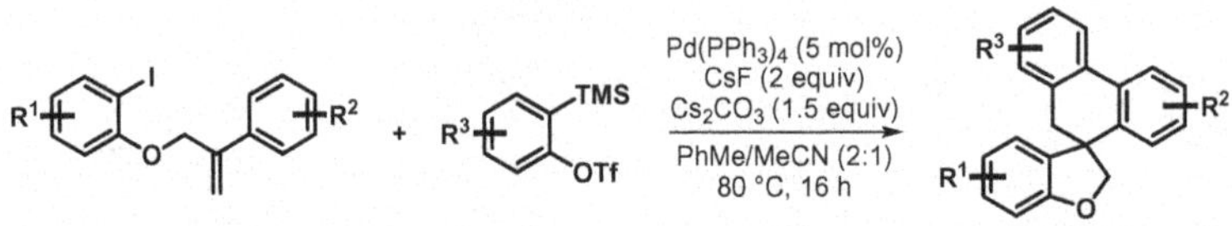


3.85 3.91 3.92

| Entry | Substrate | Product | Yield (%)[a] |
|---|---|---|---|
| 1 | 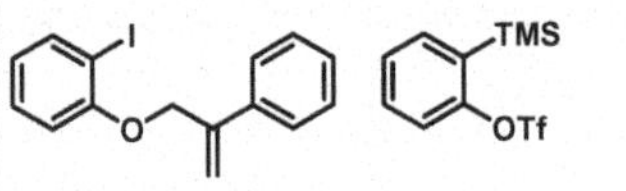 3.85a 3.91a | 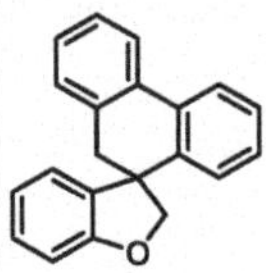 3.92a | 84<br>86[b] |
| 2 | 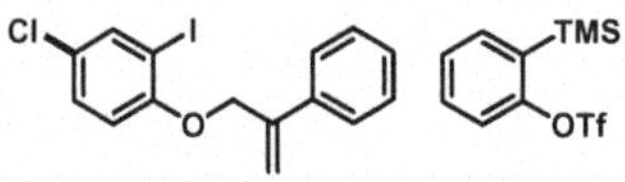 3.85b[c] 3.91a | 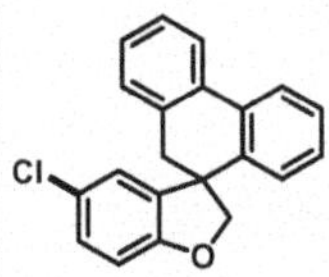 3.92b | 94 |
| 3 | 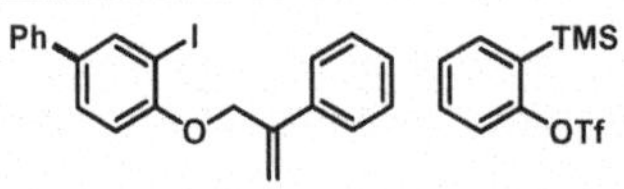 3.85c[c] 3.91a | 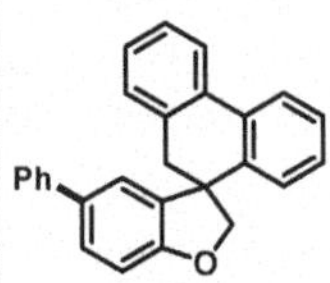 3.92c | 93 |
| 4 | 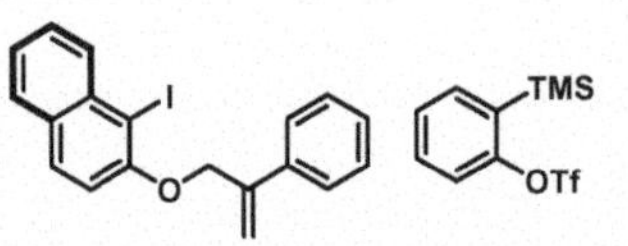 3.85d 3.91a | 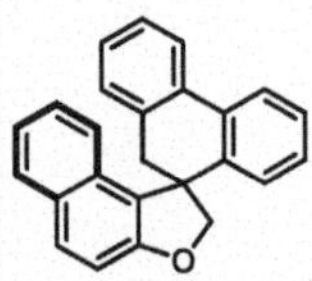 3.92d | 67 |
| 5 | 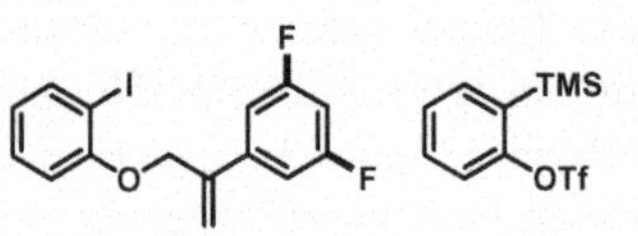 3.85e[d] 3.91a | 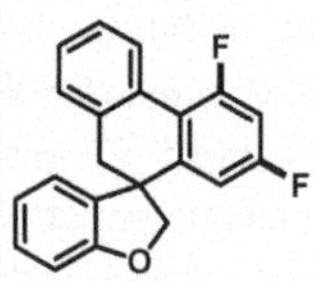 3.92e | 94 |
| 6 | 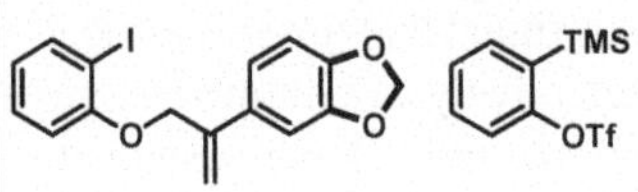 3.85f[e] 3.91a | 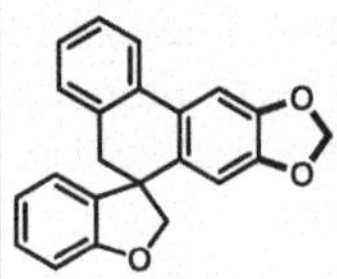 3.92f[e] Major 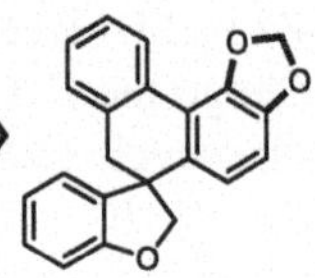 3.92f'[e] Minor | 76<br>(6.7:1 rr) |

(continued)

**Table 3.5** (continued)

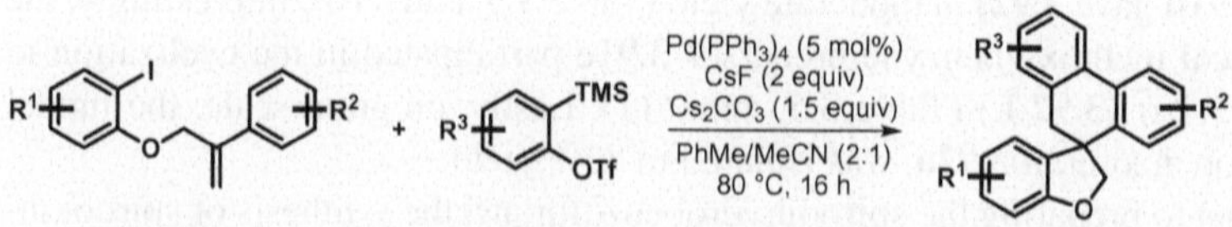


3.85 3.91 3.92

| Entry | Substrate | Product | Yield (%)[a] |
|---|---|---|---|
| 7 | 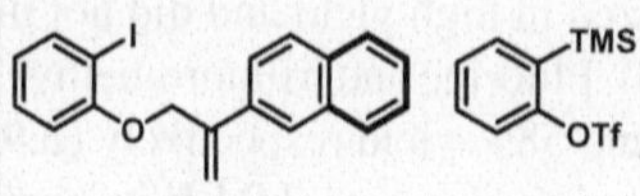 3.85g[e] 3.91a | 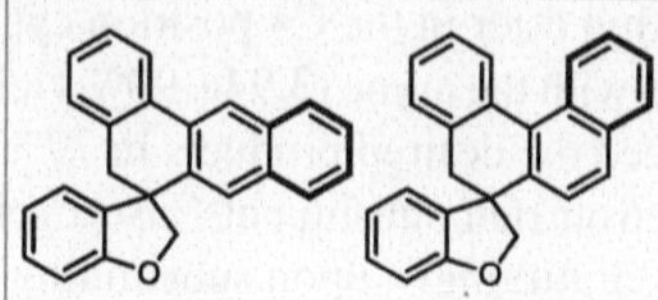 3.92g[e] Major; 3.92g'[e] Minor | 94 (9:1 rr) |
| 8 | 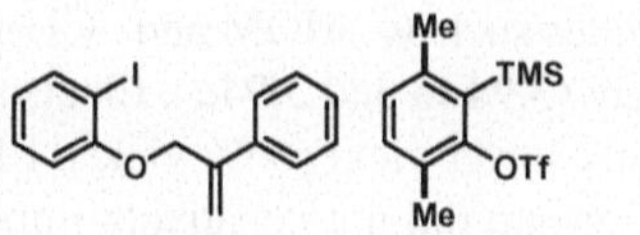 3.85a 3.91b | 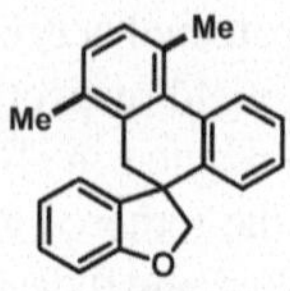 3.92h | 98 |
| 9 | 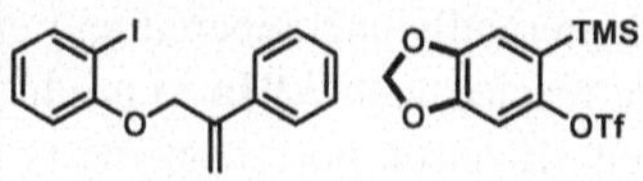 3.85a 3.91c | 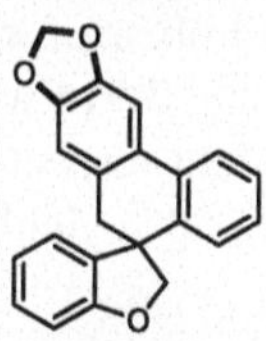 3.92i | 84 |
| 10 | 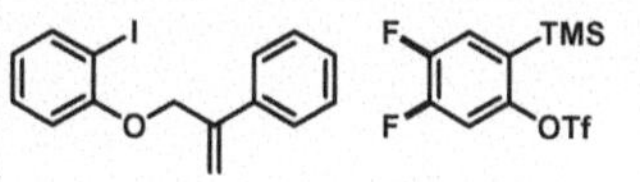 3.85a 3.91d | 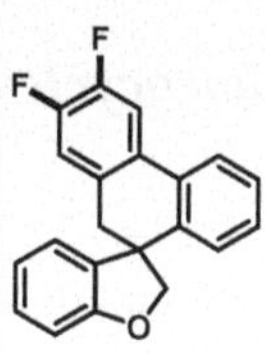 3.92j | 55 |
| 11 | 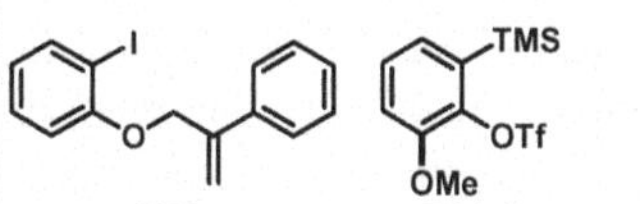 3.85a 3.91e[e] | 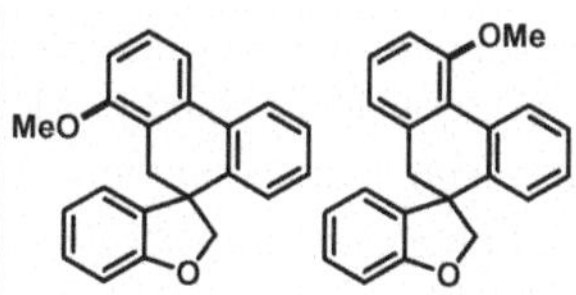 3.92k Major; 3.92k' Minor | 68 (3.3:1 rr) |

Reactions were run on 0.2 mmol scale. [a]Isolated yields. [b]Reaction run on 2.97 mmol scale. [c]This compound was prepared by Yoshifumi Yasukawa. [d]This compound was prepared by Daniel Schmidmeier. [e]This compound was prepared by Felicitas Landau. rr = Ratio of regioisomers

98% and 84% yield respectively (Table 3.5, entry 8 and 9). Electron poor benzyne precursor **3.91d** gave **3.92j** in moderate yield (Table 3.5, entry 10). Interestingly, the unsymmetrical methoxy benzyne precursor **3.91e** participated in the cyclization in a modest 3.3:1 rr (**3.92 k**) (Table 3.5, entry 11). Lastly, on gram-scale, the model spirodihydrobenzofuran **3.92a**, was isolated in 86% yield.

In addition to preparing the spirodihydrobenzofurans, the synthesis of spirooxindoles was also investigated (Table 3.6). Acrylamides bearing halogens and electron neutral substituents at 4- and 5- positions were well tolerated (**3.88b–3.88d**, 89–95% yield). A methyl ester in the C4 position cyclized in high yield and did not undergo side reactions with the aryne (**3.94e**, 90% yield). Fluorine and trifluoromethyl substitution produced the desired products in 79% and 58% yield respectively (**3.94f** and **3.94 g**). Electron rich substituents also cyclized to generate **3.94 h** in good yield (84% yield). Gratifyingly, upon substituting the tethered aryl group with a fluorine, the reactivity improved giving **3.94i** in 91% yield. *N*-benzylated acrylamides also participated in the reaction in excellent yields (**3.94j** and **3.94 k**, 85% and 87% yield respectively). Other *N*-protected acrylamides such as –MOM and –$CH_2CO_2Et$ also cyclized in 92% and 86% yield respectively (**3.94 m** and **3.94n**). Electron poor *N*-protected acrylamide **3.94o** reacted in slightly lower yield (66% yield). Electron rich and poor substituents on the tethered aryl group did not negatively impact the cyclization (**3.94p** and **3.94q**, 96% and 92% yield respectively). An ortho- substituent in the tethered aryl group produced the spirooxindole in good yield (**3.94r**, 78% yield). Similar to the spirodihydrobenzofuran, sterically encumbered and electron rich benzyne precursors **3.91b** and **3.91c** were reacted with **3.91a** to produce **3.94 s** and **3.94t** in 85% and 82% yield respectively. Electron poor benzyne precursor **3.91d** cyclized to form **3.94u** in 50% yield. On gram-scale, **3.94a** was isolated in 85% yield (Table 3.6).

### *3.5.4 Derivatization of Spirocycles*

Derivatizations of the model substrates **3.92a** and **3.94a** were explored to examine their synthetic versatility. We reasoned that benzylic bromination of the spirodihydrobenzofuran with NBS would generate the respective alkyl bromide (Scheme 3.27). The reaction was performed using NBS, DBPO, in benzene at 100 °C. Surprisingly, the desired brominated spirocycle was not observed and instead a ring expanded chromene product **3.95** was obtained in 41% yield. The product can be rationalized to have undergone a ring expansion via a Wagner-Meerwein rearrangement. Analogously, the spirooxindoles also underwent the ring expansion in 40% yield (**3.96**). We were unable to reduce the lactam functionality using standard conditions ($LiAlH_4$, $BH_3$, or DIBAL). These reactions led to recovery of starting material or complex mixtures.

**Table 3.6** Pd-catalyzed spirocyclization via C–H and benzyne insertion forming spirooxindoles. Substrate scope

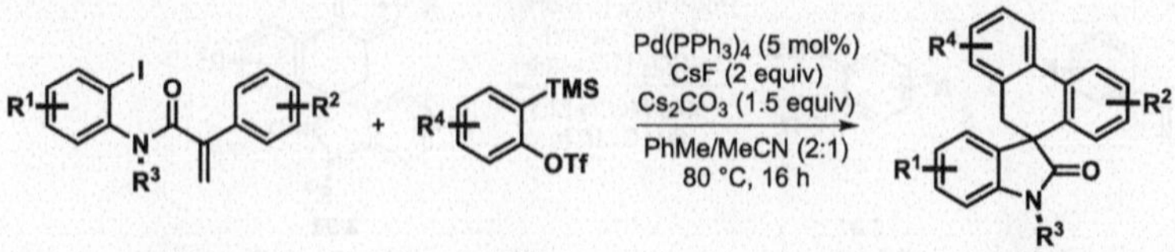

3.88 1 equiv; 3.91 1.5 equiv; 3.94 0.2 mmol scale

| Entry | Substrate | Product | Yield (%)[a] |
|---|---|---|---|
| 1 | 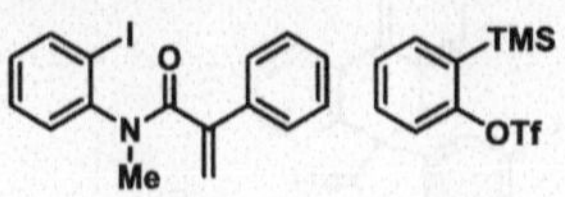<br>3.88a 3.91a | 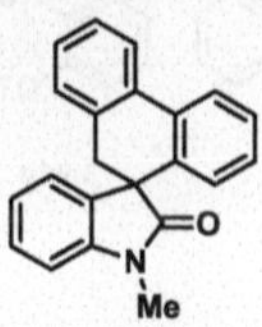<br>3.94a | 96<br>85[b] |
| 2 | 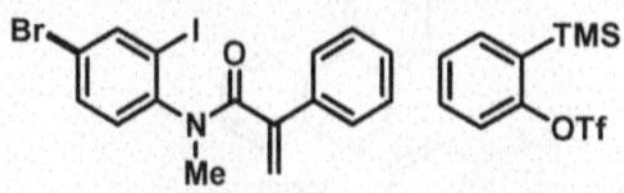<br>3.88b[c] 3.91a | 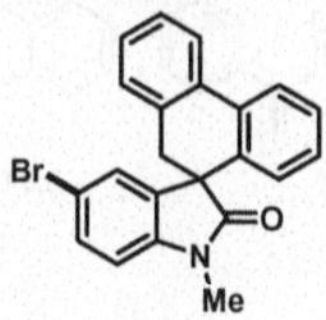<br>3.94b | 89 |
| 3 | 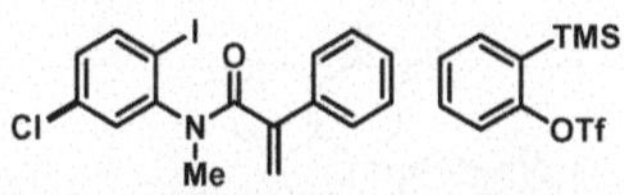<br>3.88c[c] 3.91a | 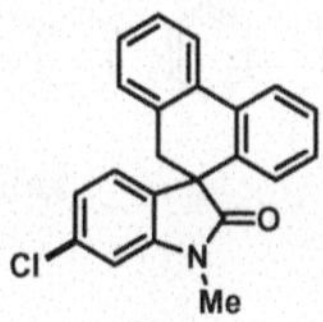<br>3.94c | 95 |
| 4 | 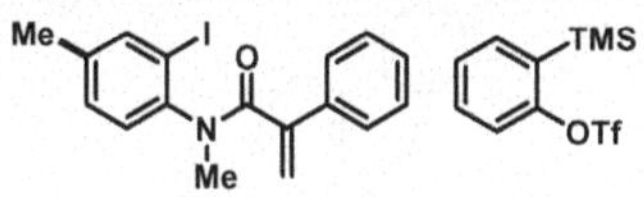<br>3.88d[c] 3.91a | 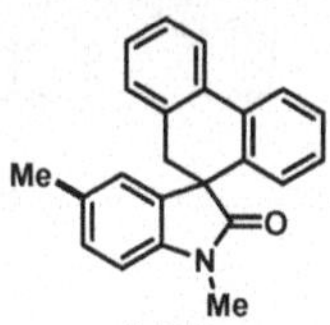<br>3.94d | 90 |
| 5 | 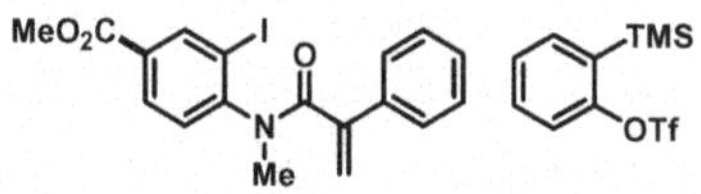<br>3.88e[c] 3.91a | 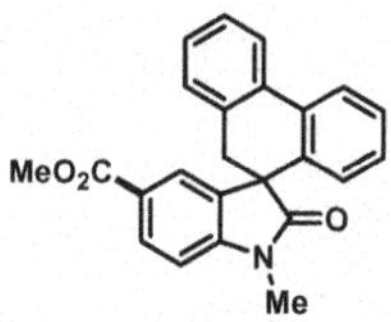<br>3.94e[c] | 90 |

(continued)

**Table 3.6** (continued)

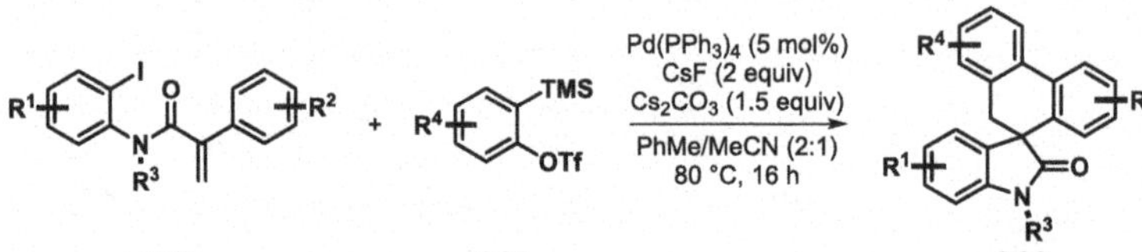


**3.88** 1 equiv | **3.91** 1.5 equiv | **3.94** 0.2 mmol scale

| Entry | Substrate | Product | Yield (%)[a] |
|---|---|---|---|
| 6 | 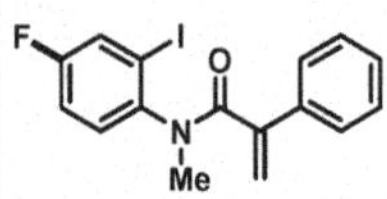 3.88f[c] 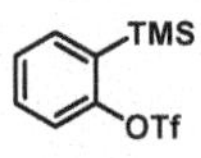 3.91a | 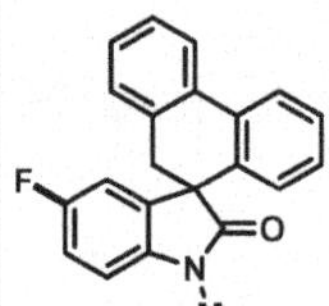 3.94f[c] | 79 |
| 7 | 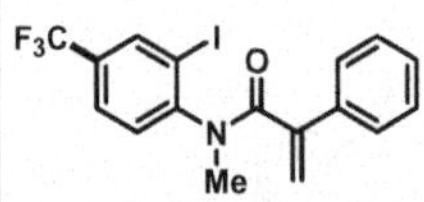 3.88g[c] 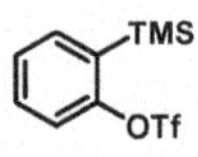 3.91a | 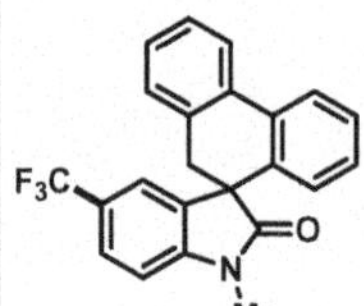 3.94g[c] | 58 |
| 8 | 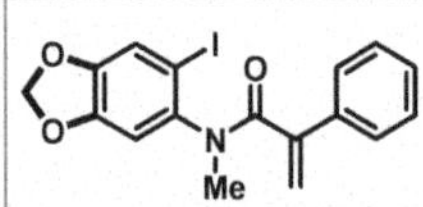 3.88h 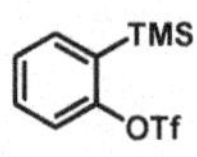 3.91a | 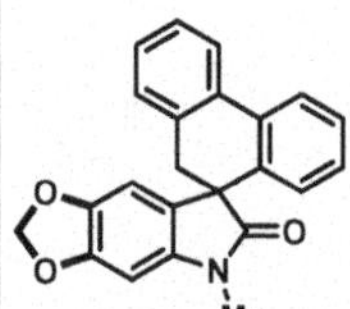 3.94h | 84 |
| 9 | 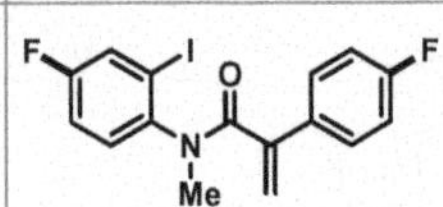 3.88i[c] 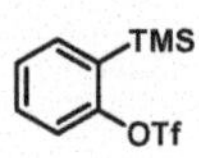 3.91a | 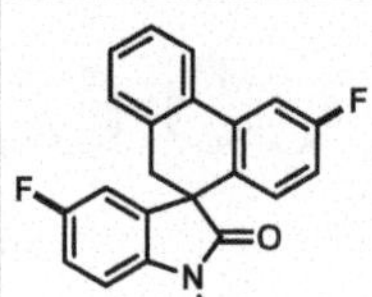 3.94i[c] | 91 |
| 10 | 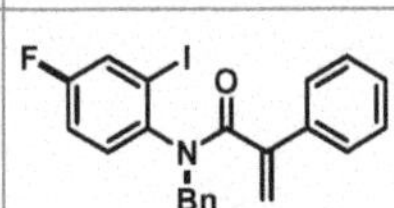 3.88j[c] 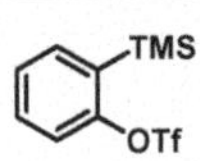 3.91a | 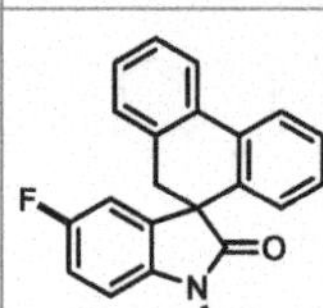 3.94j[c] | 85 |

(continued)

**Table 3.6** (continued)

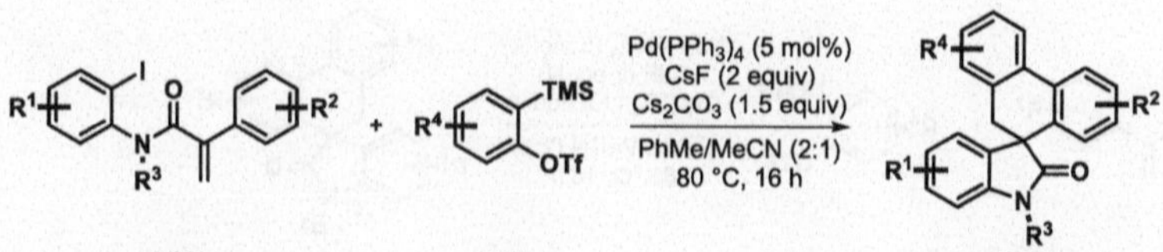


**3.88** 1 equiv    **3.91** 1.5 equiv    **3.94** 0.2 mmol scale

| Entry | Substrate | Product | Yield (%)[a] |
|---|---|---|---|
| 11 | 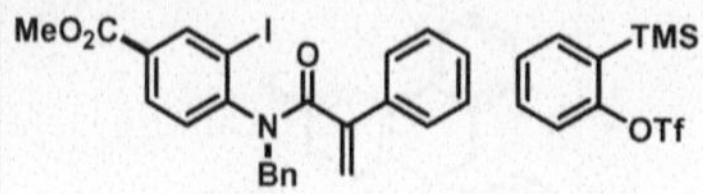 3.88k[c] 3.91a | 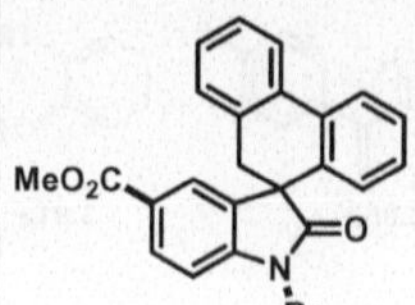 3.94k[c] | 87 |
| 12 | 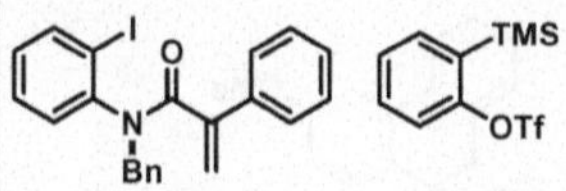 3.88l 3.91a | 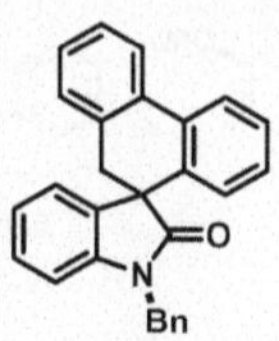 3.94l | 93 |
| 13 | 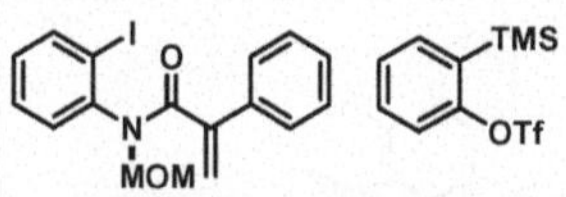 3.88m[c] 3.91a | 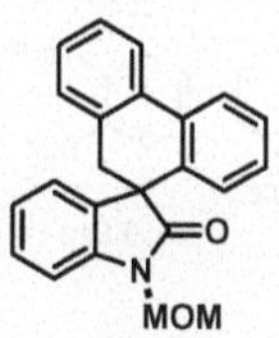 3.94m[c] | 92 |
| 14 | 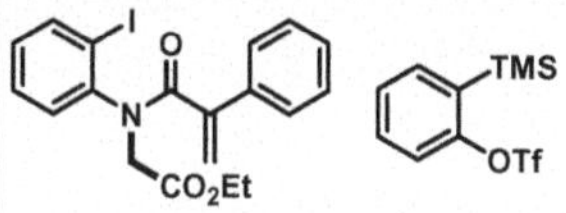 3.88n[c] 3.91a | 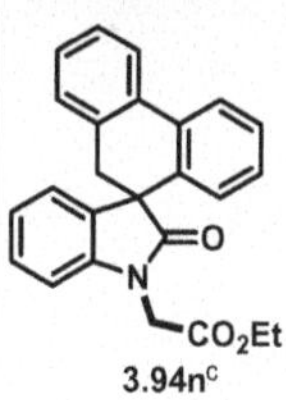 3.94n[c] | 86 |
| 15 | 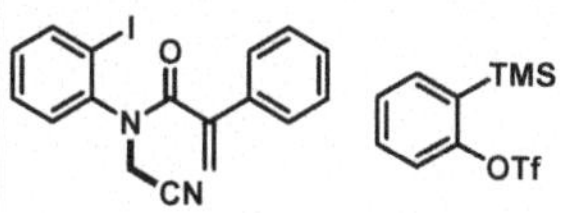 3.88o[c] 3.91a | 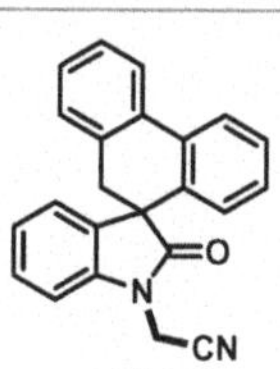 3.94o[c] | 66 |

(continued)

**Table 3.6** (continued)

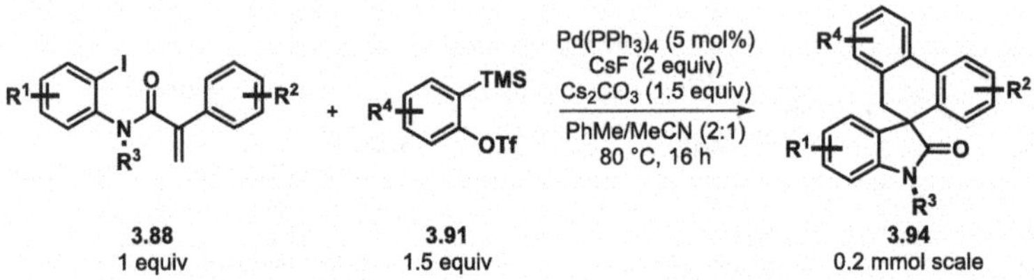


| Entry | Substrate | Product | Yield (%)[a] |
|---|---|---|---|
| 16 | 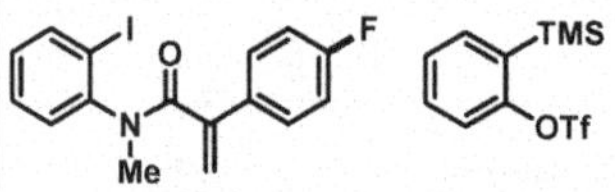 3.88p[c] 3.91a | 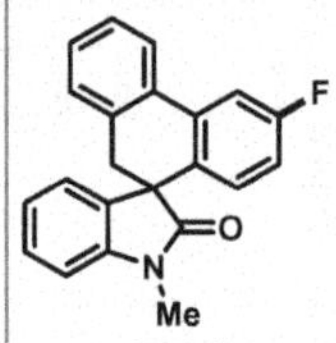 3.94p[c] | 96 |
| 17 | 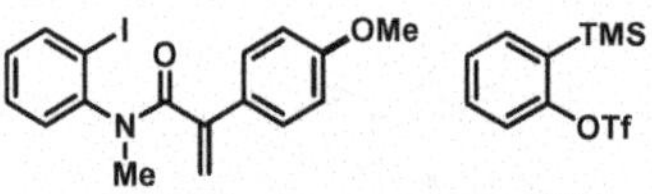 3.88q 3.91a | 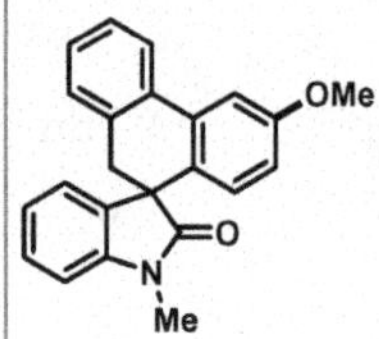 3.94q | 92 |
| 18 | 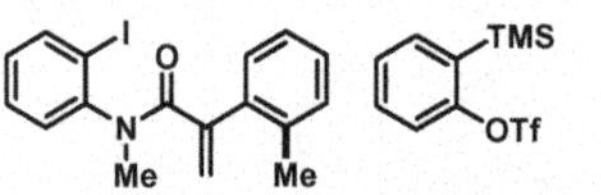 3.88r[c] 3.91a | 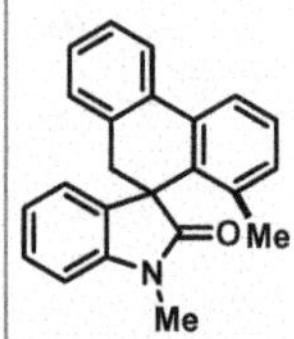 3.94r[c] | 78 |

(continued)

**Table 3.6** (continued)

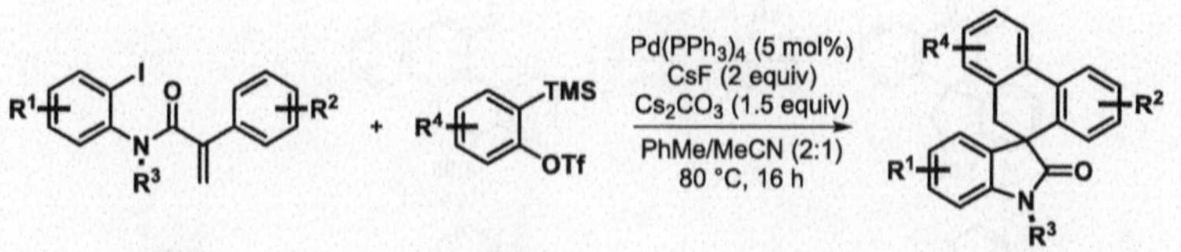


3.88 1 equiv; 3.91 1.5 equiv; 3.94 0.2 mmol scale

| Entry | Substrate | Product | Yield (%)[a] |
|---|---|---|---|
| 19 | 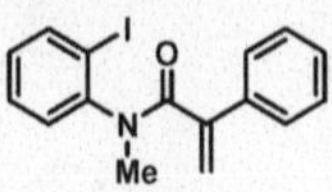 3.88a 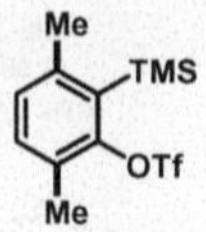 3.91b | 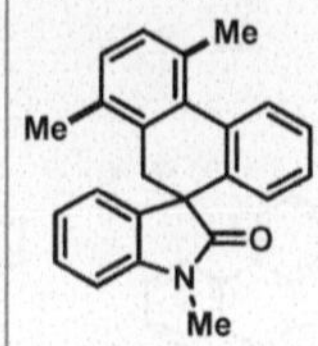 3.94s[c] | 85 |
| 20 | 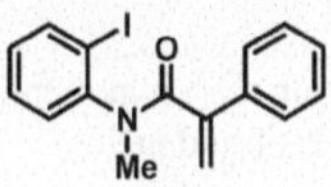 3.88a 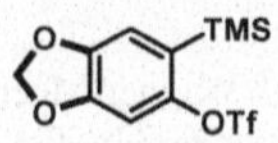 3.91c | 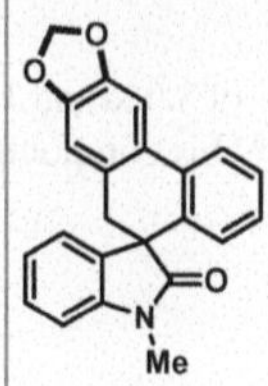 3.94t | 82 |
| 21 | 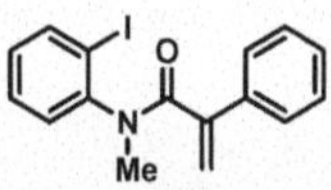 3.88a 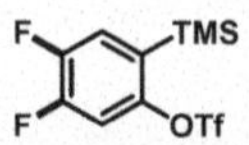 3.91d | 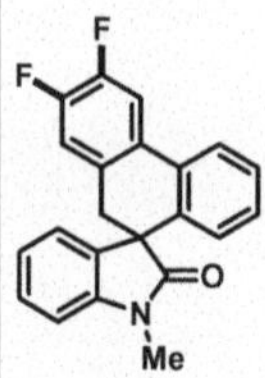 3.94u[c] | 50 |

Reactions were run on 0.2 mmol scale. [a]Isolated yields. [b]Reaction run on 2.75 mmol scale. [c]This compound was prepared by Alexis Lossouarn

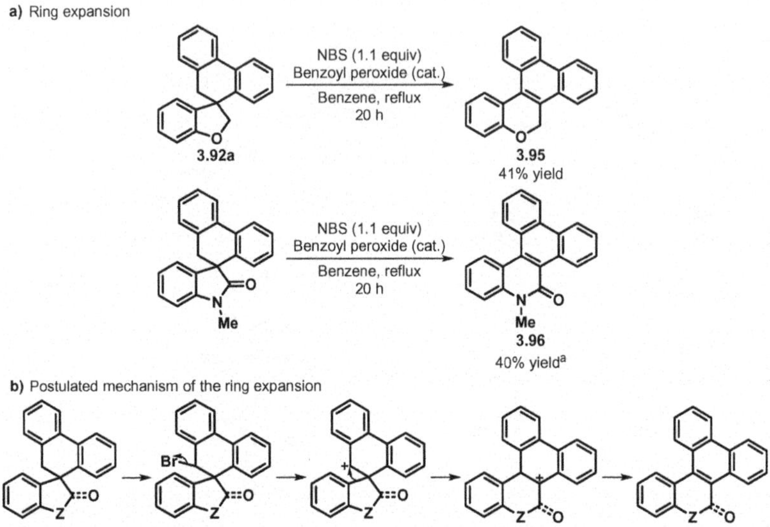


**Scheme 3.27** Ring expansion of 3.92a and 3.94a via a Wagner-Meerwein rearrangement. Reactions were run on 0.2 mmol scale. [a]This compound was prepared by Felicitas Landau

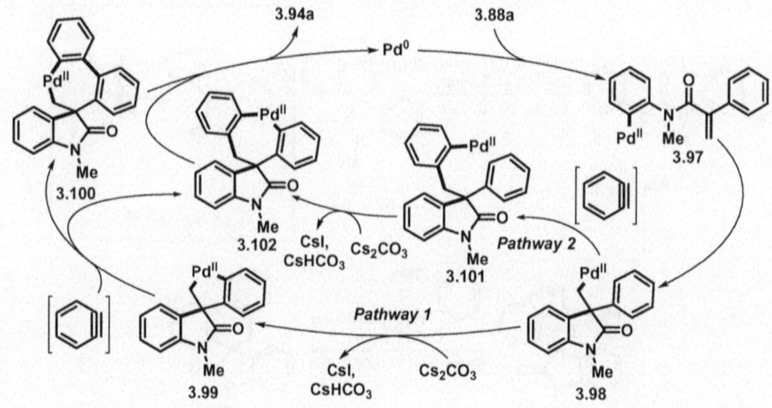


**Scheme 3.28** Postulated mechanism for the spirocyclization

## 3.5.5 *Postulated Mechanism*

The spirocyclization reaction is explained through the postulated mechanism (Scheme 3.28). Acrylamide **3.88a** undergoes a Pd-catalyzed oxidative addition followed by a carbopalladation generating the alkylpalladium (II) intermediate **3.98** where two divergent pathways are possible. In the first pathway, an intramolecular C–H activation of the pendant aryl group generates palladacycle **3.99**. An insertion of the aryne leads to intermediate **3.100** followed by reductive elimination to release **3.94a** and regenerate the catalyst. In contrast, pathway 2 involves an intermolecular insertion of the aryne to form intermediate **3.101** followed by C–H activation to yield intermediate **3.102**, and reductive elimination to form **3.94a**.

## 3.5.6 *Mechanistic Studies*

Intrigued by the two possible divergent pathways, we sought to gain insight into the C–H activation versus addition to the aryne by performing a series of mechanistic experiments. Pathway 1 involves the initial formation of the palladacycle **3.99**. To support the viability of the mechanism, **3.99** was prepared in 89% yield by reacting **3.88a** with stoichiometric quantities of $Pd(PPh_3)_4$ and excess $Cs_2CO_3$. The structure was confirmed by spectroscopic data and single X-ray crystallography. Palladacycle **3.99** was then reacted with the in situ generated benzyne to give the spirooxindole **3.94a** in 83% yield (Scheme 3.29). The formation of **3.94a** from **3.99** suggests that **3.99** is a competent intermediate in the catalytic cycle.

Pathway 2 was explored by generating **3.101** in situ from the oxidative addition into the corresponding aryl iodide **3.103**. However, under the standard reaction conditions, the cyclization from C–H activation did not occur. Instead, the arylation

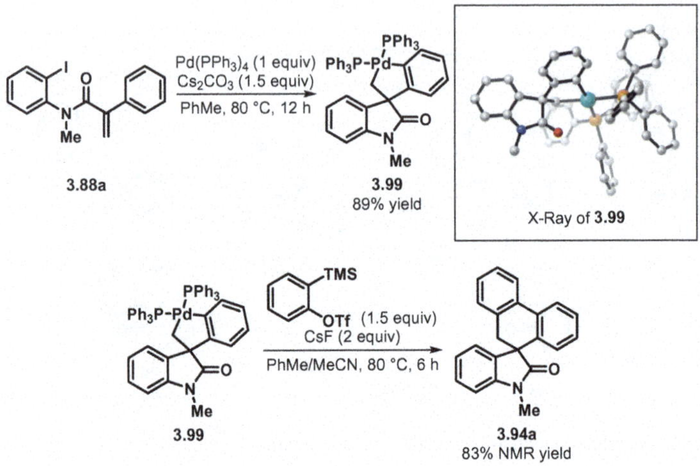


**Scheme 3.29** Mechanistic studies probing pathway 1: formation and reaction of palladacycle 3.99

of acetonitrile proceeded exclusively to give **3.104** in 52% yield and recovery of the starting aryl iodide **3.103** in 42% yield. The absence of spirooxindole **3.94a** suggests that the intramolecular C–H activation forming **3.102** after the benzyne insertion, is unfavourable under the reaction conditions. The absence of the desired product suggests that the reaction exclusively proceeds via pathway 1 (Scheme 3.30).

Lastly, to determine if the C–H activation is the rate determining step, a kinetic isotope effect was measured using parallel experiments. The initial rates of cyclization of acrylamide and the deuterated derivative were determined and a $k_H/k_D$ of 1.04 was measured. This result provides evidence that C–H activation is not the rate-limiting step in the catalytic cycle (Scheme 3.31).

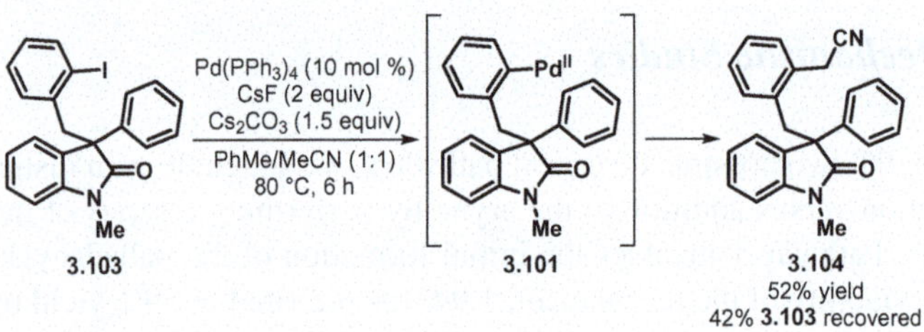


**Scheme 3.30** Mechanistic studies probing pathway 2: in situ formation 3.101 and formation of 3.104

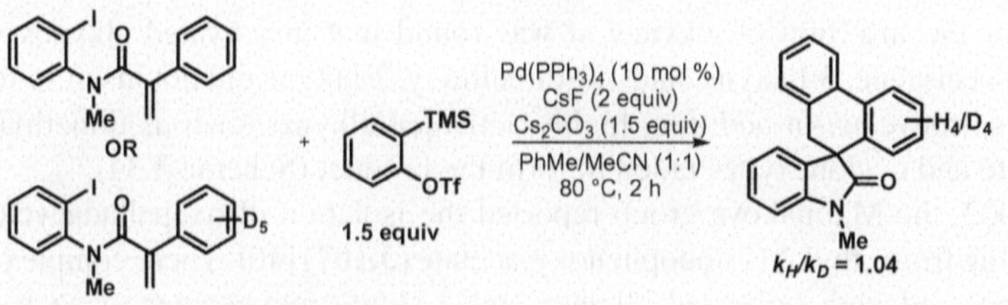


**Scheme 3.31** Parallel KIE experiment

## 3.6 Research Goal: Part 2

While exploring the scope of the aryne insertion, a series of limitations were found relating to modest regioselectivity and the inability to insert into alkynes. 3-Methoxybenzyne was reacted and it was found that the products were formed in approximately 3:1 regioselectivity for both the spirodihydrobenzofurans and oxindoles [38]. Previously, Garg and Houk have experimentally and computationally demonstrated that 3-methoxybenzyne **3.105** undergoes regioselective nucleophilic addition on C1, whereas in our reaction, we observed poor regiocontrol (Scheme 3.32) [39] (Scheme 3.33).

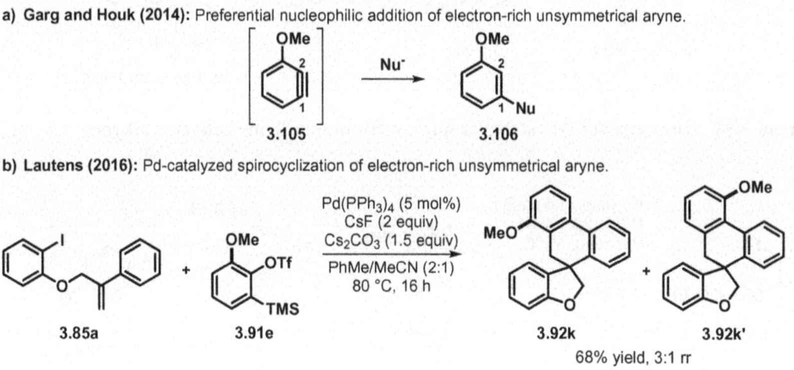


**Scheme 3.32** Nucleophilc addition and Pd-catalyzed spirocyclization of unsymmetrical arynes

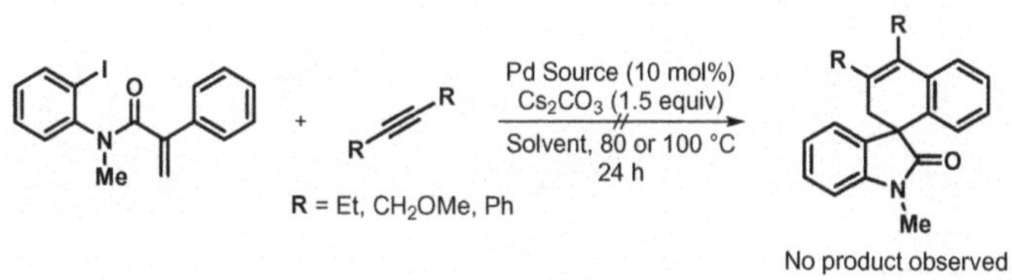


**Scheme 3.33** Unsuccessful Pd-catalyzed spirocyclization through alkyne insertion. Pd sources screened: $Pd(PPh_3)_4$, $Pd(dppf)Cl_2$ and XPhos Pd G2. Solvents screened: PhMe, MeCN, DMF

As for the insertion of alkynes, it was found that unactivated alkynes such as diphenylacetylene, 3-hexyne, and 1,4-dimethoxy-2-butyne did not insert to form the desired spirocycles. In addition, highly activated alkynes such as dimethyl but-2-ynedioate and cyclohexynes failed to form the product (Scheme 3.34).

In 2002, the Malinakova group reported the isolation of oxapalladacycle **3.108** originating from ethyl 2-(2-iodophenoxy)acetate (**3.107**) [40]. These complexes were further reacted with polarized alkynes and a single regioisomer of each desired product was prepared in good to excellent yields (**3.109**, Scheme 3.35).

As seen with the insertion of arynes, we disclosed that the generation of the palladacycle occurs prior to the insertion. Employing a similar procedure, we sought to insert polarized and unsymmetrical alkynes in a highly regioselective manner to generate spirodihydrobenzofurans and spirooxindoles (Scheme 3.36).

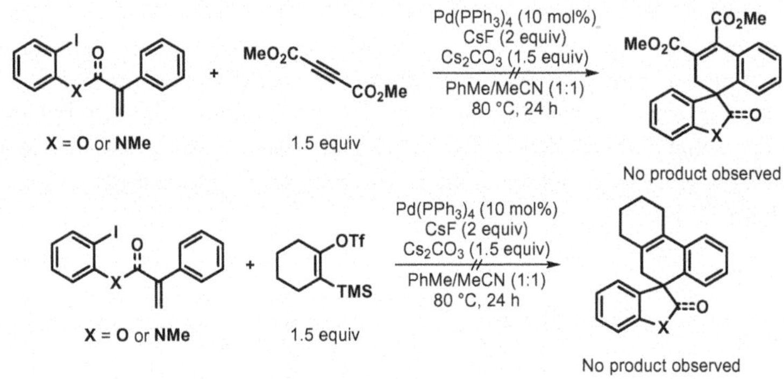


**Scheme 3.34** Unsuccessful Pd-catalyzed spirocyclization utilizing activated alkynes

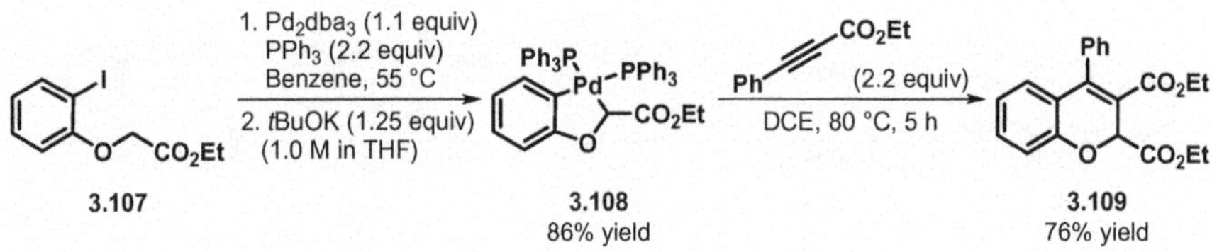


**Scheme 3.35** Synthesis of benzopyrans via palladacycles

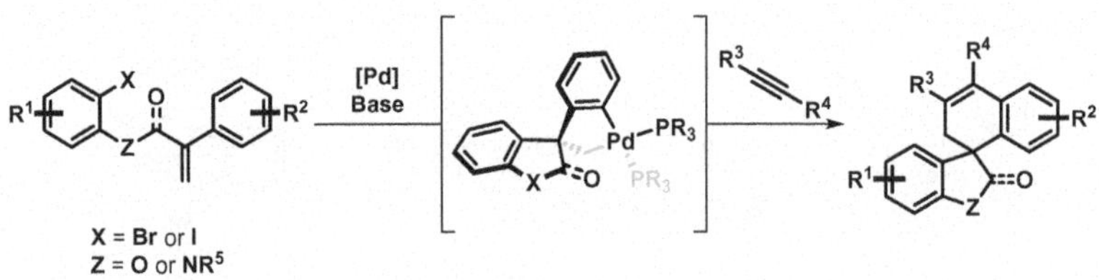


**Scheme 3.36** Proposed Pd-catalyzed spirocyclization via alkyne insertion

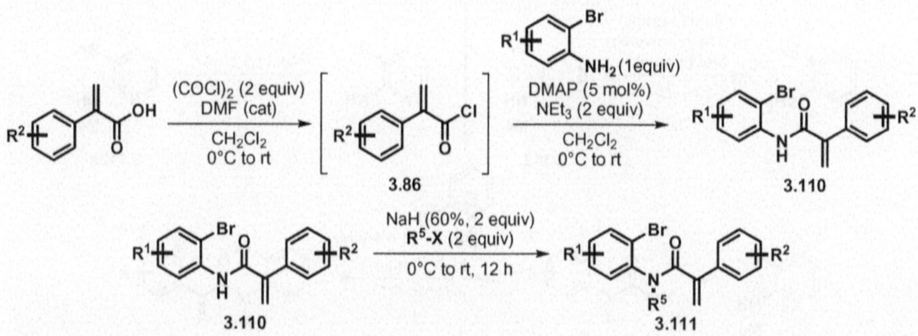


**Scheme 3.37** General method to 2-bromoacrylamides (3.111)

## 3.7 Results and Discussion: Spirocyclization via Alkyne Insertion

### *3.7.1 Starting Material Preparation*

The majority of the starting 2-bromoacrylamides were prepared in a similar fashion to the 2-iodoacrylamides (**3.88**) described in chapter 4.3.1. The substituted 2-bromoanilines were used in the first step (Scheme 3.37).

Some substrates were synthesized in an alternative route than shown in chapter 3.5.1. The alkylation of the 2-bromoaniline is performed prior to the amide bond formation via Mukaiyama coupling. In addition, some of the 2-bromoacrylamides bearing substituted pendant aromatic groups installed the olefin in the final step (Scheme 3.38).

### *3.7.2 Optimization*

Model substrates 2-iodoarylether **3.85a** and 2-iodoacrylamide **3.88a** were reacted with ethyl phenylpropiolate under the reaction conditions outlined by the Malinakova group. The respective spirocycles **3.114a** and **3.115a** were generated in 45% and 14% yield in >20:1 rr (Scheme 3.39).

The formation of the two spirocyclic scaffolds were optimized and a series of variations from these conditions were performed to summarize the findings. The standard conditions for the cyclization forming **3.114a** was subjecting model 2-iodoarylether **3.85a** and ethyl phenylpropiolate (1.1 equiv) to $Pd(PPh_3)_4$ (5 mol%) and $Cs_2CO_3$ (1.1 equiv) in PhMe at 110 °C (88% yield) (Table 3.7, entry 1). Utilizing other Pd catalysts resulted in lower yield or formation of the benzofused cyclobutane byproduct **3.74a** (Table 3.7, entries 2–4). Replacement of the cesium counterion with potassium gave significantly lower yields, likely due to the lower solubility of $K_2CO_3$ in PhMe (Table 3.7, entry 5). CsOPiv and CsOAc salts generated complex mixtures and no

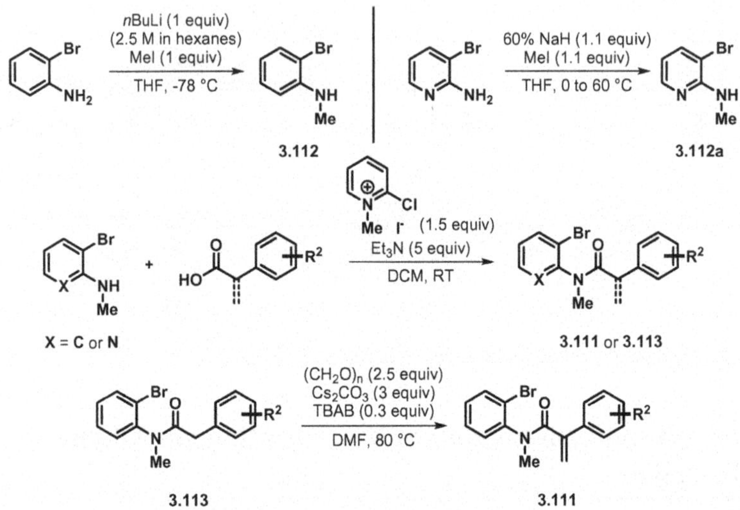


Scheme 3.38 Alternative syntheses of 2-bromoacrylamides 3.111

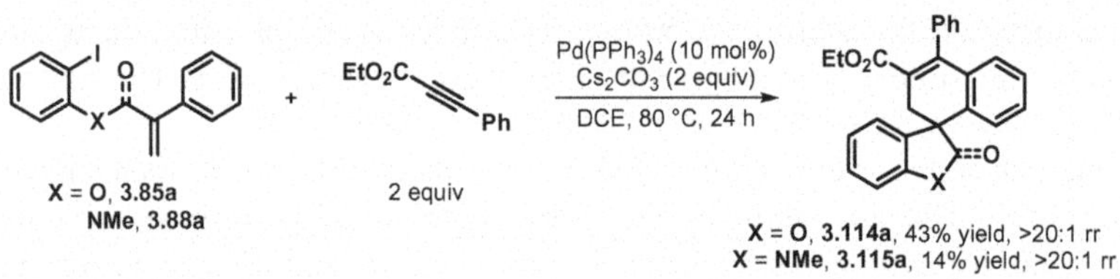


Scheme 3.39 Initial conditions

apparent formation of product via $^{1}$H NMR analysis (Table 3.7, entries 6 and 7). The reaction in dioxane proceeded in comparable yield while MeCN was found to be incompatible with the reaction (Table 3.7, entries 8 and 9). We also found 2-bromoarylethers were found to be less reactive than iodides and lower temperatures did not promote the reaction (Table 3.7, entries 10 and 11).

The optimized conditions for the synthesis of spirooxindole **3.115a** was found to employ 2-bromoacrylamide **3.111a**, 2 equivalents of ethyl phenylpropiolate, $Cs_2CO_3$ (1.5 equiv), and $Pd(PPh_3)_4$ (10 mol%) in PhMe for 24 h (Table 3.8, entry 1). Other catalyst systems were found to give lower yield or no product formation (Table 3.8, entries 2–5). Potassium carbonate gave similar yields but also gave byproduct **3.65**, arising from a 1,4-Pd shift and subsequent C–H activation, in 24% yield (Table 3.8, entry 6). In addition, CsOAc and CsOPiv exclusively gave byproduct **3.65** in good to excellent yields. Employing the bromoacrylamide was found to be essential for the reaction as the iodoacrylamide was found to undergo decomposition under high temperatures and the chloroacrylamide was found to be unreactive under the reaction conditions (Table 3.8, entries 9 and 10). In line with the reaction profile

**Table 3.7** Pd-catalyzed spirocyclization via C–H and alkyne insertion forming 3.114a: Variation from optimized reaction conditions

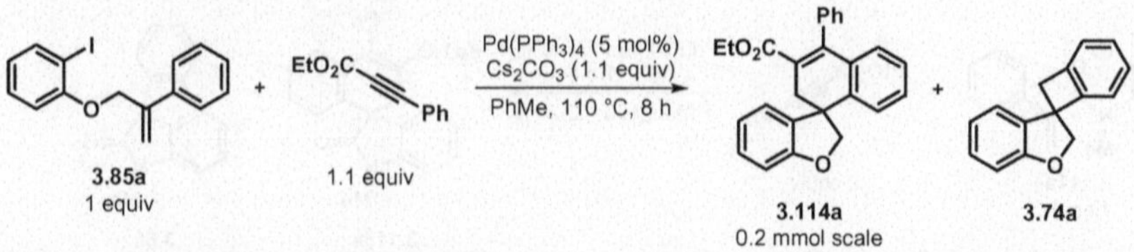


| Entry | Variation from optimized conditions | Yield 3.114a (%)[a] | Yield 3.74a (%)[a] |
|---|---|---|---|
| 1 | None | (88) | – |
| 2[c] | $Pd(dppf)Cl_2$ instead of $Pd(PPh_3)_4$ | 36 | – |
| 3[c] | Pd XPhos G2 instead of $Pd(PPh_3)_4$ | 19 | 63 |
| 4[c] | $Pd(P^tBu_3)_2$ instead of $Pd(PPh_3)_4$ | <2 | 79 |
| 5 | $K_2CO_3$ instead of $Cs_2CO_3$ | 28 | 19 |
| 6 | CsOPiv instead of $Cs_2CO_3$ | – | – |
| 7 | CsOAc instead of $Cs_2CO_3$ | – | – |
| 8 | Dioxane instead of PhMe | 83 | – |
| 9 | MeCN instead of PhMe | 19 | – |
| 10 | -Br instead of -I | 71 | – |
| 11 | 80°C instead of 110°C | 19 | – |

Reactions were run on 0.2 mmol scale. [a]Determined by $^1$H NMR analysis of the crude reaction mixtures using 1,3,5-trimethoxybenzene as internal standard. [b]Yield in parentheses are isolated yields. [c]Reaction run by Felicitas Landau

of the spirodihydrobenzofurans, other solvents and lower temperature gave lower yields (Table 3.8, entries 11–13).

With the optimized conditions in hand, the generality of the methodology forming spirodihydrobenzofurans was explored (Table 3.9). Unless stated otherwise, all spirocycles were generated in >20:1 rr. Showcasing the robustness of the reaction, model substrate **3.114a** was isolated in 94% yield with identical regioselectivity on gram-scale. In the presence of a chloride atom, the cyclization generated **3.114b** in excellent yield. The spirocyclic structure of **3.114b** was unambiguously confirmed by spectroscopic analysis and x-ray crystallography. Sterically encumbered aryl iodide **3.85d** reacted in 83% yield. Electron-rich substituents on the aryl iodide and pendant aromatic ring cyclized in good to excellent yields (**3.114d** and **3.114e**, 75% and 92% yield respectively). Electron-poor substituents on the aryl iodide or tethered aromatic ring required longer reaction times and gave moderate yields (**3.114f** and **3.114 g**, 66% and 55% yield respectively). *N*-tosyl spiroindoline was cyclized in good yield (**3.116**, 83% yield).

Internal alkynes other than ethyl phenylpropiolate were also investigated (Table 3.9). The insertion of ethyl 2-butynoate generated **3.114 h** in 83% yield. The phenyl ketone substituted alkyne **3.117a** inserted to form **3.115i** in good yield. The indole

**Table 3.8** Pd-catalyzed spirocyclization via C–H and alkyne insertion forming 3.115a: Variation from optimized reaction conditions

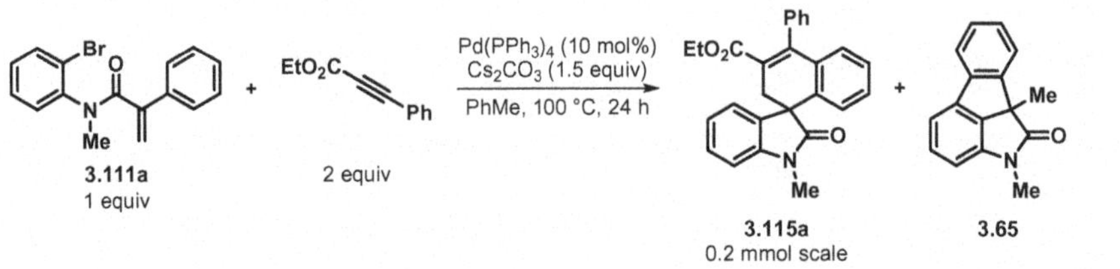


| Entry | Variation from optimized conditions | Yield 3.115a (%)[a] | Yield 3.65 (%)[a] |
|---|---|---|---|
| 1 | None | (79) | – |
| 2 | $Pd(dppf)Cl_2$ instead of $Pd(PPh_3)_4$ | 50 | – |
| 3 | Pd XPhos G2 instead of $Pd(PPh_3)_4$ | – | – |
| 4 | $Pd(P\mathit{t}Bu_3)_2$ instead of $Pd(PPh_3)_4$ | – | – |
| 5[c] | $Pd(OAc)_2$ and BINAP instead of $Pd(PPh_3)_4$ | – | – |
| 6 | $K_2CO_3$ instead of $Cs_2CO_3$ | 70 | 24 |
| 7 | CsOAc instead of $Cs_2CO_3$ | – | 92 |
| 8 | CsOPiv instead of $Cs_2CO_3$ | – | 66 |
| 9[d] | – I instead of –Br | 52 | – |
| 10[d] | – Cl instead of –Br | – | – |
| 11[d] | Dioxane instead of PhMe | 61 | – |
| 12[d] | MeCN instead of PhMe | 8 | – |
| 13[d] | 80 °C instead of 100 °C | 12 | – |

Reactions were run on 0.2 mmol scale. [a]Determined by $^1H$ NMR analysis of the crude reaction mixtures using 1,3,5-trimethoxybenzene as internal standard. [b]Yield in parentheses are isolated yields. [c]$Pd(OAc)_2$ (10 mol%) and BINAP (12 mol%) were prestirred for 15 min prior to the addition of other reagents. [d]Reaction run by Martin Rölz

derived alkyne **3.117b** and the Weinreb amide **3.117c** inserted in excellent yields. Phenylpropionitrile and the diaryl alkyne **3.117e** reacted in good yield. The 5-trifluoromethyl-2-pyridine bearing alkyne **3.117f** also provided the desired spirodihydrobenzofuran but required extended reaction time. A potential explanation for the lower yield can be attributed to the less reactive nature of the alkyne.

**Table 3.9** Pd-catalyzed spirocyclization via C–H and alkyne insertion forming spirodihydrobenzofurans. Substrate scope

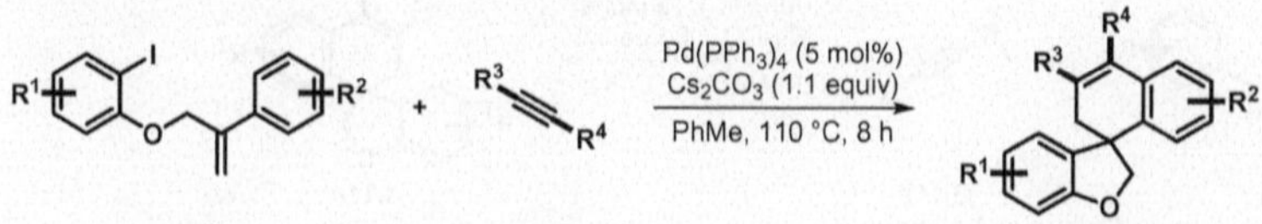


**3.85** 1 equiv; **3.117** 1.1 equiv; **3.114** >20:1 r.r. 0.2 mmol scale

| Entry | Substrate | Product | Yield (%)[a] |
|---|---|---|---|
| 1 | 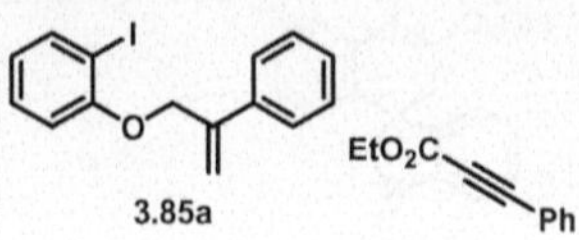 3.85a | 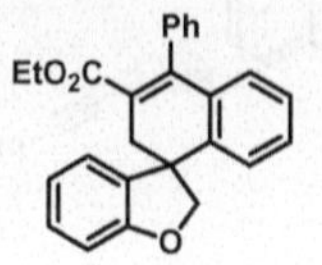 3.114a | 88<br>94[b] |
| 2 | 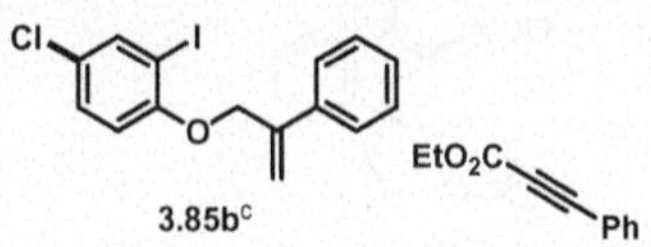 3.85b[c] | 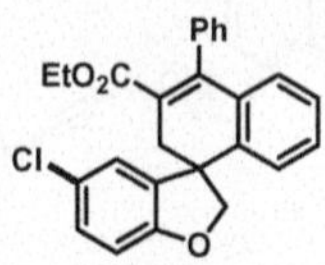 3.114b[c] | 88 |
| 3 | 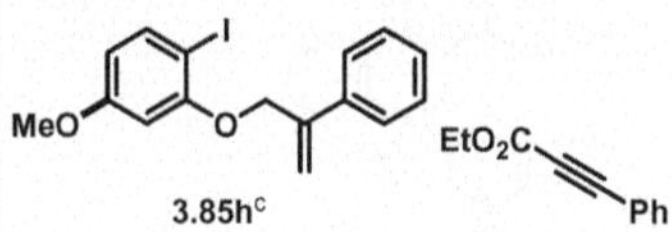 3.85h[c] | 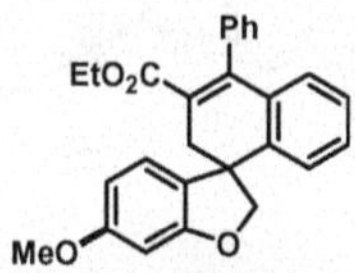 3.114d | 83 |
| 4 | 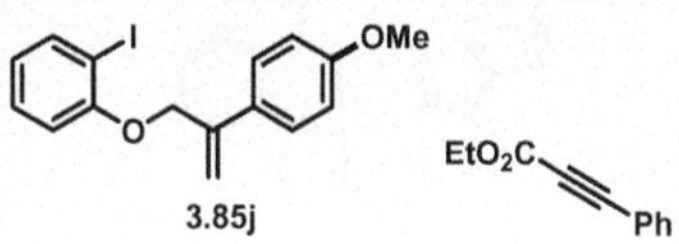 3.85j | 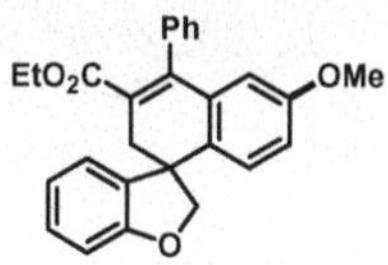 3.114e | 75 |
| 5 | 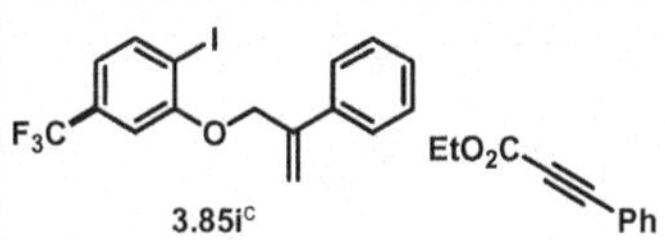 3.85i[c] | 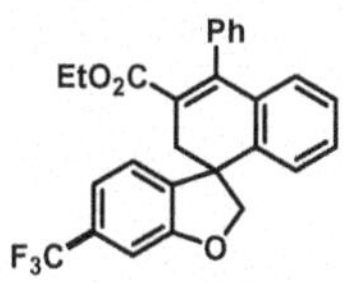 3.114f[c] | 92 |
| 6 | 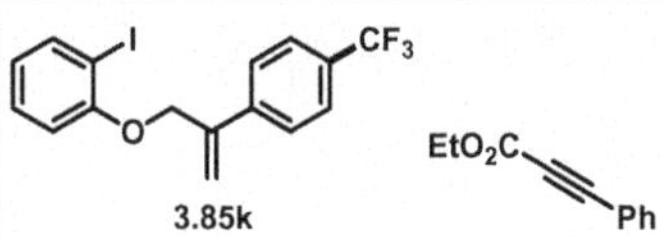 3.85k | 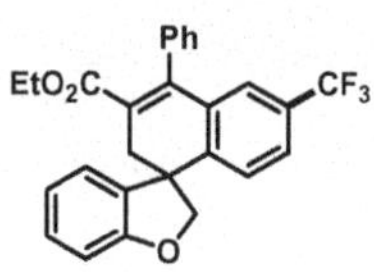 3.114g | 66 |

(continued)

**Table 3.9** (continued)

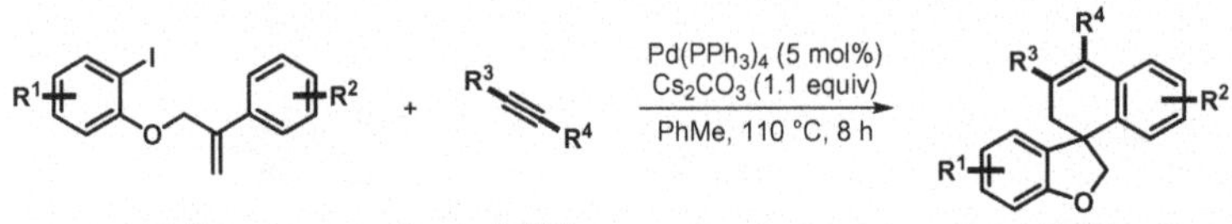

3.85 1 equiv; 3.117 1.1 equiv; 3.114 >20:1 r.r. 0.2 mmol scale

| Entry | Substrate | Product | Yield (%)[a] |
|---|---|---|---|
| 7 | 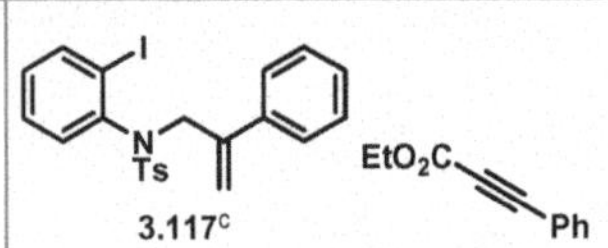 3.117[c] | 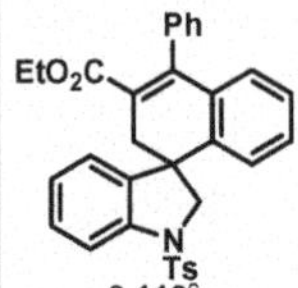 3.116[c] | 55 |
| 8 | 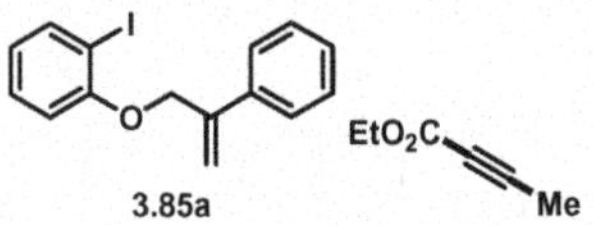 3.85a | 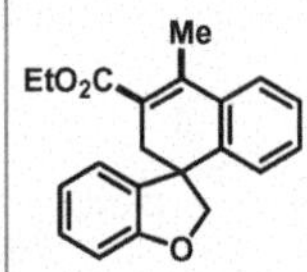 3.114h | 83 |
| 9 | 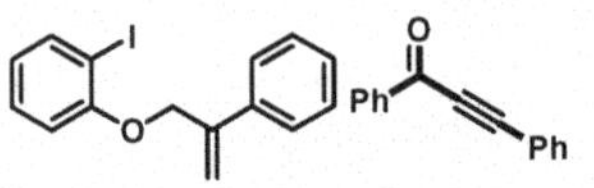 3.85a 3.117a | 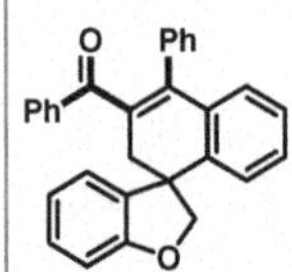 3.114i | 83 |
| 10 | 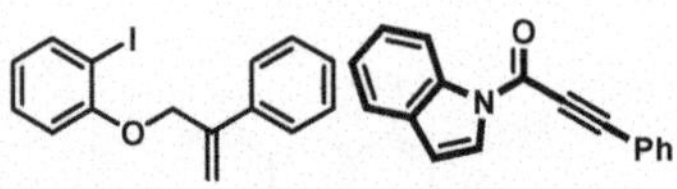 3.85a 3.117b | 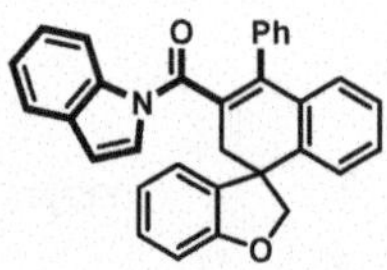 3.114j | 84 |
| 11 | 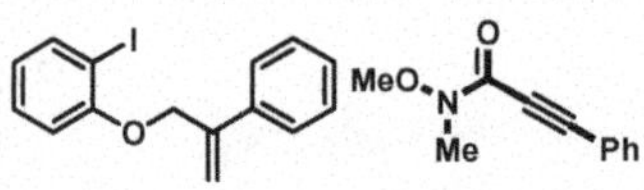 3.85a 3.117c[d] | 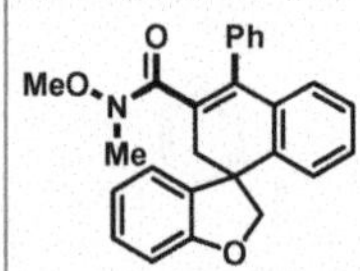 3.114k | 90 |
| 12 | 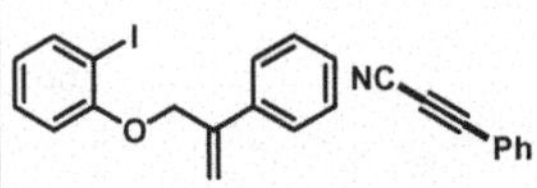 3.85a 3.117d | 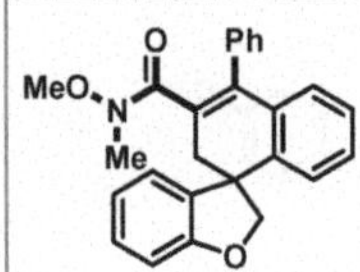 3.114k | 86 |

(continued)

**Table 3.9** (continued)

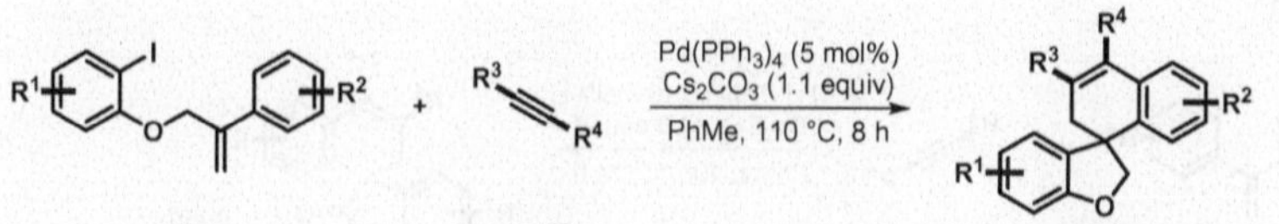


3.85, 1 equiv; 3.117, 1.1 equiv; 3.114, >20:1 r.r., 0.2 mmol scale

| Entry | Substrate | Product | Yield (%)[a] |
|---|---|---|---|
| 13 | 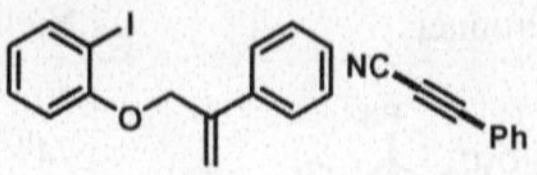 3.85a 3.117d | 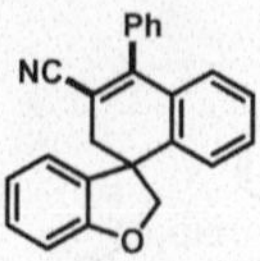 3.114l | 76 |
| 14 | 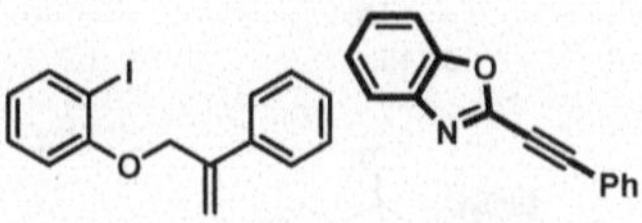 3.85a 3.117e[d] | 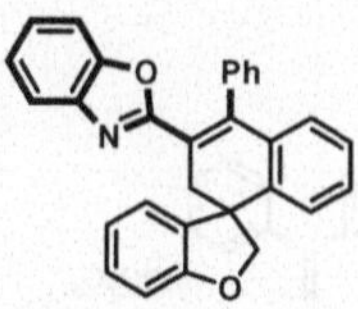 | 83 |
| 15 | 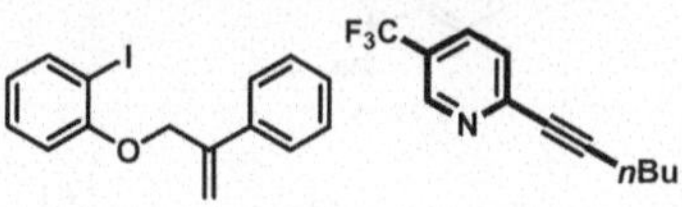 3.85a 3.117f[d] | 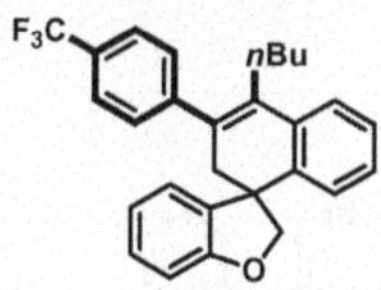 3.114n | 53 |

Reactions were run on 0.2 mmol scale. [a]Isolated yields. [b]Reaction run on 2.97 mmol scale. [c]Reaction run for 24 h. [c]This compound was prepared by Felicitas Landau. [d]This compound was prepared by Martin Rölz

Following the success of the spirodihydrobenzofurans, the spirooxindoles were targeted (Table 3.10). On gram scale, **3.115a** was synthesized in 74% yield; however, a longer reaction time was necessary for the full consumption of **3.111a**. Fluorinated acrylamide **3.111b** reacted to give **3.115b** in 74% yield. Electron-rich substituents were tolerated and gave **3.115c** and **3.115d** in moderate yields. Other nitrogen protecting groups such as –Bn and –MOM were tolerated in the cyclization. Substitution by a pyridinylbromide did not impede the generation of **3.115 g**. Electron-rich substituents on the pendant aromatic ring aided in the cyclization forming **3.115 h** in 88% yield. Substitution to a tethered thiophene group reacted to give **3.115i** in moderate yield.

**Table 3.10** Pd-catalyzed spirocyclization via C–H and alkyne insertion forming spirooxindoles. Substrate scope

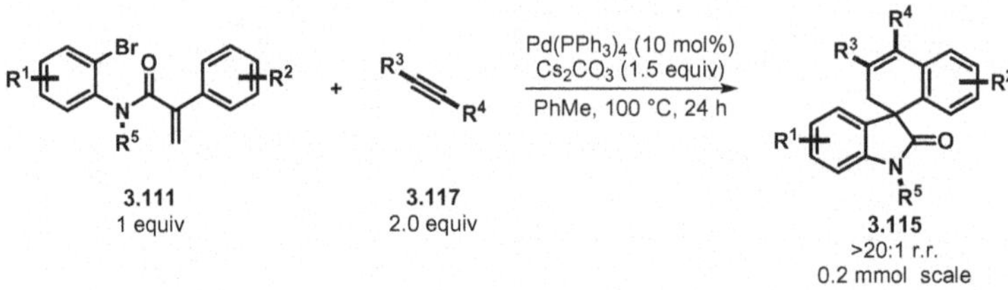


| Entry | Substrate | Product | Yield (%)[a] |
|---|---|---|---|
| 1 | 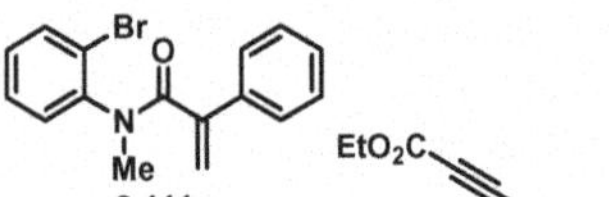 3.111a | 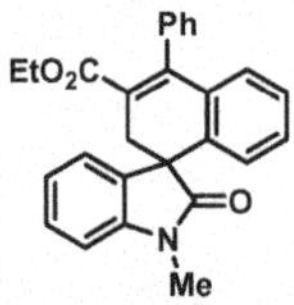 3.115a | 79<br>74[b,c] |
| 2 | 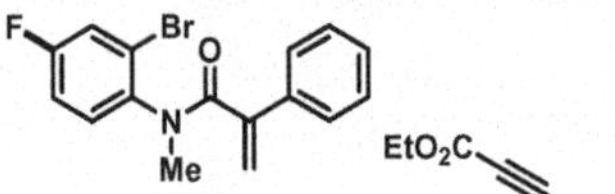 3.111b | 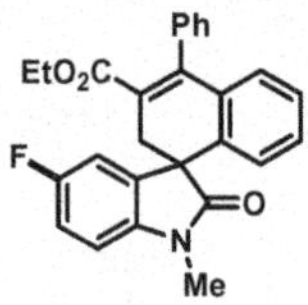 3.115b | 74 |
| 3 | 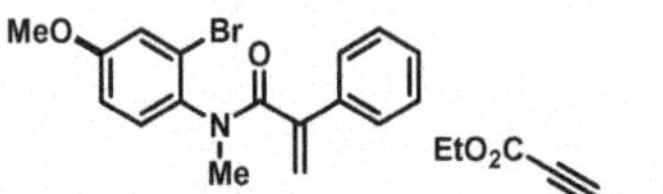 3.111c | 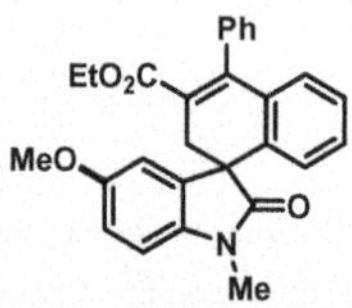 3.115c | 61 |
| 4 | 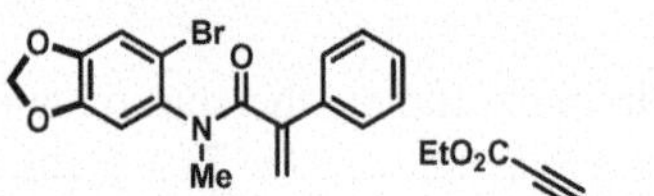 3.111d | 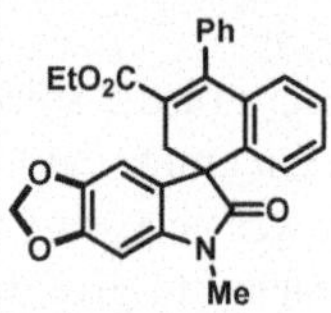 3.115d | 62 |
| 5 | 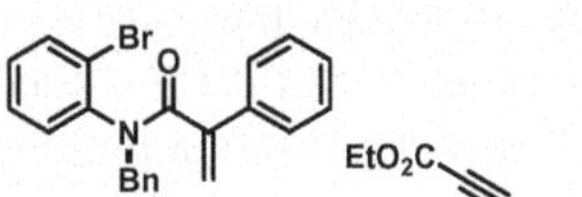 3.111e | 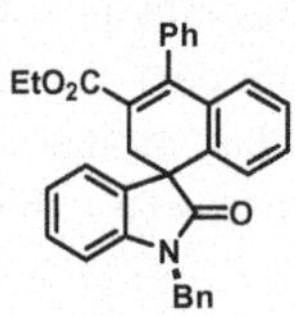 3.115e | 73 |

(continued)

**Table 3.10** (continued)

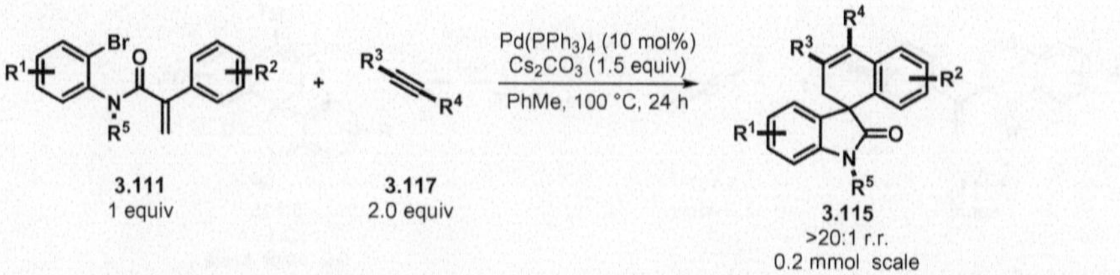


| Entry | Substrate | Product | Yield (%)[a] |
|---|---|---|---|
| 6 | 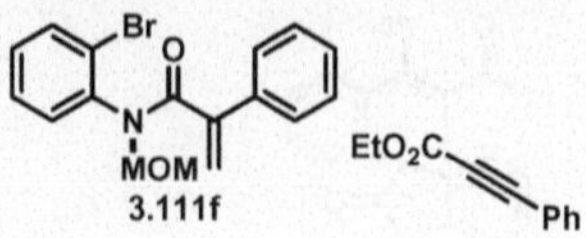 3.111f | 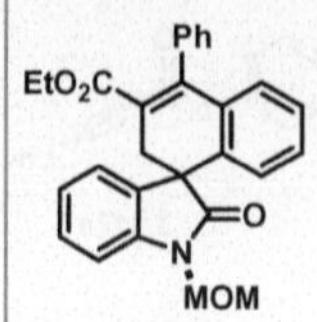 3.115f | 92 |
| 7 | 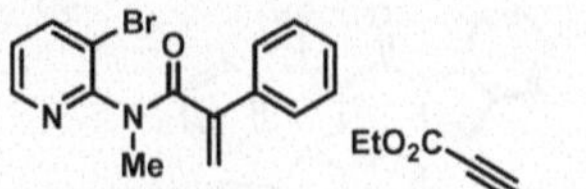 3.111g | 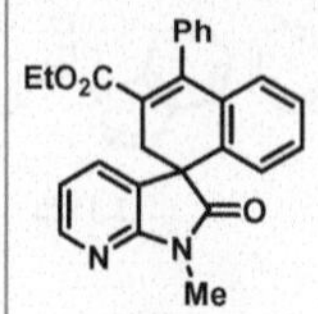 3.115g | 67 |
| 8 | 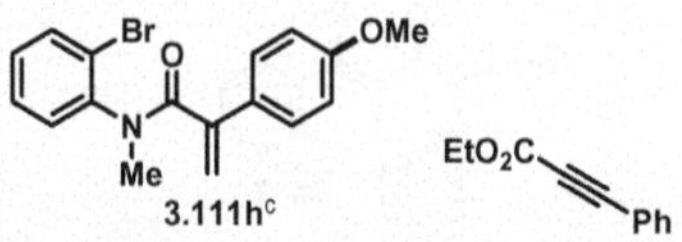 3.111h[c] | 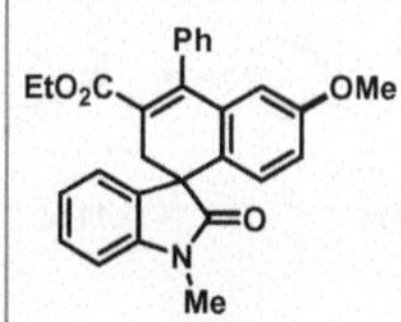 3.115h[c] | 88 |
| 9 | 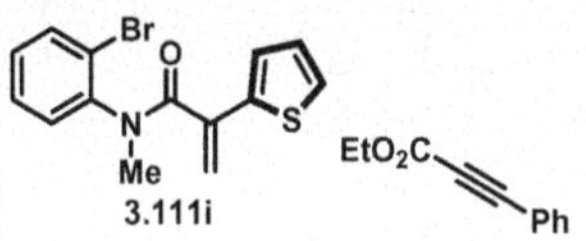 3.111i | 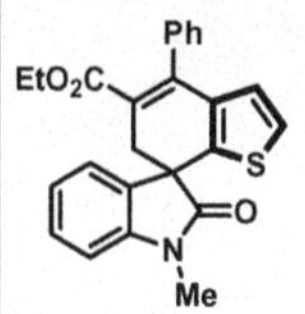 3.115i | 56 |
| 9 | 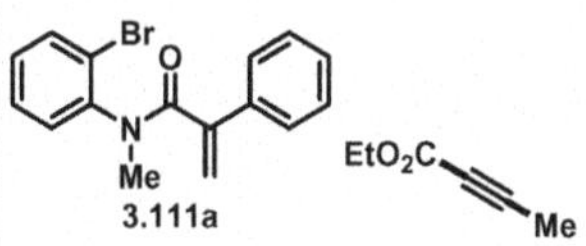 3.111a | 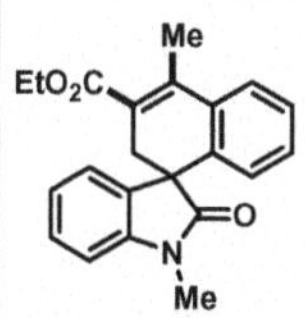 3.115j | 85 |

(continued)

**Table 3.10** (continued)

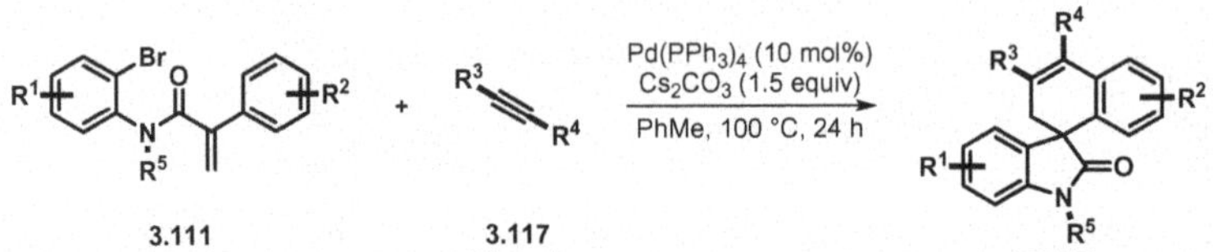


3.111 1 equiv; 3.117 2.0 equiv; 3.115 >20:1 r.r. 0.2 mmol scale

| Entry | Substrate | Product | Yield (%)[a] |
|---|---|---|---|
| 10 | 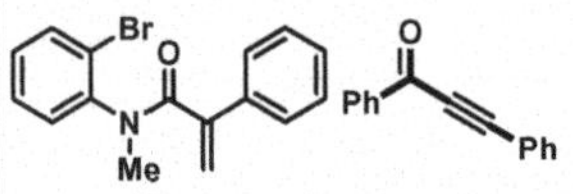 3.111a 3.117a | 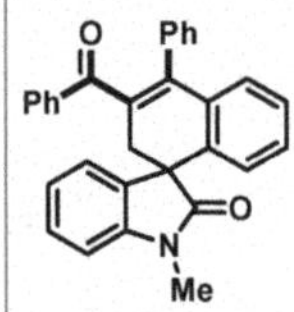 3.115k | 79 |
| 11 | 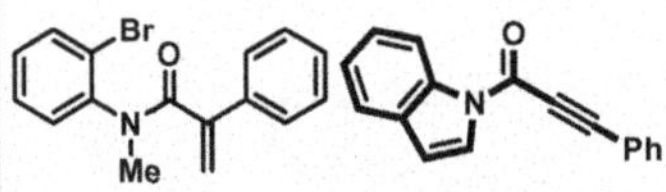 3.111a 3.117b[c] | 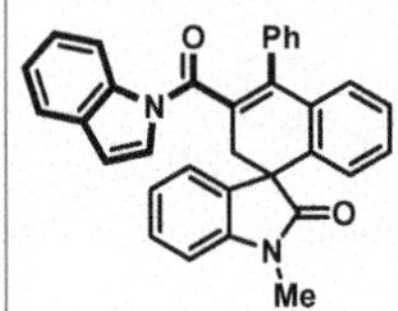 3.115l | 90 |
| 12 | 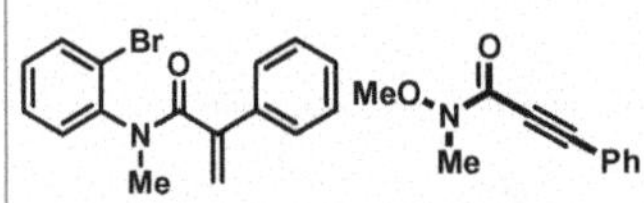 3.111a 3.117c | 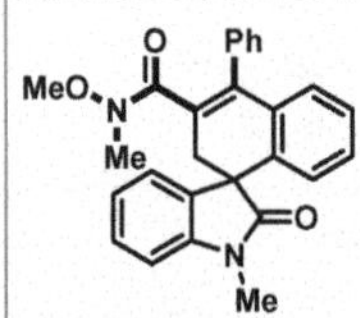 3.115m | 78 |
| 13 | 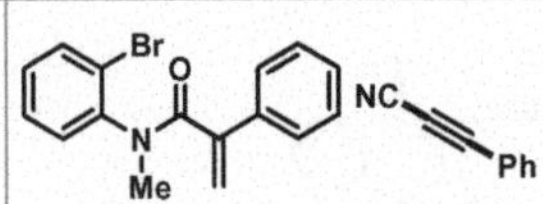 3.111a 3.117d | 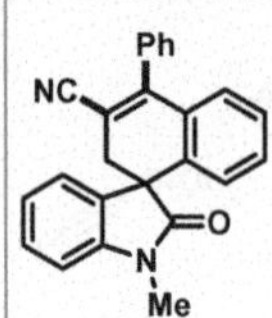 3.115n | 84 (9:1 r.r.) |

(continued)

**Table 3.10** (continued)

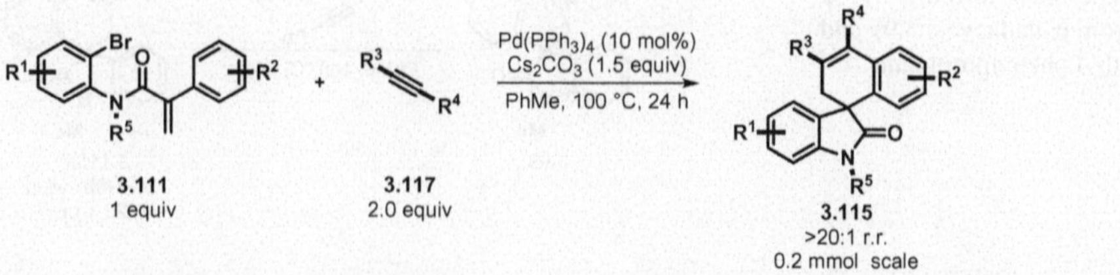


| Entry | Substrate | Product | Yield (%)[a] |
|---|---|---|---|
| 14 | 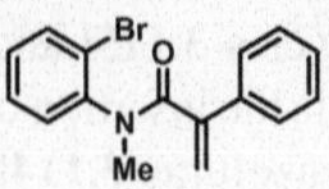 3.111a 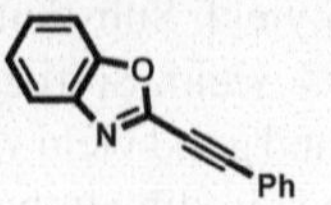 3.117e | 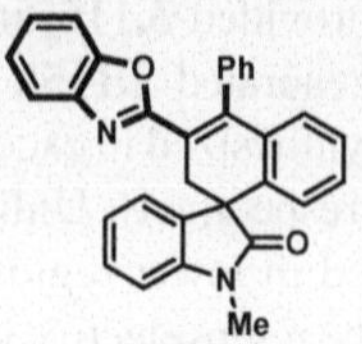 3.115o | 63 |
| 15 | 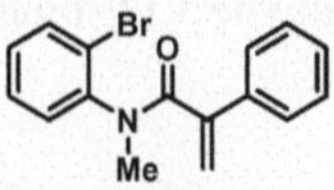 3.111a 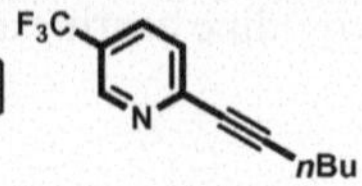 3.117f | 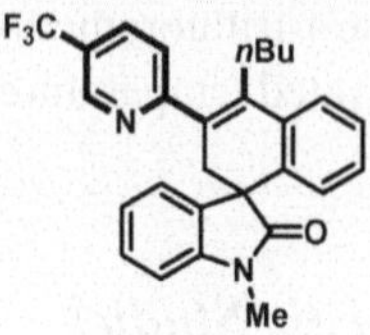 3.115p | 68 |

Reactions were run on 0.2 mmol scale. [a]Isolated yields. [b]Reaction run on 3.16 mmol scale. [c]Reaction run for 48 h. [c]This compound was prepared by Hyung Yoon

**Scheme 3.40** Stoichiometric experiment forming 3.115a from palladacycle 3.99 and ethyl phenylpropiolate

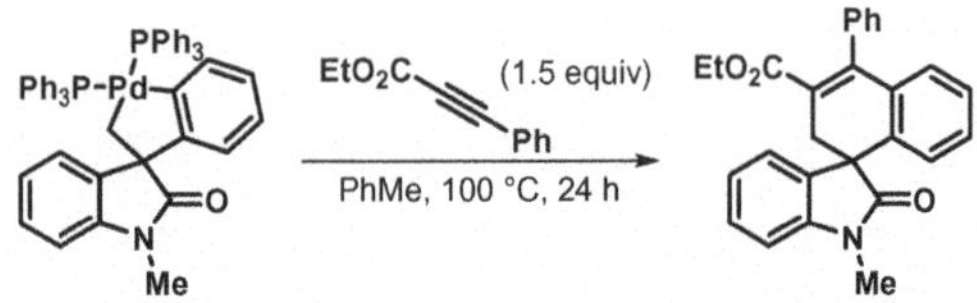


3.99

3.115a
75% NMR yield
>20:1 r.r.

The internal alkynes were screened with 2-bromoacrylamide **3.111a** (Table 3.10). Ethyl 2-butynoate provided **3.115j** in 85% yield. Substitution of the ester to a non-enolizable ketone generated **3.115k** in 79% yield. Spirocycles **3.115l** and **3.115m** were successfully synthesized in excellent and good yield via the insertion of alkynes **3.117b** and **3.117c** respectively. Unlike the spirodihydrobenzofuran **3.114l**, Phenylpropionitrile inserted in lower regioselectivity (**3.115n**, 9:1 r.r.). The diaryl alkyne **3.117e** underwent the regioselective insertion at elevated temperatures to give **3.115o** in 63% yield. The connectivity was confirmed by spectroscopic analysis and X-ray crystallography. The 4-trifluoromethyl-2-pyridine bearing alkyne **3.117f** gave 3.115p in 68% yield at elevated temperature.

### *3.7.3 Mechanistic Studies*

Alluding back to chapter 3.5.5 and 3.5.6, the postulated mechanism proceeds via the C–H activation prior to the insertion of the benzyne. Similarly, we propose that the C–H activation occurs prior to the alkyne insertion. To support these claims, a stoichiometric study involving palladacycle 3.99 and ethyl phenylpropiolate was performed. Spirooxindole 3.115a was formed 75% yield with consistent regioselectivity (Scheme 3.40).

### *3.7.4 Limitations*

Despite the wide range of products generated, some substrates were unreactive under the described reaction conditions. A major limitation exhibited in both classes of scaffolds is that substrates bearing highly electron deficient pendant aromatics result in lower yield or no product formation. Substrate 3.85e with a 3,5-difluorinated pendant aromatic ring failed to give the respective product (Scheme 3.41a). In addition, 2-bromoacrylamide bearing a *p*-$CF_3$ group on the pendant aromatic ring was also found to be unreactive. Gaining further insight to the reaction profile, palladacycle **3.118** was independently synthesized. Under similar reaction conditions to synthesize palladacycle **3.99**, palladacycle **3.118** was formed in good yield but required significantly longer reaction times. The subsequent reaction with the model alkyne

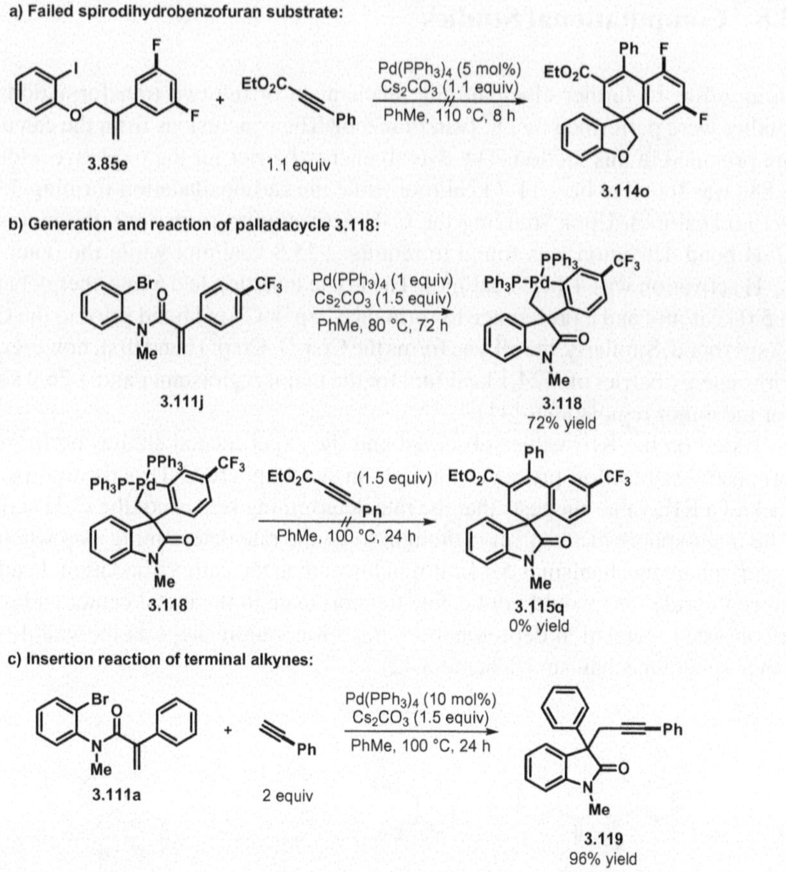


**Scheme 3.41** Limitations in the Pd-catalyzed spirocyclization via alkyne insertion

led to full consumption of **3.118**; however, no desired spirooxindole was formed. This suggests that the insertion of the alkyne is kinetically disfavoured with respect to the decomposition pathway (Scheme 3.41b). Lastly, it was found that in the presence of terminal alkynes, the reaction gives the direct Sonogashira coupling of the generated alkylpalladium species as opposed to the intramolecular C–H activation (Scheme 3.41c).

## 3.8 Computational Studies

In an effort to further elucidate the mechanism of the two transformations, DFT studies were performed by Dr. Ivan Franzoni. The conclusions from the calculations are provided in this section. The overall energy barrier for the oxidative addition of 3.88a was found to be +11.7 kcal/mol while the carbopalladation forming 3.98 was +19.6 kcal/mol. Upon studying the C–H activation to form 3.99, the inner-sphere C–H bond activation was found to require +25.8 kcal/mol while the outer-sphere C–H activation was +14.2 kcal/mol. The aryne insertion had a low energy barrier of +5.0 kcal/mol and a preference to form the $C(sp^2)$–$C(sp^2)$ bond prior to the $C(sp^3)$–$C(sp^2)$ bond. Similarly, the alkyne forms the $C(sp^2)$–$C(sp^2)$ bond first; however, with a large energy barrier of +24.1 kcal/mol for the major regioisomer and +26.9 kcal/mol for the minor regioisomer [41].

Based on the KIE values observed and the experimental studies performed, we propose that reaction proceeds through an outer-sphere C–H activation [42]. The lack of a KIE value suggests that the rate determining step is not the C–H activation. The inner-sphere mechanism is thought to be the rate determining step whereas the outer-sphere mechanism is 5.4 kcal/mol lower than the carbopalladation. In addition, since fluoride ions would not be able to coordinate to the metal center and promote a concerted metalation deprotonation, this observation suggests the viability of an outer-sphere mechanism (Scheme 3.42).

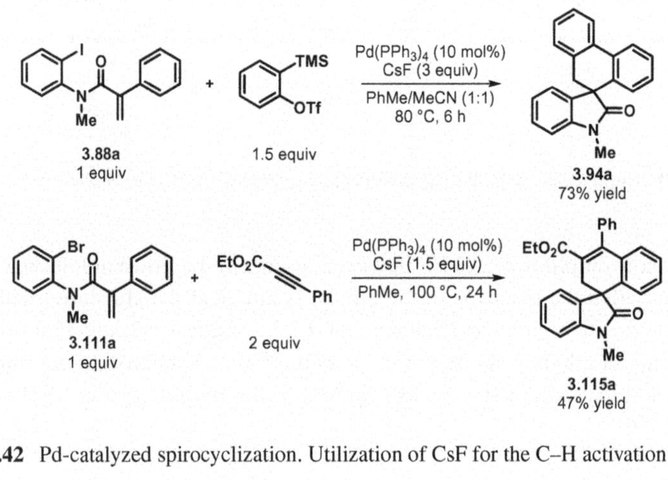


**Scheme 3.42** Pd-catalyzed spirocyclization. Utilization of CsF for the C–H activation

## 3.9 Recent Advances

Concurrently to the two spirocyclization reactions via aryne and alkyne insertion, examples of α-diazocarbonyl and $CH_2Br_2$ insertion were reported by the García-López and Zhang group respectively [43]. The insertion of the α-diazocarbonyl compound yielded 3.122 in excellent yield albeit in low diastereomeric control. In addition, spirooxindole 3.123 was synthesized in excellent yield and the insertion of $CH_2Br_2$ into palladacycle 3.99 was postulated to proceed through a $Pd^{IV}$ intermediate (Scheme 3.43).

Recently, our group disclosed the synthesis of dihydrobenzoindolones via the insertion of polarized alkynes in excellent yields and regioselectivities (Scheme 3.44) [44]. Interestingly, in contrast to the spirocyclization process, CsOPiv was essential for the C–H activation *ortho* to the arylhalide.

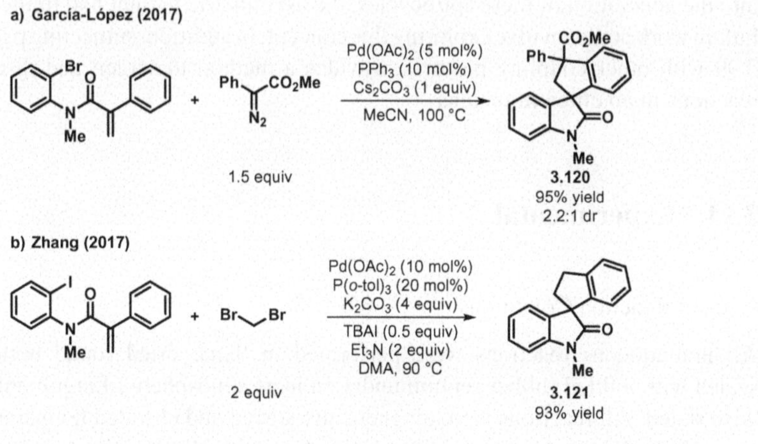


**Scheme 3.43** Pd-catalyzed spirocyclization reactions via diazocarbonyl and $CH_2Br_2$ insertion

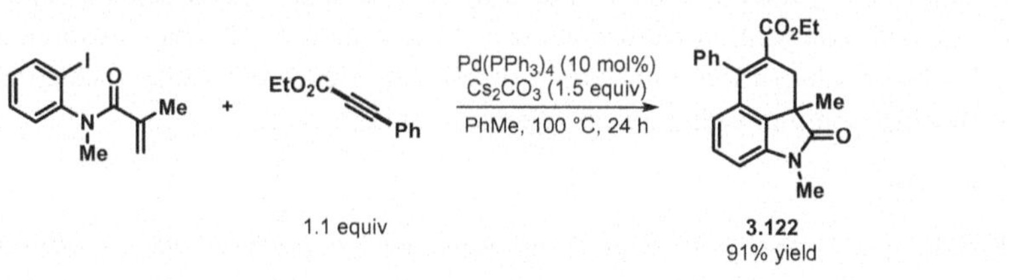


**Scheme 3.44** Pd-catalyzed cascade reaction forming dihydrobenzoindolones via C–H activation and alkyne insertion

## 3.10 Conclusion

The use of C–H functionalization has enabled the synthesis of novel scaffolds with improved step- and atom- economy. In particular, the use of C–H functionalization in a domino fashion has extended the versatility and applicability of the reaction as it enables the activation of remote bonds that may otherwise be considered unreactive. We described the Pd-catalyzed spirocyclization via C–H and $\pi$-system insertion to furnish spirodihydrobenzofurans and spirooxindoles in good to excellent yields. A range of arynes were successfully incorporated; however, modest regioselectivities were observed with unsymmetrical arynes. Alternatively, high regioselectivities were observed when unsymmetrical alkynes were employed. On the basis of experimental and computational mechanistic studies, we believe that the reaction proceeds via an outer-sphere C–H activation prior to the insertion of the $\pi$-system.

Although we have gained significant insight experimentally and computationally into the generation of these spirocycles, the asymmetric variant had limited success. Future work could involve exploring this concept. In addition, subjecting palladacycle 3.99 with other coupling partners provides a method to screen and discover new reactions in an efficient manner.

## 3.11 Experimental

General Reaction Conditions

All non-aqueous reactions were performed in flame dried round bottom flasks sealed with a fitted rubber septum under an inert atmosphere of argon unless otherwise stated. All reactions were magnetically stirred and elevated temperatures were reported as the temperature of the surrounding oil bath. Reactions were monitored by thin layer chromatography (TLC) or by crude $^{1}$H-NMR analysis of a worked up aliquot. TLC visualization was performed under a UV lamp or $KMnO_4$/CAM stain developed with heat. Solvent evaporation was conducted by rotary evaporation at the appropriate temperature and pressure. All reported yields reflect spectroscopically ($^{1}$H-NMR) pure material unless otherwise stated.

Materials

Unless stated otherwise, all reagents were used as received and the following reaction solvents were distilled under anhydrous conditions over the appropriate drying agent and transferred under argon via a syringe. Dichloromethane was distilled over $CaH_2$, tetrahydrofuran was distilled over Na (1% w:v) and benzophenone (1% w:v), 1,4-dioxane was distilled over Na (1% w:v) and benzophenone (1% w:v), and triethylamine was distilled over KOH. Dimethylformamide was distilled over 5Å molecular sieves and stored over 5Å molecular sieves (water content was kept lower than

50 ppm). 1,3,5-trimethoxybenzene was crushed into a fine powder by a mortar and pestle, dried overnight *in vacuo* and stored in a desiccator.

Analysis

$^{1}$H-NMR and $^{13}$C-NMR spectra of catalytic starting material and products were obtained on the Agilent DD2 500 equipped with a 5 mm Xsens Cold Probe. The starting material precursor $^{1}$H-NMR and $^{13}$C-NMR spectra were obtained from one of the following spectrometers: Varian NMR system 400, Bruker Avance III 400, Varian Mercury 400 or Varian Mercury 300. All $^{19}$F-NMR spectra were obtained on the Varian Mercury 300 and Varian Mercury 400. Measurements were carried out at 23 °C and chemical shifts (δ) are reported as parts per million (ppm). The solvent resonance was used as the internal standard for $^{1}$H-NMR (Chloroform at 7.26 ppm) and $^{13}$C-NMR (Chloroform at 77.0 ppm). The *J* values are reported in hertz (Hz) and are rounded off to the nearest 0.5 Hz. Multiplicities are abbreviated as follows: singlet (s), doublet (d), triplet (t), quartet (q), multiplet (m) and broad (br). All accurate mass values were obtained from the following spectrometers: Agilent 6538 Q-TOF (ESI) and JEOL AccuTOF-DART. Melting points were obtained on a Fisher-Johns Melting Point Apparatus and uncorrected. Infrared (IR) spectra were obtained as a neat film or dissolved in $CHCl_3$ on a NaCl disk using a Shimadzu FTIR-8400S FT-IR spectrometer.

General Procedures

General procedure 1 (GP1):

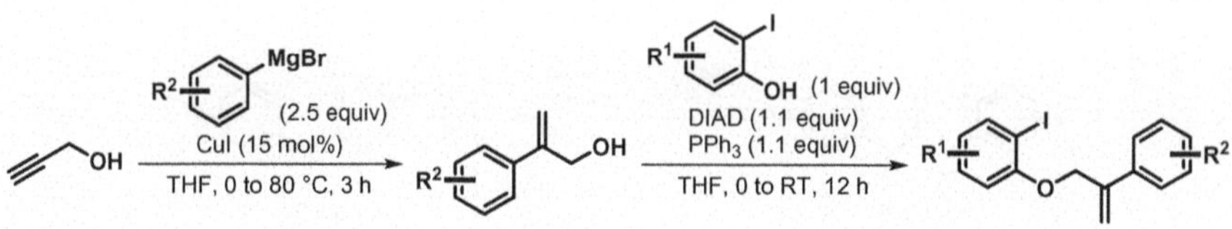


3.84 3.85

To a solution of substituted phenylmagnesium bromide (2.5 equiv) and CuI (15 mol%) in THF at 0 °C, a solution of propargyl alcohol (1.0 equiv) in THF (1.0 M) was added dropwise. After 30 min, the reaction was warmed to 80 °C for 4 h. The reaction was cooled to 0 °C and was quenched with a saturated solution of $NH_4Cl$ and extracted with EtOAc (3x). The combined organic layers were washed with brine, dried over $Na_2SO_4$ and concentrated *in vacuo*.

A solution of the substituted allylic alcohol (1.0 equiv), substituted iodophenol (1.0 equiv) and $PPh_3$ (1.1 equiv) in THF (0.66 M) was prepared and cooled to 0 °C. DIAD (1.1 equiv) was added dropwise and after 30 min, the reaction was allowed to warm to room temperature and was stirred overnight. The reaction was quenched with cold water and extracted with EtOAc (3x). The combined organic layers were washed with brine, dried with $Na_2SO_4$ and concentrated *in vacuo*. The reaction was purified via Si gel flash chromatography using the indicated mobile phase.

General procedure 2 (GP2):

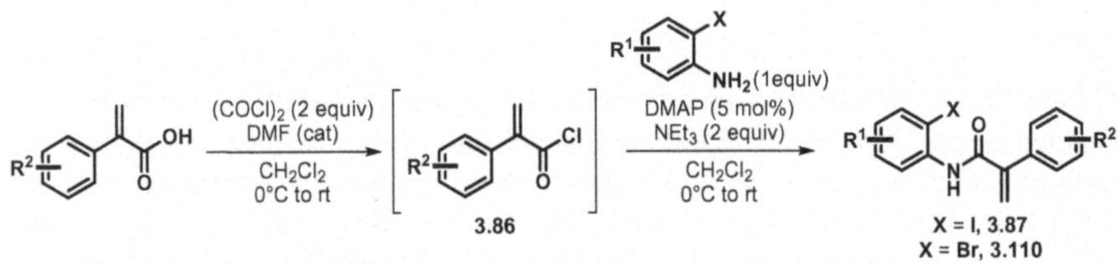

A solution of substituted 2-phenylacrylic acid (1.0 equiv) and DMF (4 drops) in $CH_2Cl_2$ (0.40 M) was prepared and cooled to 0 °C. A bubbler was attached to the vessel and $(COCl)_2$ (2 equiv) was added dropwise. After 5 min, the reaction was allowed to warm to room temperature and was stirred for 1 h. The acyl chloride was concentrated *in vacuo* and redissolved in $CH_2Cl_2$. A solution of the substituted 2-iodoaniline (1.0 equiv), DMAP (0.05 equiv) and $NEt_3$ (2.0 equiv) was prepared in $CH_2Cl_2$ (0.50 M) and cooled to 0 °C. The acyl chloride solution was added dropwise into the vessel containing the substituted 2-iodoaniline. After 5 min, the reaction was allowed to warm to room temperature and was stirred overnight. The reaction was quenched with a saturated $NaHCO_3$ solution and extracted with EtOAc (3x). The combined organic layers were washed with brine, dried over $Na_2SO_4$ and concentrated *in vacuo*. The crude unsubstituted acrylamide was passed through a plug of silica gel prior to proceeding to GP3.

General procedure 3 (GP3):

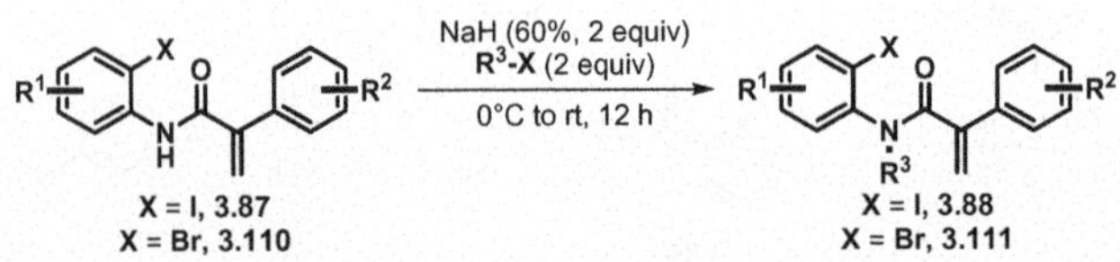

A solution of the unsubstituted acrylamide (1.0 equiv) in THF (0.20 M) was prepared and cooled to 0 °C. NaH (60%w/w, 2.0 equiv) was added to the solution and the mixture was stirred for 15 min before adding $R^3$-X (2 equiv) dropwise. The reaction was allowed to warm at room temperature after 10 min and was stirred for 2 h to overnight. The reaction was quenched with cold water and extracted with EtOAc (2x). The combined organic layers were washed with brine, dried with $Na_2SO_4$ and concentrated *in vacuo*. The reaction was purified via Si gel flash chromatography using the indicated mobile phase.

General procedure 4 (GP4):

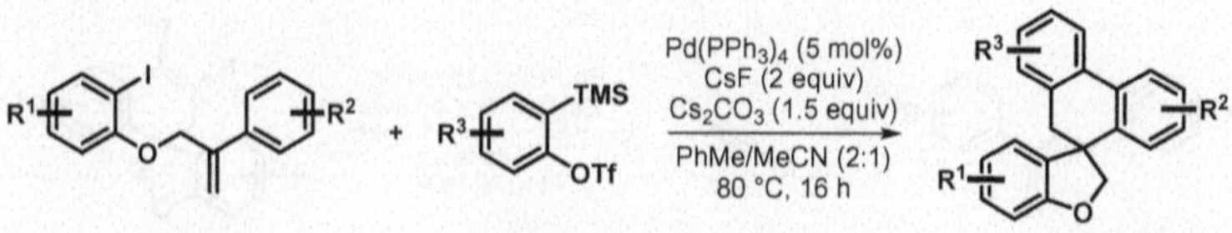


3.85 1 equiv | 3.91 1.5 equiv | 3.92 0.2 mmol scale

To a flame dried 2 dram vial cooled under argon, the 2-iodoarylether (1 equiv), $Pd(PPh_3)_4$ (5 mol% Pd), CsF (2 equiv), and $Cs_2CO_3$ (1.5 equiv) were added and allowed to purge for 10 min. PhMe/MeCN (2:1) (0.1 M) was added. The aryne precursor (1.5 equiv) was added via a microsyringe. A Teflon line screw cap was fitted on the two dram vial, sealed with Teflon tape and place in a preheated oil bath at 80 °C for 20 h. The reaction mixture was cooled to room temperature. Once cooled, the reaction was passed through a plug of silica gel using EtOAc. The pure spirodihydrobenzofurans were obtained via silica gel flash column chromatography using the indicated mobile phase.

General Procedure 5 (GP5):

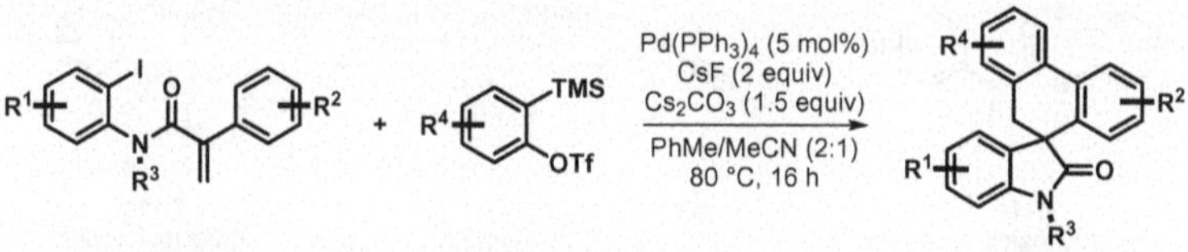


3.88 1 equiv | 3.91 1.5 equiv | 3.94 0.2 mmol scale

To a flame dried 2 dram vial cooled under argon, the 2-iodoacrylamide (1 equiv), $Pd(PPh_3)_4$ (10 mol% Pd), CsF (2 equiv), and $Cs_2CO_3$ (1.5 equiv) were added and allowed to purge for 10 min. PhMe/MeCN (1:1) (0.1 M) was added. The aryne precursor (1.5 equiv) was added via a microsyringe. A Teflon line screw cap was fitted on the two dram vial, sealed with Teflon tape and place in a preheated oil bath at 80 °C for 6 h. The reaction mixture was cooled to room temperature. Once cooled, the reaction was passed through a plug of silica gel using EtOAc. The pure spirooxindoles were obtained via silica gel flash column chromatography using the indicated mobile phase.

General Procedure 6 (GP6):

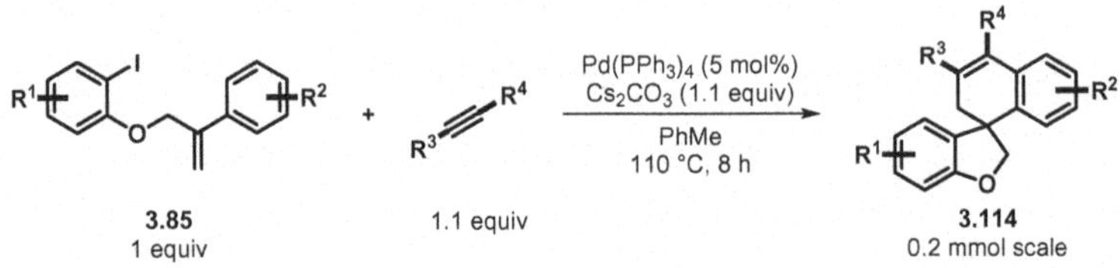


To a flame dried two dram vial under argon atmosphere, aryl iodide 3 (1 equiv), $Pd(PPh_3)_4$ (5 mol% Pd), and $Cs_2CO_3$ (1.1 equiv) were added. PhMe (0.1 M) the alkyne (1.1 equiv) were added in sequence. A Teflon line screw cap was fitted on the two dram vial. The vial was sealed with Teflon tape and placed in a preheated oil bath at 110 °C for 8 h. The reaction mixture was cooled down to room temperature and was filtered through a plug of silica gel using EtOAc. The pure spirooxindoles and spirodihydrobenzofurans were obtained via silica gel flash column chromatography using the indicated mobile phase.

General Procedure 7 (GP7):

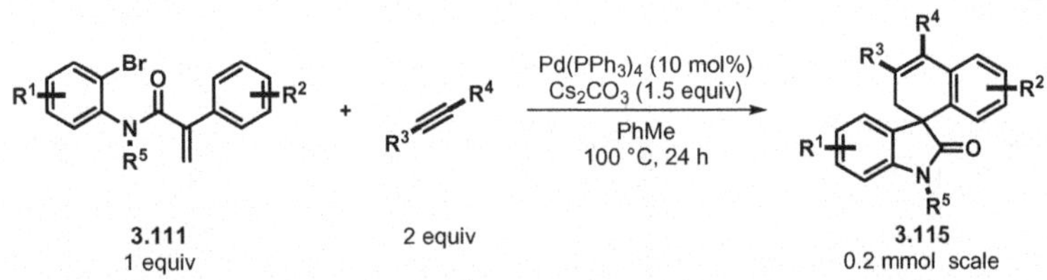


To a flame dried two dram vial under argon atmosphere, bromoacrylamide 1 (1 equiv), $Pd(PPh_3)_4$ (10 mol% Pd), and $Cs_2CO_3$ (1.5 equiv) were added. PhMe (0.1 M) and the alkyne (2 equiv) were added in sequence. A Teflon line screw cap was fitted on the two dram vial. The vial was sealed with Teflon tape and placed in a preheated oil bath at 100 °C for 24 h. The reaction mixture was cooled down to room temperature and was filtered through a plug of silica gel using EtOAc. The pure spirooxindoles were obtained via silica gel flash column chromatography using the indicated mobile phase.

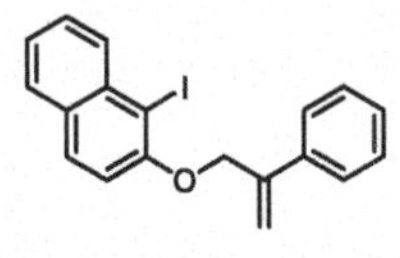


1-iodo-2-((2-phenylallyl)oxy)naphthalene (3.85d)

Prepared according to GP1 with an overall yield of 54%. The aryl iodide was purified by flash column chromatography using hexanes:DCM (10:1 v:v) and was obtained as a beige solid (M.P.: 68–69 °C).

$^1$H NMR (500 MHz, CDCl$_3$) δ 8.24–8.11 (m, 1H), 7.80 (d, $J$ = 9.0 Hz, 1H), 7.78 7.72 (m, 1H), 7.59–7.50 (m, 3H), 7.45–7.31 (m, 4H), 7.22 (d, $J$ = 9.0 Hz, 1H), 5.75–5.61 (m, 2H), 5.11 (t, $J$ = 1.5 Hz, 2H). $^{13}$C NMR (126 MHz, CDCl$_3$) δ 155.7, 142.7, 138.3, 135.7, 131.3, 130.2, 130.1, 128.6, 128.2, 128.1, 128.1, 126.2, 124.5, 115.0, 114.4, 88.7, 71.5. IR (neat film, cm$^{-1}$) 3055, 2893, 2859, 1622, 1593, 1501, 1344, 1329, 1265, 1240, 1150, 1065, 1024, 909, 799, 779, 764, 748, 706. HRMS (DART) calculated 387.02458 m/z (found 387.02389 m/z for $C_{19}H_{16}IO$).

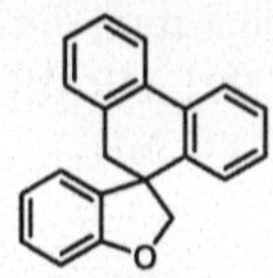


2*H*,10'*H*-spiro[benzofuran-3,9'-phenanthrene] (3.92a)

Prepared according to GP4 using 1-iodo-2-((2-phenylallyl)oxy)benzene (67 mg, 0.2 mmol) and 2-(trimethylsilyl)phenyl trifluoromethanesulfonate (89 mg, 0.3 mmol). The spirodihydrobenzofuran was purified by flash column chromatography using hexanes:DCM (3:1 v:v) and was obtained as a white solid (48 mg, 0.168 mmol, 84%, M.P.: 147–149 °C).

$^1$H NMR (500 MHz, CDCl$_3$) δ 7.89–7.84 (m, 1H), 7.84–7.80 (m, 1H), 7.42–7.34 (m, 2H), 7.32–7.21 (m, 4H), 7.17 (ddd, $J$ = 7.5, 1.5, 0.5 Hz, 1H), 7.09 (ddd, $J$ = 7.5, 1.5, 0.5 Hz, 1H), 6.98 (td, $J$ = 7.5, 1.0 Hz, 1H), 6.95–6.91 (m, 1H), 4.51 (d, $J$ = 9.0 Hz, 1H), 4.23 (dd, $J$ = 9.0, 1.0 Hz, 1H), 3.36 (d, $J$ = 15.5 Hz, 1H), 3.07 (d, $J$ = 15.5 Hz, 1H). $^{13}$C NMR (126 MHz, CDCl$_3$) δ 160.5, 140.9, 134.3, 134.0, 133.9, 132.8, 129.1, 128.9, 128.3, 127.9, 127.8, 127.6, 126.7, 124.5, 124.1, 123.9, 121.1, 110.1, 81.7, 50.5, 41.0. IR (neat film, cm$^{-1}$) 3065, 3030, 3021, 2931, 2880, 1597, 1481, 1460, 1450, 1430, 1319, 1277, 1217, 1017, 988. HRMS (DART) calculated 285.12794 m/z (found 285.12723 m/z for $C_{21}H_{17}O$).

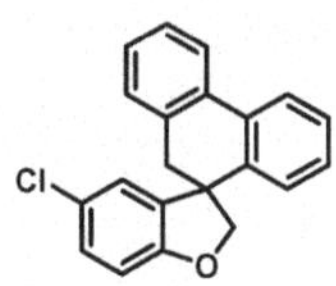


5-chloro-2*H*,10'*H*-spiro[benzofuran-3,9'-phenanthrene] (3.92b)

Prepared according to GP4 using 4-chloro-2-iodo-1-((2-phenylallyl)oxy)benzene (74 mg, 0.2 mmol) and 2-(trimethylsilyl)phenyl trifluoromethanesulfonate (89 mg, 0.3 mmol). The spirodihydrobenzofuran was purified by flash column chromatography using hexanes:DCM (3:1 v:v) and was obtained as a white solid (60 mg, 0.188 mmol, 94%, M.P.: 134–135 °C).

$^1$H NMR (500 MHz, CDCl$_3$) δ 7.85 (dd, $J$ = 8.0, 1.0 Hz, 1H), 7.81 (d, $J$ = 7.5 Hz, 1H), 7.38 (td, $J$ = 7.5, 1.5 Hz, 2H), 7.32–7.22 (m, 3H), 7.21 (dd, $J$ = 8.5, 2.5 Hz, 1H), 7.11 (d, $J$ = 2.5 Hz, 1H), 7.08 (dd, $J$ = 7.5, 1.0 Hz, 1H), 6.84 (d, $J$ = 8.5 Hz, 1H), 4.52 (d, $J$ = 9.0 Hz, 1H), 4.24 (dd, $J$ = 9.0, 1.0 Hz, 1H), 3.32 (d, $J$ = 15.0 Hz, 1H), 3.05 (d, $J$ = 15.0 Hz, 1H). $^{13}$C NMR (126 MHz, CDCl$_3$) δ 159.1, 140.0, 134.9, 134.1, 133.9, 133.5, 129.1, 128.9, 128.4, 128.0, 128.0, 127.8, 126.5, 125.7, 124.6, 124.2, 123.9, 111.1, 82.2, 50.8, 40.9. IR (neat film, cm$^{-1}$) 3067, 3021, 2930, 2882, 1480, 1453, 1256, 1219, 1159, 1090, 1071, 991. HRMS (DART) calculated 336.11552 m/z (found 336.11506 m/z for $C_{21}H_{16}ClO + NH_3$).

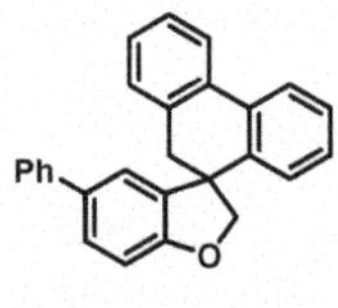


5-phenyl-2*H*,10'*H*-spiro[benzofuran-3,9'-phenanthrene] (3.93c)

Prepared according to GP4 using 3-iodo-4-((2-phenylallyl)oxy)-1,1'-biphenyl (83 mg, 0.2 mmol) and 2-(trimethylsilyl)phenyl trifluoromethanesulfonate (89 mg, 0.3 mmol). The spirodihydrobenzofuran was purified by flash column chromatography using hexanes:DCM (3:1 v:v) and was obtained as a white solid (67 mg, 0.186 mmol, 93%, M.P.: 145–147 °C).

$^1$H NMR (500 MHz, CDCl$_3$) δ 7.88–7.85 (m, 1H), 7.83 (d, $J$ = 7.7 Hz, 1H), 7.57–7.53 (m, 2H), 7.52 (dd, $J$ = 8.5, 2.0 Hz, 1H), 7.44–7.35 (m, 5H), 7.34–7.23 (m, 4H), 7.17 (ddd, $J$ = 7.5, 1.5, 0.5 Hz, 1H), 7.00 (dd, $J$ = 8.5, 0.5 Hz, 1H), 4.57 (d, $J$ = 9.0 Hz, 1H), 4.27 (dd, $J$ = 9.0, 1.0 Hz, 1H), 3.50–3.32 (m, 1H), 3.12 (d, $J$ = 15.5 Hz, 1H). $^{13}$C NMR (126 MHz, CDCl$_3$) δ 160.2, 141.1, 140.6, 134.6, 134.3, 134.0, 134.0, 133.5, 129.1, 128.7, 128.3, 128.0, 127.9, 127.9, 127.7, 126.8, 126.8, 126.6, 124.1, 123.9, 123.2, 110.2, 82.1, 50.7, 41.1. IR (neat film, cm$^{-1}$) 3065, 3030, 2932, 2880, 1611, 1601, 1480, 1452, 1256, 1219, 1123, 1088, 1038, 990. HRMS (DART) calculated 378.18579 m/z (found 378.18564 m/z for $C_{27}H_{21}O + NH_3$).

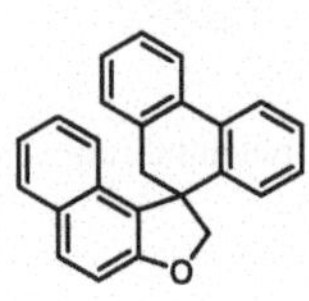


2*H*,10'*H*-spiro[naphtho[2,1-*b*]furan-1,9'-phenanthrene] (3.92d)

Prepared according to GP4 using 1-iodo-2-((2-phenylallyl)oxy)naphthalene (77 mg, 0.2 mmol) and 2-(trimethylsilyl)phenyl trifluoromethanesulfonate (89 mg, 0.3 mmol). The spirodihydrobenzofuran was purified by flash column chromatography using hexanes:DCM (3:1 v:v) and was obtained as a white solid (45 mg, 0.134 mmol, 67%, M.P.: 182–183 °C).

$^1$H NMR (500 MHz, $CDCl_3$) δ 7.94–7.90 (m, 1H), 7.90–7.87 (m, 1H), 7.87–7.84 (m, 1H), 7.84–7.81 (m, 1H), 7.50–7.45 (m, 1H), 7.42–7.35 (m, 2H), 7.31–7.25 (m, 3H), 7.24–7.20 (m, 2H), 7.20–7.15 (m, 1H), 7.12 (ddd, $J$ = 8.0, 1.5, 0.5 Hz, 1H), 4.73 (d, $J$ = 9.0 Hz, 1H), 4.25 (dd, $J$ = 9.0, 1.5 Hz, 1H), 3.93–3.80 (m, 1H), 3.06 (d, $J$ = 15.5 Hz, 1H). $^{13}$C NMR (126 MHz, $CDCl_3$) δ 158.5, 139.8, 134.3, 134.1, 133.8, 130.7, 130.4, 130.1, 129.4, 129.4, 128.3, 128.1, 127.9, 127.8, 127.6, 126.3, 124.0, 123.9, 123.8, 122.7, 122.0, 112.5, 81.9, 52.2, 39.0. IR (neat film, $cm^{-1}$) 3067, 3021, 2930, 2882, 1480, 1453, 1256, 1219, 1159, 1090, 1071, 991. HRMS (DART) calculated 335.14359 m/z (found 335.14353 m/z for $C_{25}H_{19}O$).

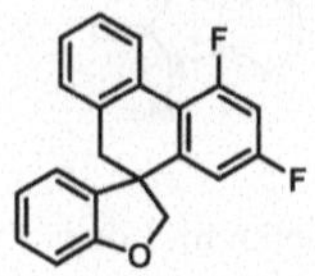


5',7'-difluoro-2*H*,10'*H*-spiro[benzofuran-3,9'-phenanthrene] (3.92e)

Prepared according to GP4 using 1,3-difluoro-5-(3-(2-iodophenoxy)prop-1-en-2-yl)benzene (74 mg, 0.2 mmol) and 2-(trimethylsilyl)phenyl trifluoromethanesulfonate (89 mg, 0.3 mmol). The spirodihydrobenzofuran was purified by flash column chromatography using hexanes:DCM (2:1 v:v) and was obtained as a white solid (60 mg, 0.188 mmol, 94%, M.P.: 98–100 °C).

$^1$H NMR (500 MHz, $CDCl_3$) δ 8.04 (d, $J$ = 8.0 Hz, 1H), 7.42–7.33 (m, 1H), 7.33–7.23 (m, 3H), 7.16–7.08 (m, 1H), 7.02–6.96 (m, 1H), 6.95–6.89 (m, 1H), 6.84 (ddd, $J$ = 11.0, 8.5, 2.5 Hz, 1H), 6.59 (ddd, $J$ = 9.0, 2.5, 1.0 Hz, 1H), 4.41 (d, $J$ = 9.0 Hz, 1H), 4.18 (dd, $J$ = 9.0, 0.5 Hz, 1H), 3.35 (d, $J$ = 15.0 Hz, 1H), 3.02 (d, $J$ = 15.0 Hz, 1H). $^{13}$C NMR (126 MHz, $CDCl_3$) δ 161.7 (dd, $J$ = 250.0, 13.5 Hz), 160.7 (dd, $J$ = 254.0, 12.0 Hz), 160.4, 146.2 (dd, $J$ = 8.0, 4.5 Hz), 133.9 (d, $J$ = 1.0 Hz), 131.4, 129.9 (dd, $J$ = 3.5, 1.0 Hz), 129.4, 129.0, 128.1, 127.7 (d, $J$ = 10.0 Hz), 127.6 (d, $J$ = 7.0 Hz), 124.4, 121.3, 118.5 (dd, $J$ = 10.5, 4.0 Hz), 110.36, 109.8 (dd, $J$ = 22.5, 3.5 Hz), 103.8 (dd, $J$ = 28.0, 25.0 Hz), 81.1, 50.9, 40.8. $^{19}$F NMR (470 MHz, $CDCl_3$) δ -109.76 (q, $J$ = 8.5 Hz), -110.47–-110.57 (m). IR (neat film, $cm^{-1}$) 3084, 3067, 3034, 2932, 2882, 1620, 1593, 1481, 1458, 1449, 1427, 1323, 1246, 1213, 1117, 1028, 976. HRMS (DART) calculated 321.10910 m/z (found 321.10864 m/z for $C_{21}H_{15}F_2O$).

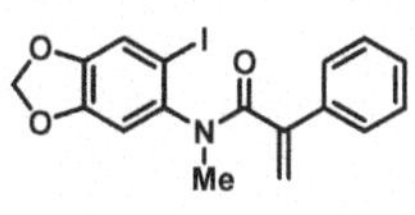


*N*-(6-iodobenzo[*d*][1, 3]dioxol-5-yl)-*N*-methyl-2-phenylacrylamide (3.88 h)

Prepared according to GP2 and GP3 with an overall yield of 13%. The *N*-substituted acrylamide was purified by flash column chromatography using hexanes:EtOAc (2:1 v:v) and was obtained as a clear yellow oil.

*Three rotamers were observed in a 5:1:1 ratio. The major rotamer is reported below.*

$^1$H NMR (500 MHz, $CDCl_3$) δ 7.25–7.20 (m, 3H), 7.17–7.11 (m, 2H), 7.09 (s, 1H), 6.18 (s, 1H), 5.94–5.81 (m, 2H), 5.66 (d, *J* = 0.5 Hz, 1H), 5.38 (d, *J* = 0.5 Hz, 1H), 3.22 (s, 3H). $^{13}$C NMR (126 MHz, $CDCl_3$) δ 170.4, 148.2, 147.7, 145.7, 139.1, 137.2, 128.3, 128.0, 126.1, 117.9, 117.8, 109.8, 102.2, 88.3, 36.3. IR (neat film, cm$^{-1}$) 3007, 2903, 1636, 1507, 1476, 1368, 1225, 1140, 1225, 1036, 930, 862, 756. HRMS (DART) calculated 408.00897 m/z (found 408.00966 m/z for $C_{17}H_{15}INO_3$).

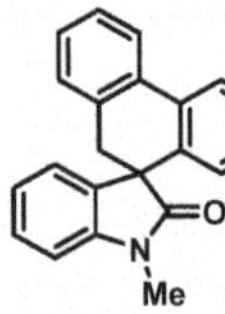


1-methyl-10'*H*-spiro[indoline-3,9'-phenanthren]-2-one (3.94a)

Prepared according to GP5 using *N*-(2-iodophenyl)-*N*-methyl-2-phenylacrylamide (73 mg, 0.2 mmol) and 2-(trimethylsilyl)phenyl trifluoromethane-sulfonate (89 mg, 0.3 mmol). The spirooxindole was purified by flash column chromatography using hexanes:DCM:EtOAc (10:5:1 v:v:v) and was obtained as a white solid (60 mg, 0.193 mmol, 96%, M.P.: 186–187 °C).

$^1$H NMR (500 MHz, $CDCl_3$) δ 7.93–7.90 (m, 1H), 7.90–7.86 (m, 1H), 7.45–7.39 (m, 1H), 7.35 (ddd, *J* = 8.0, 7.5, 1.5 Hz, 1H), 7.27 (td, *J* = 7.5, 1.0 Hz, 1H), 7.21 (td, *J* = 7.5, 1.5 Hz, 1H), 7.18 (td, *J* = 7.5, 1.5 Hz, 1H), 7.16–7.14 (m, 1H), 6.94–6.85 (m, 2H), 6.74 (td, *J* = 7.5, 1.0 Hz, 1H), 6.49 (ddd, *J* = 7.5, 1.5, 0.5 Hz, 1H), 3.68 (d, *J* = 15.0 Hz, 1H), 3.38 (s, 3H), 2.90 (d, *J* = 15.0 Hz, 1H). $^{13}$C NMR (126 MHz, $CDCl_3$) δ 179.2, 141.9, 136.8, 134.3, 134.1, 133.1, 133.0, 129.2, 128.4, 128.3, 128.2, 127.9, 127.7, 126.3, 124.4, 123.6, 123.5, 122.7, 108.3, 52.9, 38.5, 26.6. IR (neat film, cm$^{-1}$) 3069, 3051, 2957, 2929, 1705, 1607, 1490, 1469, 1373, 1343, 1249, 1091, 956. HRMS (DART) calculated 312.13884 m/z (found 312.13809 m/z for $C_{22}H_{18}$-NO).

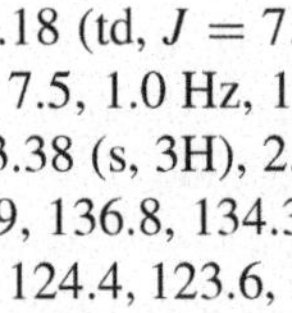


5-bromo-1-methyl-10'*H*-spiro[indoline-3,9'-phenanthren]-2-one (3.94b)

Prepared according to GP5 using *N*-(4-bromo-2-iodophenyl)-*N*-methyl-2-phenylacrylamide (88 mg, 0.2 mmol) and 2-(trimethylsilyl)phenyl trifluoromethanesulfonate (89 mg, 0.3 mmol). The spirooxindole was purified by flash column chromatography using hexanes:DCM:EtOAc (14:6:1 v:v:v) and was obtained as a white solid (69 mg, 0.178 mmol, 89%, M.P.: 170–172 °C).

$^1$H NMR (500 MHz, $CDCl_3$) δ 7.94–7.90 (m, 1H), 7.90–7.86 (m, 1H), 7.46–7.41 (m, 1H), 7.38 (ddd, *J* = 8.0, 7.5, 1.5 Hz, 1H), 7.34 (dd, *J* = 8.5, 2.0 Hz, 1H), 7.29 (td, *J* = 7.5, 1.0 Hz, 1H), 7.22–7.17 (m, 1H), 7.17–7.14 (m, 1H), 6.88 (ddd, *J* = 8.0, 1.5, 0.5 Hz, 1H), 6.77 (d, *J* = 8.5 Hz, 1H), 6.58 (dd, *J* = 2.0, 0.5 Hz, 1H), 3.65 (d, *J* = 15.0 Hz, 1H), 3.34 (s, 3H), 2.89 (d, *J* = 15.0 Hz, 1H). $^{13}$C NMR (126 MHz, $CDCl_3$) δ 178.6, 141.1, 136.0, 135.0, 134.1, 134.0, 132.3, 131.1, 129.2, 128.6, 128.4, 128.2, 128.0, 126.8, 126.2, 124.6, 123.8, 115.3, 109.7, 53.0, 38.5, 26.7. IR (neat film, $cm^{-1}$) 3110, 3079, 3063, 3027, 2932, 2854, 1712, 1604, 1483, 1341, 1266, 1099, 959, 806, 770, 743, 727. HRMS (DART) calculated 390.04935 m/z (found 390.04842 m/z for $C_{22}H_{18} \cdot BrNO$).

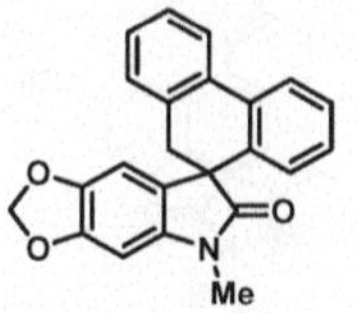


5'-methyl-10*H*-spiro[phenanthrene-9,7'-[1, 3]dioxolo[4,5-*f*]indol]-6'(5'*H*)-one (3.94 h)

Prepared according to GP5 using *N*-(6-iodobenzo[d][1, 3]dioxol-5-yl)-*N*-methyl-2-phenylacrylamide (81 mg, 0.2 mmol) and 2-(trimethylsilyl)phenyl trifluoromethanesulfonate (89 mg, 0.3 mmol). The spirooxindole was purified by flash column chromatography using hexanes:DCM:EtOAc (5:5:1 v:v:v) and was obtained as a white solid (60 mg, 0.168 mmol, 84%, M.P.:163–164 °C).

$^1$H NMR (500 MHz, $CDCl_3$) δ 7.92–7.88 (m, 1H), 7.88–7.83 (m, 1H), 7.43–7.38 (m, 1H), 7.35 (ddd, *J* = 8.0, 7.5, 1.5 Hz, 1H), 7.27 (td, *J* = 7.5, 1.0 Hz, 1H), 7.19 (td, *J* = 7.5, 1.5 Hz, 1H), 7.17–7.14 (m, 1H), 6.91 (ddd, *J* = 7.5, 1.5, 0.5 Hz, 1H), 6.54–6.47 (m, 1H), 5.99 (d, *J* = 0.5 Hz, 1H), 5.79 (dd, *J* = 13.5, 1.5 Hz, 2H), 3.66 (d, *J* = 15.0 Hz, 1H), 3.33 (s, 3H), 2.85 (d, *J* = 15.0 Hz, 1H). $^{13}$C NMR (126 MHz, $CDCl_3$) δ 179.5, 147.4, 142.9, 137.0, 136.3, 134.1, 133.9, 132.9, 129.2, 128.5, 128.3, 128.0, 127.8, 126.2, 125.1, 124.4, 123.7, 105.4, 100.9, 92.1, 53.1, 38.7, 26.8. IR (neat film, $cm^{-1}$) 3067, 2928, 2890, 1711, 1625, 1474, 1422, 1379, 1337, 1229, 1120, 1029, 923. HRMS (DART) calculated 356.12867 m/z (found 356.12865 m/z for $C_{23}H_{18}NO_3$).

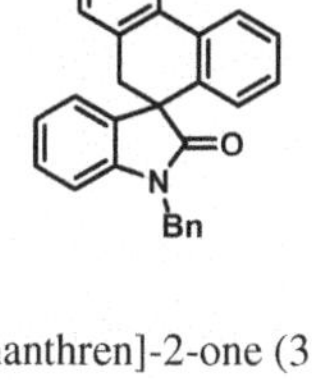


1-benzyl-10'*H*-spiro[indoline-3,9'-phenanthren]-2-one (3.94 l)

Prepared according to GP5 using *N*-benzyl-*N*-(2-iodophenyl)-2-phenylacrylamide (78.6 mg, 0.2 mmol) and 2-(trimethylsilyl)phenyl trifluoromethanesulfonate (89.05 mg, 0.3 mmol). The spirooxindole was purified by flash column chromatography using hexanes:DCM:EtOAc (10:10:2 v:v:v) and was obtained as a white solid (72 mg, 0.186 mmol, 93%, M.P.: 175–176 °C).

$^1$H NMR (500 MHz, $CDCl_3$) δ 7.95–7.92 (m, 1H), 7.92–7.87 (m, 1H), 7.47–7.35 (m, 6H), 7.35–7.27 (m, 2H), 7.24–7.15 (m, 2H), 7.09 (td, $J$ = 8.0, 1.5 Hz, 1H), 6.97 (ddd, $J$ = 7.5, 1.5, 0.5 Hz, 1H), 6.85–6.77 (m, 1H), 6.71 (td, $J$ = 7.5, 1.0 Hz, 1H), 6.51 (ddd, $J$ = 7.5, 1.5, 0.5 Hz, 1H), 5.22–4.99 (m, 2H), 3.75 (d, $J$ = 15.0 Hz, 1H), 2.97 (d, $J$ = 15.0 Hz, 1H). $^{13}$C NMR (126 MHz, $CDCl_3$) δ 179.3, 141.1, 136.9, 136.0, 134.3, 134.1, 133.1, 132.9, 129.2, 128.9, 128.5, 128.3, 128.1, 128.0, 127.8, 127.7, 127.4, 126.4, 124.5, 123.7, 123.6, 122.7, 109.4, 52.9, 44.0, 38.9. IR (neat film, $cm^{-1}$) 3060, 3020, 2923, 2888, 2864, 2831, 1708, 1605, 1484, 1466, 1347, 1197, 1184, 1014, 946. HRMS (DART) calculated 388.17014 m/z (found 388.17053 m/z for $C_{28}H_{22}NO$).

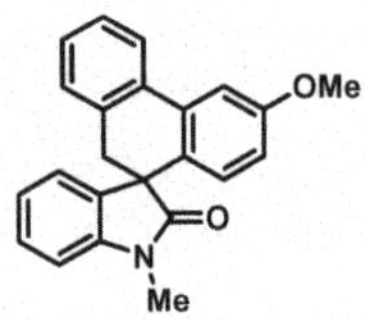


6'-methoxy-1-methyl-10'*H*-spiro[indoline-3,9'-phenanthren]-2-one (3.94q)

Prepared according to GP5 using *N*-(2-iodophenyl)-2-(4-methoxyphenyl)-*N*-methylacrylamide (79 mg, 0.2 mmol) and 2-(trimethylsilyl)phenyl trifluoromethanesulfonate (89 mg, 0.3 mmol). The spirooxindole was purified by flash column chromatography using hexanes:DCM:EtOAc (10:10:1 v:v:v) and was obtained as a white solid (63 mg, 0.184 mmol, 92%, M.P.: 83–84 °C).

$^1$H NMR (500 MHz, $CDCl_3$) δ 7.90–7.81 (m, 1H), 7.48–7.38 (m, 2H), 7.32–7.24 (m, 1H), 7.20 (td, $J$ = 7.5, 1.5 Hz, 1H), 7.17–7.13 (m, 1H), 6.91–6.86 (m, 1H), 6.85–6.81 (m, 1H), 6.79–6.71 (m, 2H), 6.49 (ddd, $J$ = 7.5, 1.5, 0.5 Hz, 1H), 3.85 (s, 3H), 3.63 (d, $J$ = 15.0 Hz, 1H), 3.36 (s, 3H), 2.90 (d, $J$ = 15.0 Hz, 1H). $^{13}$C NMR (126 MHz, $CDCl_3$) δ 179.4, 159.6, 141.9, 135.4, 134.3, 133.4, 133.3, 129.2, 129.2, 128.1 (2 overlapping signals), 127.7, 127.4, 123.6, 123.5, 122.7, 113.6, 110.2, 108.3, 55.4, 52.3, 38.9, 26.6. IR (neat film, $cm^{-1}$) 3059, 2936, 2833, 1706, 1606, 1565, 1491, 1468, 1371, 1345, 1299, 1219, 1127, 1087, 1021, 810. HRMS (DART) calculated 342.14940 m/z (found 342.14937 m/z for $C_{23}H_{20}NO_2$).

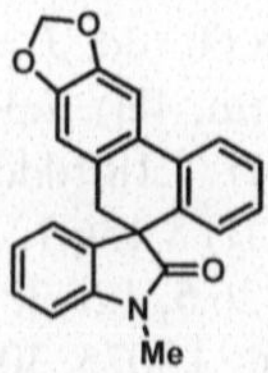

1-methyl-6'*H*-spiro[indoline-3,5'-phenanthro[2,3-*d*][1, 3]dioxol]-2-one (3.94t)

Prepared according to GP5 using *N*-(2-iodophenyl)-*N*-methyl-2-phenylacrylamide (79 mg, 0.2 mmol) and 6-(trimethylsilyl)benzo[*d*][1, 3]dioxol-5-yl trifluoromethanesulfonate (103 mg, 0.3 mmol). The spirooxindole was purified by flash column chromatography using hexanes:DCM:EtOAc (10:10:2 v:v:v) and was obtained as a white solid (58 mg, 0.164 mmol, 82%, M.P.: 176–178 °C).

$^1$H NMR (500 MHz, $CDCl_3$) δ 7.76–7.68 (m, 1H), 7.36 (s, 1H), 7.31 (ddd, *J* = 8.0, 7.5, 1.5 Hz, 1H), 7.22 (td, *J* = 7.5, 1.5 Hz, 1H), 7.12 (td, *J* = 7.5, 1.5 Hz, 1H), 6.92–6.88 (m, 1H), 6.87 (ddd, *J* = 7.5 1.5, 0.5 Hz, 1H), 6.79 (td, *J* = 7.5, 1.0 Hz, 1H), 6.68–6.57 (m, 2H), 6.03 (d, *J* = 1.5 Hz, 1H), 5.99 (d, *J* = 1.5 Hz, 1H), 3.57 (dd, *J* = 15.0, 1.0 Hz, 1H), 3.36 (s, 3H), 2.79 (d, *J* = 15.0 Hz, 1H). $^{13}$C NMR (126 MHz, $CDCl_3$) δ 179.2, 147.6, 147.2, 142.0, 136.1, 134.2, 133.1, 128.3, 128.2, 128.2, 127.7, 127.0, 126.2, 123.9, 123.6, 122.7, 109.5, 108.3, 104.3, 101.1, 52.9, 38.6, 26.6. IR (neat film, $cm^{-1}$) 3055, 3011, 2891, 2359, 2342, 2332, 1705, 1609, 1483, 1373, 1221, 1167, 1132, 1090, 1038, 934, 752. HRMS (DART) calculated 356.12867 m/z (found 356.12875 m/z for $C_{23}H_{18}NO_3$).

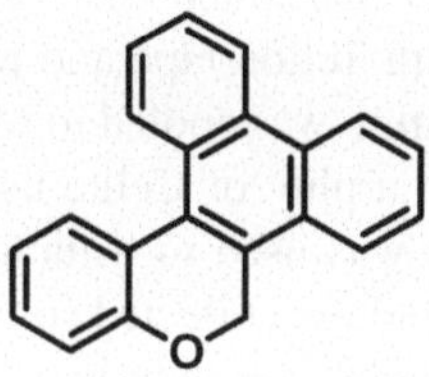

5*H*-phenanthro[9,10-*c*]chromene (3.95)

To a flame dried 1 dram vial cooled under argon, 3.92a (57 mg, 1 equiv), *N*-bromosuccinimide (39 mg, 1.1 equiv) and dibenzoylperoxide (~5 mg) were added and allowed to purge for 10 min. Benzene (0.4 mL, 0.5 M) was added. The aryne precursor (1.5 equiv) was added via a microsyringe. A Teflon line screw cap was fitted on the two dram vial, sealed with Teflon tape and place in a preheated oil bath at 100 °C for 16 h. The reaction mixture was cooled to room temperature. Once cooled, the reaction was passed through a plug of silica gel using EtOAc. 3.95 was purified by flash column chromatography using hexanes:$Et_2O$ (10:1 v:v) and was obtained as a white solid (23 mg, 0.08 mmol, 41%, M.P.: 78–79 °C).

$^{1}$H NMR (500 MHz, $CDCl_3$) δ 8.60 (dd, $J$ = 8.0, 1.5 Hz, 2H), 8.01–7.93 (m, 1H), 7.92–7.84 (m, 1H), 7.75–7.58 (m, 4H), 7.34 (ddd, $J$ = 8.0, 7.5, 1.5 Hz, 1H), 7.23 (ddd, $J$ = 8.0, 1.5, 0.5 Hz, 1H), 7.18 (ddd, $J$ = 8.0, 7.5, 1.5 Hz, 1H), 5.53 (s, 2H). $^{13}$C NMR (126 MHz, $CDCl_3$) δ 156.1, 131.1, 129.7, 129.7, 128.6, 128.2, 127.9, 127.8, 127.2, 126.8, 126.7, 126.5, 126.3, 125.8, 124.2, 123.3, 123.2, 122.5, 121.7, 117.4, 66.6. IR (neat film, cm$^{-1}$) 3075, 3030, 2837, 1595, 1481, 1460, 1265, 1244, 1215, 1113, 1062, 1049, 1036, 1001, 812, 752, 723. HRMS (DART) calculated 283.11229 m/z (found 283.11154 m/z for $C_{21}H_{15}O$).

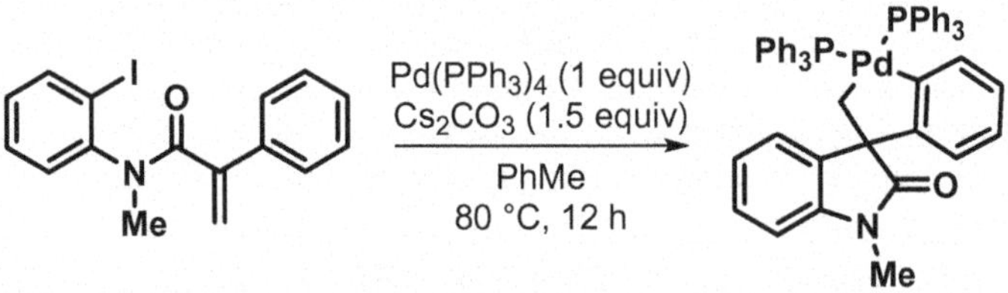


Palladacycle (3.99)

**Note:** All DCM and $CHCl_3$/$CDCl_3$ were passed through a plug of basic alumina prior to use.

To a flame dried 2 dram vial cooled under argon, *N*-(2-iodophenyl)-*N*-methyl-2-phenylacrylamide (3.88a) (73 mg, 0.2 mmol, 1 equiv), $Pd(PPh_3)_4$ (231 mg, 0.2 mmol, 1 equiv), and $Cs_2CO_3$ (98 mg, 0.3 mmol, 1.5 equiv) were added and allowed to purge for 10 min. PhMe (2 mL, 0.1 M) was added and a Teflon line screw cap was fitted on the two dram vial, sealed with Teflon tape and place in a preheated oil bath at 80 °C for 12 h. The reaction mixture was cooled to room temperature. Once cooled, the reaction was passed through a plug of Celite using DCM and concentrated *in vacuo*. Once solidified, hexanes was used to triturate the compound. The mixture was passed through glass wool and the collected solid was redissolved in DCM and concentrated *in vacuo*. The palladacycle then recrystallized $Et_2O$ and hexanes to obtain a pale yellow solid (87 mg, 0.1 mmol, 50%).

$^1$H NMR (500 MHz, $CDCl_3$) δ 7.65–7.57 (m, 7H), 7.34–7.27 (m, 6H), 7.27–7.17 (m, 7H), 7.17–7.10 (m, 6H), 7.09–7.03 (m, 6H), 7.01 (td, $J$ = 7.5, 1.0 Hz, 1H), 6.88–6.83 (m, 1H), 6.82–6.79 (m, 1H), 6.65 (td, $J$ = 7.5, 1.0 Hz, 1H), 6.44 (ddd, $J$ = 7.5, 2.5, 1.5 Hz, 1H), 6.40–6.33 (m, 1H), 3.26 (s, 3H), 2.19–2.06 (m, 2H). $^{13}$C NMR (126 MHz, $CDCl_3$) δ 181.6 (dd, $J$ = 5.0, 1.0 Hz), 170.5 (dd, $J$ = 113.0, 11.5 Hz), 158.8 (t, $J$ = 3.5 Hz), 143.0, 140.7 (dd, $J$ = 11.0, 4.0 Hz), 138.8 (dd, $J$ = 5.0, 0.5 Hz), 134.8 (d, $J$ = 13.5 Hz), 134.4 (dd, $J$ = 30.5, 2.0 Hz), 134.2 (d, $J$ = 12.5 Hz), 133.4 (dd, $J$ = 32.5, 2.0 Hz), 129.3 (d, $J$ = 2.0 Hz), 129.2 (d, $J$ = 2.0 Hz), 127.9 (d, $J$ = 9.5 Hz), 127.8 (d, $J$ = 9.5 Hz), 126.5, 124.5, 124.0 (dd, $J$ = 8.5, 3.0 Hz), 122.9, 122.4 (d, $J$ = 3.0 Hz), 121.8, 107.1, 68.9 (dd, $J$ = 7.5, 5.0 Hz), 47.5 (dd, $J$ = 90.5, 8.0 Hz), 26.3. $^{31}$P NMR (202 MHz, $CDCl_3$) δ 25.15 (dd, $J$ = 178.0, 22.5 Hz). IR (neat film, $cm^{-1}$) 3051, 3003, 2886, 1699, 1653, 1608, 1479, 1470, 1435, 1343, 1217, 1096, 752. HRMS (DART) calculated 862.1927 m/z (found 862.1949 m/z for $C_{52}H_{44}NOP_2Pd$).

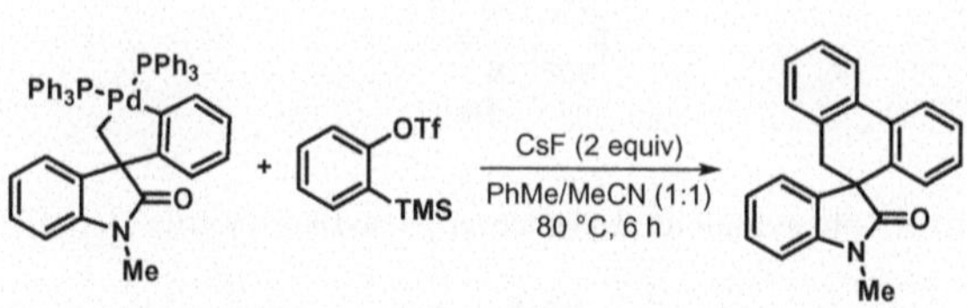


Mechanistic Study Probing Pathway 1

To a flame dried 2 dram vial cooled under argon, palladacycle 3.99 (43 mg, 0.05 mmol, 1 equiv), CsF (15 mg, 0.3 mmol, 1.5 equiv) were added and allowed to purge for 10 min. PhMe:MeCN (1:1) (2 mL, 0.1 M) and 2-(trimethylsilyl)phenyl trifluoromethanesulfonate (22 mg, 0.075 mmol) was added and a Teflon line screw cap was fitted on the two dram vial, sealed with Teflon tape and place in a preheated oil bath at 80 °C for 6 h. The reaction mixture was cooled to room temperature. Once cooled, the reaction was passed through a plug of Celite using DCM and concentrated *in vacuo*. Once solidified, hexanes was used to triturate the compound. The reaction mixture was cooled to room temperature. Once cooled, the reaction was passed through a plug of silica gel using EtOAc. NMR yield was taken using 1,3,5-trimethoxybenzene as the internal standard.

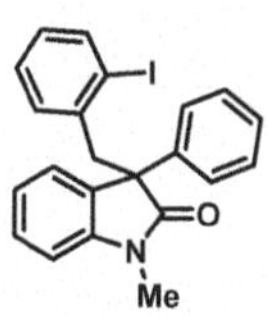


3-(2-iodobenzyl)-1-methyl-3-phenylindolin-2-one (3.103)

*Two rotamers were observed in a 12.1:1 ratio. The major rotamer is reported below.*

$^1$H NMR (500 MHz, $CDCl_3$) δ 7.65 (dd, $J$ = 8.0, 1.5 Hz, 1H), 7.57–7.52 (m, 2H), 7.39–7.33 (m, 2H), 7.32–7.27 (m, 1H), 7.24 (td, $J$ = 7.5, 1.5 Hz, 1H), 7.18 (ddd, $J$ = 7.5, 1.0, 0.5 Hz, 1H), 7.06–6.96 (m, 2H), 6.91 (dd, $J$ = 8.0, 1.5 Hz, 1H), 6.75 (ddd, $J$ = 8.0, 7.5, 1.5 Hz, 1H), 6.73–6.70 (m, 1H), 3.96–3.76 (m, 2H), 3.15 (s, 3H). $^{13}$C NMR (126 MHz, $CDCl_3$) δ 178.0, 143.5, 139.5, 139.4, 139.4, 129.7, 129.2, 128.6, 128.6, 128.4, 128.3, 127.6, 127.5, 127.4, 127.3, 127.0, 122.1, 122.1, 107.9, 103.7, 57.6, 47.3, 26.4. IR (neat film, cm$^{-1}$) 3055, 3015, 2932, 1713, 1699, 1611, 1495, 1373, 1352, 1258, 1130, 1092, 1080, 1013, 750, 729. HRMS (DART) calculated 440.05113 m/z (found 440.05038 m/z for $C_{22}H_{19}INO$). M.P.: 70–72 °C

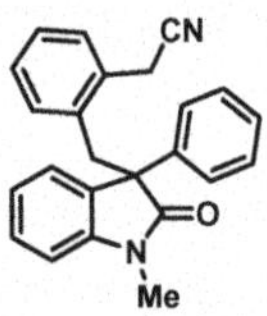


2-(2-((1-methyl-2-oxo-3-phenylindolin-3-yl)methyl)phenyl)acetonitrile (3.104)

Prepared according to GP5 using 3-(2-iodobenzyl)-1-methyl-3-phenylindolin-2-one (88 mg, 0.2 mmol). The product was purified by flash column chromatography using hexanes:EtOAc (4:1 v:v) and was obtained as a white foam (37 mg, 0.102 mmol, 52%).

$^1$H NMR (500 MHz, $CDCl_3$) δ 7.56–7.48 (m, 2H), 7.37 (m, 2H), 7.33–7.27 (m, 3H), 7.18 (m, 1H), 7.16–7.09 (m, 2H), 6.92 (m, 1H), 6.72 (d, $J$ = 8.0 Hz, 1H), 6.58 (d, $J$ = 8.0 Hz, 1H), 3.78 (d, $J$ = 18.5 Hz, 1H), 3.71–3.67 (m, 2H), 3.56 (d, $J$ = 14.0 Hz, 1H), 2.94 (s, 3H). $^{13}$C NMR (126 MHz, $CDCl_3$) δ 177.8, 143.7, 139.3, 133.9, 130.9, 130.8, 129.7, 128.8, 128.8, 128.8, 127.7, 127.7, 127.2, 127.2, 125.4, 122.6, 118.4, 108.5, 57.9, 39.6, 26.2, 21.8. IR (neat film, cm$^{-1}$) 3057, 3024, 2934, 2247, 1701, 1611, 1493, 1373, 1352, 1258, 1130, 1080, 756. HRMS (DART) calculated 353.1648 m/z (found 353.1646 m/z for $C_{24}H_{21}N_2O$).

KIE Experiment:

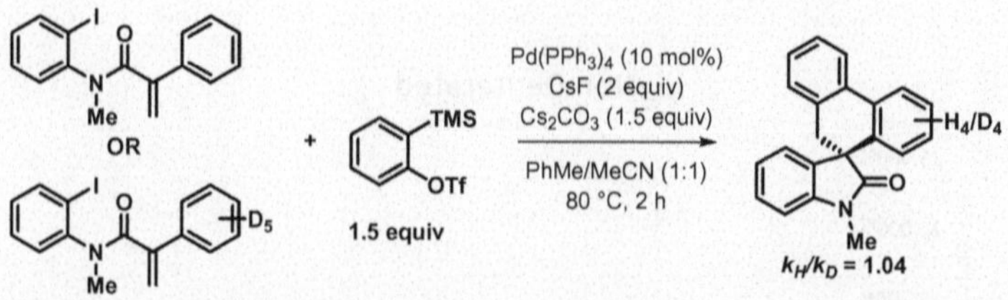


The reaction was run on 0.4 mmol scale. To a flame dried 3 dram vial cooled under argon, 3.88a or 3.88a'($D_5$) (1 equiv), $Pd(PPh_3)_4$ (10 mol% Pd), CsF (2 equiv), $Cs_2CO_3$ (1.5 equiv), and 1,3,5-trimethoxybenzene (~10 mg) were added and allowed to purge for 10 min. PhMe/MeCN (1:1) (0.1 M) was added. The aryne precursor (1.5 equiv) was added via a microsyringe. A septa with an argon balloon was fitted on the 3 dram vial. The vial was placed in a preheated oil bath at 80 °C for 2 h. Every 10 min during the first hour, approximately 100 μL aliquots were removed via a long needle and syringe. Every 15 min during the second hour, approximately 100 μL aliquots were removed via a long needle and syringe. The aliquots quenched with water extracted with EtOAc (3x). The combined organic layers were washed with brine, dried over $Na_2SO_4$ and concentrated *in vacuo*.

**Note:** An induction period was observed in both the deuterated and the non-deuterated substrate.

| Datapoints for 3.94a (Non deuterated) | |
|---|---|
| Time (min) | Yield of product (%) |
| 10 | 0 |
| 20 | 0 |
| 30 | 2.475 |
| 40 | 3.960 |
| 50 | 5.445 |
| 60 | 7.920 |
| 75 | 11.384 |
| 90 | 15.344 |
| 105 | 18.314 |
| 120 | 22.769 |

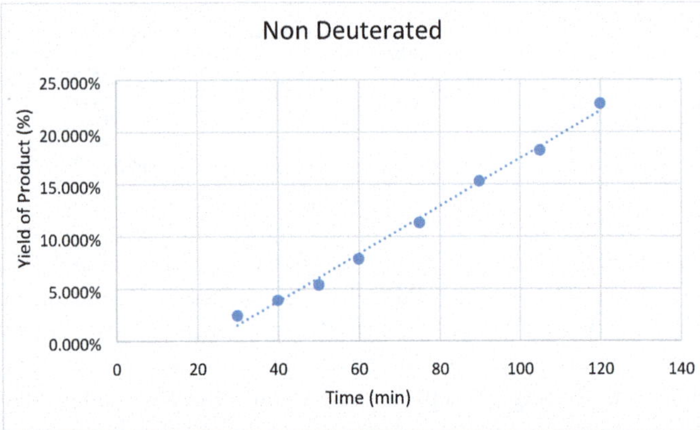


| Datapoints for 3.94a' ($D_5$) | |
|---|---|
| Time (min) | Yield of product (%) |
| 10 | 0 |
| 20 | 0 |
| 30 | 3.971 |
| 40 | 5.991 |
| 50 | 8.430 |
| 60 | 11.363 |
| 75 | 14.185 |
| 90 | 17.510 |
| 105 | 20.564 |
| 120 | 24.126 |

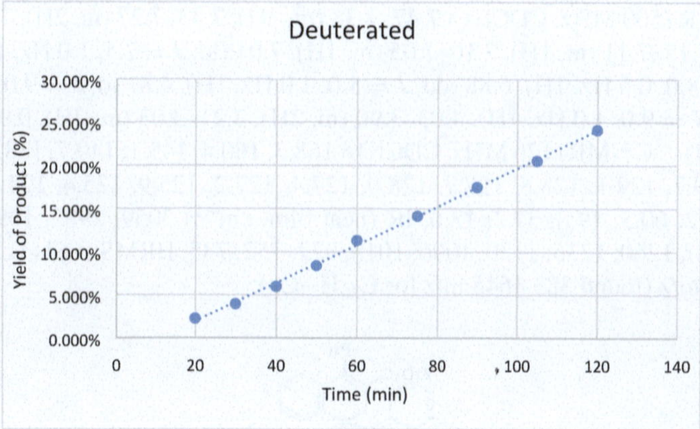


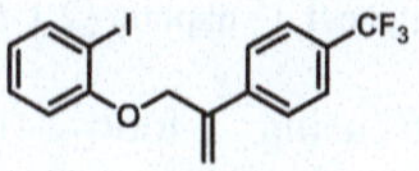

1-iodo-2-((2-(4-(trifluoromethyl)phenyl)allyl)oxy)benzene (3.85 k)

Prepared according to GP1. The aryl iodide was purified by flash column chromatography using hexanes:DCM (3:1 v:v) and was obtained as clear colourless oil (531 mg, 1.31 mmol, 40%).

$^1$H NMR (500 MHz, $CDCl_3$) δ 7.80 (dd, $J = 8.0$, 1.5 Hz, 1H), 7.68–7.56 (m, 4H), 7.31 (ddd, $J = 8.0, 7.5, 1.5$ Hz, 1H), 6.88 (dd, $J = 8.0$, 1.5 Hz, 1H), 6.75 (td, $J = 7.5$, 1.5 Hz, 1H), 5.74 (s, 1H), 5.72 (s, 1H), 4.94 (s, 3H). $^{13}$C NMR (126 MHz, $CDCl_3$) δ 156.9, 141.8 (q, $J$ = 1.5 Hz), 141.5, 139.7, 130.0 (q, $J$ = 32.5 Hz), 129.4, 126.6, 124.2 (q, $J$ = 272.0 Hz, overlapping 125.4), 125.4 (q, $J$ = 4.0 Hz), 123.1, 117.2, 112.5, 86.6, 70.4. $^{19}$F NMR (470 MHz, $CDCl_3$) δ -62.6. IR (neat film, cm$^{-1}$) 3061, 2931, 2870, 1619, 1582, 1570, 1470, 1439, 1327, 1246, 1167, 1121, 1053, 1015. HRMS (DART) calculated 404.99632 m/z (found 404.99589 m/z for $C_{16}H_{13}F_3IO$).

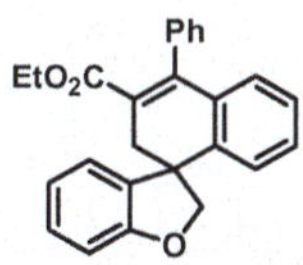

Ethyl 4'-phenyl-2*H*,2'*H*-spiro[benzofuran-3,1'-naphthalene]-3'-carboxylate (3.114a)

Prepared according to GP6 using 1-iodo-2-((2-phenylallyl)oxy)benzene (67 mg, 0.2 mmol) and ethyl 3-phenylpropiolate (38 mg, 0.22 mmol). The spirodihydrobenzofuran was purified by flash column chromatography using hexanes:DCM (1:6 v:v) and was obtained as a clear colourless oil (67 mg, 0.176 mmol, 88%, >20:1 rr).

$^{1}$H NMR (500 MHz, $CDCl_3$) δ 7.47–7.38 (m, 3H), 7.33–7.27 (m, 2H), 7.27–7.20 (m, 3H), 7.17–7.11 (m, 1H), 7.10–7.05 (m, 1H), 7.01 (td, $J$ = 7.5, 1.0 Hz, 1H), 6.94 (ddd, $J$ = 8.0, 0.5 Hz, 1H), 6.88 (dd, $J$ = 8.0, 1.0 Hz, 1H), 4.81 (d, $J$ = 9.0 Hz, 1H), 4.40 (dd, $J$ = 9.0, 1.0 Hz, 1H), 4.03–3.90 (m, 2H), 3.21–3.03 (m, 2H), 0.93 (t, $J$ = 7.0 Hz, 3H). $^{13}$C NMR (126 MHz, $CDCl_3$) δ 168.3, 160.4, 145.1, 140.7, 138.7, 134.7, 132.4, 129.7, 129.1, 128.8, 128.7, 128.0, 127.4, 127.2, 125.9, 125.4, 124.4, 121.2, 110.1, 82.2, 60.3, 49.8, 37.2, 13.6. IR (neat film, $cm^{-1}$) 3059, 2982, 1695, 1479, 1460, 1370, 1260, 1236, 1130, 1096, 1018, 972, 752, 735. HRMS (ESI+) calculated 383.1642 m/z (found 383.1645 m/z for $C_{26}H_{23}O_3$).

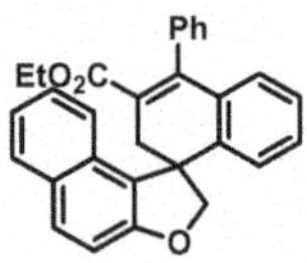


Ethyl 4-phenyl-2*H*,2'*H*-spiro[naphthalene-1,1'-naphtho[2,1-*b*]furan]-3-carboxylate (3.114c)

Prepared according to GP6 using 1-iodo-2-((2-phenylallyl)oxy)naphthalene (77 mg, 0.2 mmol) and ethyl 3-phenylpropiolate (38 mg, 0.22 mmol). The spirodihydrobenzofuran was purified by flash column chromatography using hexanes:EtOAc (10:1 v:v) and was obtained as a white powder (70 mg, 0.166 mmol, 83%, >20:1 rr, M.P.: 66–67 °C).

$^{1}$H NMR (500 MHz, $CDCl_3$) δ 7.92–7.86 (m, 1H), 7.83 (d, $J$ = 8.5 Hz, 1H), 7.58–7.50 (m, 1H), 7.49–7.38 (m, 3H), 7.37–7.24 (m, 4H), 7.22 (d, $J$ = 8.5 Hz, 1H), 7.20–7.13 (m, 2H), 7.13–7.08 (m, 1H), 6.97–6.88 (m, 1H), 5.10 (d, $J$ = 9.0 Hz, 1H), 4.37 (dd, $J$ = 9.0, 1.5 Hz, 1H), 4.03–3.87 (m, 2H), 3.54 (dd, $J$ = 17.0, 1.0 Hz, 1H), 3.08 (d, $J$ = 17.0 Hz, 1H), 0.92 (t, $J$ = 7.0 Hz, 3H). $^{13}$C NMR (126 MHz, $CDCl_3$) δ 168.4, 158.6, 145.0, 139.7, 138.8, 134.7, 130.8, 130.5, 129.9, 129.7, 129.4, 128.9, 128.7, 128.1, 127.5, 127.3, 127.3, 126.3, 125.8, 123.8, 122.7, 121.4, 112.5, 82.4, 60.3, 51.5, 35.4, 13.6. IR (neat film, $cm^{-1}$) 3061, 3023, 2978, 2933, 2883, 1701, 1626, 1597, 1580, 1518, 1418, 1443, 1370, 1344, 1301, 1260, 1198. HRMS (DART) calculated 433.18037 m/z (found 433.18072 m/z for $C_{30}H_{25}O_3$).

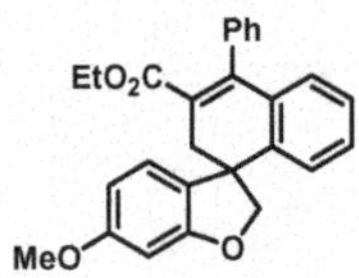


Ethyl 6-methoxy-4'-phenyl-2*H*,2'*H*-spiro[benzofuran-3,1'-naphthalene]-3'-carboxylate (3.114d)

Prepared according to GP6 using 1-iodo-4-methoxy-2-((2-phenylallyl)oxy)benzene (73 mg, 0.2 mmol) and ethyl 3-phenylpropiolate (38 mg, 0.22 mmol). The spirodihydrobenzofuran was purified by flash column chromatography using hexanes:DCM (1:1 v:v) and was obtained as a pale yellow powder (62 mg, 0.15 mmol, 75%, >20:1 rr, M.P.: 132–133 °C).

$^{1}H$ NMR (500 MHz, $CDCl_3$) δ 7.46–7.36 (m, 3H), 7.25–7.17 (m, 3H), 7.16–7.09 (m, 2H), 7.09–7.04 (m, 1H), 6.84 (d, $J = 8.0$ Hz, 1H), 6.55 (dd, $J = 8.5, 2.5$ Hz, 1H), 6.51 (d, $J = 2.5$ Hz, 1H), 4.79 (d, $J = 9.0$ Hz, 1H), 4.39 (d, $J = 9.0$ Hz, 1H), 4.03–3.89 (m, 2H), 3.83 (s, 3H), 3.06 (q, $J = 17.0$ Hz, 2H), 0.91 (t, $J = 7.0$ Hz, 3H). $^{13}C$ NMR (126 MHz, $CDCl_3$) δ 168.4, 161.7, 161.1, 145.1, 141.1, 138.8, 134.7, 129.7, 128.8, 128.7, 128.1, 127.4, 127.2, 125.8, 125.6, 124.5, 124.3, 107.1, 96.5, 83.1, 60.3, 55.5, 49.3, 37.3, 13.6, 13.6. IR (neat film, $cm^{-1}$) 2980, 1699, 1622, 1595, 1497, 1445, 1370, 1277, 1260, 1194, 1148, 1126, 1098, 760. HRMS (DART) calculated 413.17528 m/z (found 413.17523 m/z for $C_{27}H_{25}O_4$).

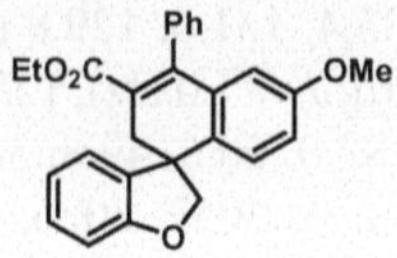


Ethyl 6'-methoxy-4'-phenyl-2*H*,2'*H*-spiro[benzofuran-3,1'-naphthalene]-3'-carboxylate (3.114e)

Prepared according to GP6 using 1-iodo-2-((2-(4-methoxyphenyl)allyl)oxy)benzene (73 mg, 0.2 mmol) and ethyl 3-phenylpropiolate (38 mg, 0.22 mmol). The spirodihydrobenzofuran was purified by flash column chromatography using hexanes:EtOAc (10:1 v:v) and was obtained as a clear colourless oil (76 mg, 0.184 mmol, 92%, >20:1 rr).

$^{1}H$ NMR (500 MHz, $CDCl_3$) δ 7.45–7.36 (m, 3H), 7.29–7.23 (m, 2H), 7.23–7.18 (m, 2H), 7.03–6.96 (m, 2H), 6.92 (d, $J = 8.0$ Hz, 1H), 6.75 (dd, $J = 8.5, 2.5$ Hz, 1H), 6.42 (d, $J = 2.5$ Hz, 1H), 4.76 (d, $J = 8.5$ Hz, 1H), 4.34 (dd, $J = 8.5, 0.5$ Hz, 1H), 4.04–3.87 (m, 2H), 3.63 (s, 3H), 3.09 (d, $J = 17.0$ Hz, 1H), 3.04 (d, $J = 17.0$ Hz, 1H), 0.91 (t, $J = 7.0$ Hz, 3H). $^{13}C$ NMR (126 MHz, $CDCl_3$) δ 168.4, 160.4, 158.6, 144.9, 138.6, 136.1, 132.9, 132.7, 129.0, 128.8, 128.1, 127.5, 127.1, 126.2, 124.4, 121.2, 115.2, 114.0, 110.1, 82.4, 60.4, 55.2, 49.3, 37.5, 13.6. IR (neat film, $cm^{-1}$) 3053, 2978, 2938, 2903, 1699, 1601, 1568, 1479, 1458, 1370, 1302, 1259, 1225, 1169, 1136, 1092. HRMS (DART) calculated 413.17528 m/z (found 413.17575 m/z for $C_{27}H_{25}O_4$).

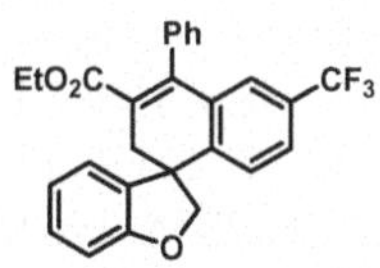


Ethyl 4'-phenyl-6'-(trifluoromethyl)-2*H*,2'*H*-spiro[benzofuran-3,1'-naphthalene]-3'-carboxylate (3.114 g)

Prepared according to GP5 using 1-iodo-2-((2-(4-(trifluoromethyl)phenyl)allyl)oxy)benzene (81 mg, 0.2 mmol) and ethyl 3-phenylpropiolate (38 mg, 0.22 mmol). The spirodihydrobenzofuran was purified by flash column chromatography using hexanes:DCM (1.5:1 v:v) and was obtained as a clear colourless gel (50 mg, 0.110 mmol, 55%, >20:1 rr).

$^{1}$H NMR (500 MHz, $CDCl_3$) δ 7.48–7.40 (m, 4H), 7.31–7.28 (m, 1H), 7.28–7.24 (m, 1H), 7.22–7.18 (m, 2H), 7.16 (d, $J$ = 8.0 Hz, 1H), 7.09 (d, $J$ = 1.5 Hz, 1H), 7.04–6.98 (m, 1H), 6.96–6.93 (m, 1H), 4.78 (d, $J$ = 9.0 Hz, 1H), 4.39 (dd, $J$ = 9.0, 0.5 Hz, 1H), 4.09–3.80 (m, 2H), 3.15 (d, $J$ = 17.0 Hz, 1H), 3.08 (d, $J$ = 17.0 Hz, 1H), 0.91 (t, $J$ = 7.1 Hz, 3H). $^{13}$C NMR (126 MHz, $CDCl_3$) δ 168.1, 160.4, 144.4 (q, $J$ = 1.0 Hz), 143.5, 137.6, 135.4, 131.6, 129.8 (q, $J$ = 32.5 Hz), 129.5, 128.8, 128.4, 128.0, 127.3, 126.6, 126.2 (q, $J$ = 4.0 Hz), 125.1 (q, $J$ = 4.0 Hz), 124.2, 123.7 (q, $J$ = 272.5 Hz), 121.5, 110.4, 82.0, 60.6, 49.8, 36.8, 13.6. $^{19}$F NMR (376 MHz, $CDCl_3$) δ-62.83. IR (neat film, $cm^{-1}$) 3057, 3026, 2982, 2937, 1701, 1613, 1597, 1481, 1460, 1416, 1371, 1333, 1302, 1263, 1169, 1128, 1094, 1018. HRMS (DART) calculated 451.15210 m/z (found 451.15173 m/z for $C_{27}H_{22}F_3O_3$).

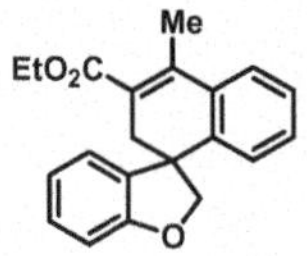


Ethyl 4'-methyl-2*H*,2'*H*-spiro[benzofuran-3,1'-naphthalene]-3'-carboxylate (3.114 h)

Prepared according to GP6 using 1-iodo-2-((2-phenylallyl)oxy)benzene (67 mg, 0.2 mmol) and ethyl but-2-ynoate (25 mg, 0.22 mmol). The spirodihydrobenzofuran was purified by flash column chromatography using hexanes:EtOAc (15:1 v:v) and was obtained as a pale yellow solid (53 mg, 0.166 mmol, 83%, >20:1 rr, M.P.: 95–97 °C).

$^{1}$H NMR (500 MHz, $CDCl_3$) δ 7.56 (dd, $J$ = 8.0, 1.0 Hz, 1H), 7.34–7.29 (m, 1H), 7.27–7.22 (m, 2H), 7.19 (ddd, $J$ = 7.5, 1.5, 0.5 Hz, 1H), 7.01–6.94 (m, 2H), 6.90 (ddd, $J$ = 8.0, 1.0, 0.5 Hz, 1H), 4.67 (d, $J$ = 9.0 Hz, 1H), 4.25 (qd, $J$ = 7.0, 0.5 Hz, 2H), 4.20 (dd, $J$ = 9.0, 1.0 Hz, 1H), 3.00 (dd, $J$ = 16.5, 1.0 Hz, 1H), 2.85 (dq, $J$ = 16.5, 2.5 Hz, 1H), 2.51 (dd, $J$ = 2.5, 1.0 Hz, 3H), 1.32 (t, $J$ = 7.0 Hz, 3H). $^{13}$C NMR (126 MHz, $CDCl_3$) δ 168.4, 160.5, 141.5, 140.8, 135.3, 132.5, 129.4, 129.0, 127.4, 126.0, 125.6, 124.5, 124.1, 121.1, 110.0, 82.1, 60.6, 49.7, 36.9, 16.2, 14.3. IR (neat film, $cm^{-1}$) 3067, 2978, 2955, 1931, 2878, 1705, 1613, 1595, 1480, 1458, 1279, 1234, 1196, 1134, 1040, 986. HRMS (DART) calculated 338.17562 m/z (found 338.17518 m/z for $C_{21}H_{21}O_3 + NH_3$).

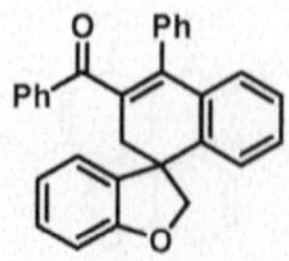


Phenyl(4'-phenyl-2*H*,2'*H*-spiro[benzofuran-3,1'-naphthalen]-3'-yl)methanone (3.114i)

Prepared according to GP6 using 1-iodo-2-((2-phenylallyl)oxy)benzene (67 mg, 0.2 mmol) and 1,3-diphenylprop-2-yn-1-one (45 mg, 0.22 mmol). The spirodihydrobenzofuran was purified by flash column chromatography using hexanes:EtOAc (20:1 v:v) and was obtained as a pale yellow solid (70 mg, 0.169 mmol, 84%, >20:1 rr, M.P.: 135–137 °C).

$^1$H NMR (500 MHz, $CDCl_3$) δ 7.73–7.67 (m, 2H), 7.38 (ddt, $J$ = 8.0, 7.0, 1.5 Hz, 1H), 7.30 (ddd, $J$ = 7.5, 1.5, 0.5 Hz, 1H), 7.29–7.23 (m, 4H), 7.21–7.12 (m, 6H), 7.08–7.05 (m, 1H), 7.03–6.97 (m, 2H), 6.94–6.88 (m, 1H), 4.90 (d, $J$ = 8.5 Hz, 1H), 4.51 (dd, $J$ = 8.5, 1.0 Hz, 1H), 3.32 (d, $J$ = 16.5 Hz, 1H), 2.85 (d, $J$ = 16.5 Hz, 1H). $^{13}$C NMR (126 MHz, $CDCl_3$) δ 160.4, 153.7, 140.6, 136.0, 132.9, 131.2, 131.1, 129.6, 129.4, 129.3, 128.9, 128.7, 127.6, 126.4, 124.3, 121.5, 119.0, 110.5, 105.2, 81.8, 49.6, 37.4. IR (neat film, $cm^{-1}$) 3059, 3026, 2930, 2878, 1647, 1597, 1480, 1460, 1449, 1346, 1312, 1260, 1236, 1217, 1175. HRMS (DART) calculated 415.16980 m/z (found 415.16969 m/z for $C_{30}H_{23}O_3$).

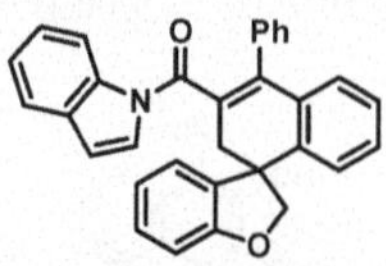


(1*H*-indol-1-yl)(4'-phenyl-2*H*,2'*H*-spiro[benzofuran-3,1'-naphthalen]-3'-yl)methanone (3.114j)

Prepared according to GP6 using 1-iodo-2-((2-phenylallyl)oxy)benzene (67 mg, 0.2 mmol) and 1-(1*H*-indol-1-yl)-3-phenylprop-2-yn-1-one (54 mg, 0.22 mmol). The spirodihydrobenzofuran was purified by flash column chromatography using DCM:hexanes (3:1 v:v) and was obtained as a yellow solid (82 mg, 0.18 mmol, 90%, >20:1 rr, M.P.: 97–99 °C).

$^1$H NMR (600 MHz, DMSO-$d_6$ at 115 °C) δ 8.19–8.12 (m, 1H), 7.59 (d, $J$ = 4.0 Hz, 1H), 7.51–7.47 (m, 1H), 7.43–7.36 (m, 2H), 7.32–7.15 (m, 9H), 7.02–6.95 (m, 2H), 6.94–6.87 (m, 2H), 6.57 (d, $J$ = 3.5 Hz, 1H), 4.82 (d, $J$ = 9.0 Hz, 1H), 4.63 (d, $J$ = 9.0 Hz, 1H), 3.37 (d, $J$ = 16.5 Hz, 1H), 2.94 (d, $J$ = 16.5 Hz, 1H). $^{13}$C NMR (126 MHz, $CDCl_3$) δ 168.8, 160.5, 140.7, 140.3, 136.5, 134.9, 133.8, 131.8, 130.7, 129.6, 129.3, 129.2 (broad), 128.7 (broad), 128.5 (broad), 128.3, 128.1, 127.5, 126.3, 125.7, 124.9, 124.5, 123.9, 121.4, 120.7, 116.2, 110.2, 109.2, 82.5, 77.3, 50.2, 38.6. IR (neat film, $cm^{-1}$) 3057, 3026, 2928, 2878, 1684, 1597, 1533, 1480, 1451, 1379, 1352, 1329, 1234, 1207, 1150. HRMS (DART) calculated 454.18070 m/z (found 454.17948 m/z for $C_{32}H_{24}NO_2$).

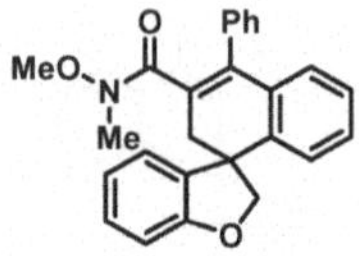

*N*-methoxy-*N*-methyl-4'-phenyl-2*H*,2'*H*-spiro[benzofuran-3,1'-naphthalene]-3'-carboxamide (3.114 k)

Prepared according to GP6 using 1-iodo-2-((2-phenylallyl)oxy)benzene (67 mg, 0.2 mmol) and N-methoxy-N-methyl-3-phenylpropiolamide (42 mg, 0.22 mmol). The spirodihydrobenzofuran was purified by flash column chromatography using hexanes:DCM:EtOAc (12:8:2 v:v) and was obtained as a white solid (68 mg, 0.172 mmol, 86%, >20:1 rr, M.P.: 56–57 °C).

$^1$H NMR (600 MHz, DMSO-$d_6$ at 115 °C) δ 7.44–7.39 (m, 2H), 7.39–7.35 (m, 1H), 7.35–7.33 (m, 1H), 7.32–7.27 (m, 2H), 7.26–7.21 (m, 1H), 7.19 (td, $J$ = 7.5, 1.5 Hz, 1H), 7.16–7.12 (m, 1H), 6.98–6.94 (m, 1H), 6.92–6.87 (m, 2H), 6.81 (dd, $J$ = 7.5, 1.5 Hz, 1H), 4.74 (d, $J$ = 9.0 Hz, 1H), 4.47 (d, $J$ = 9.0 Hz, 1H), 3.48 (s, 3H), 3.11 (d, $J$ = 16.5 Hz, 1H), 2.97 (s, 3H), 2.71 (d, $J$ = 16.5 Hz, 1H). $^{13}$C NMR (151 MHz, DMSO-$d_6$ at 115 °C) δ 160.5, 140.4, 137.7, 136.7, 134.5, 132.9, 130.6, 129.8, 129.4, 128.9, 128.4, 128.1, 127.5, 127.2, 125.8, 124.9, 121.5, 110.2, 105.0, 83.0, 61.2, 49.7, 37.8. IR (neat film, cm$^{-1}$) 3059, 3007, 2936, 2878, 1653, 1597, 1480, 1460, 1443, 1379, 1321, 1236, 1217, 1179, 1109. HRMS (DART) calculated 398.17562 m/z (found 398.17546 m/z for $C_{26}H_{24}NO_3$).

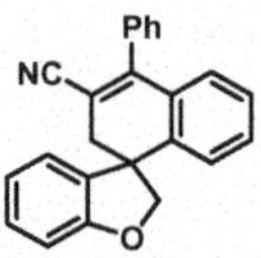

4'-phenyl-2*H*,2'*H*-spiro[benzofuran-3,1'-naphthalene]-3'-carbonitrile (3.114 l)

Prepared according to GP6 using 1-iodo-2-((2-phenylallyl)oxy)benzene (67 mg, 0.2 mmol) and 3-phenylpropiolonitrile (28 mg, 0.22 mmol). The spirodihydrobenzofuran was purified by flash column chromatography using hexanes:EtOAc (15:1 v:v) and was obtained as a white solid (51 mg, 0.152 mmol, 76%, >20:1 rr, M.P.: 60–61 °C).

$^1$H NMR (500 MHz, $CDCl_3$) δ 7.55–7.48 (m, 3H), 7.41 (s, broad, 2H), 7.34–7.27 (m, 2H), 7.25–7.18 (m, 2H), 7.12–7.08 (m, 1H), 7.03–6.97 (m, 2H), 6.97–6.93 (m, 1H), 4.78 (d, $J$ = 9.0 Hz, 1H), 4.37 (dd, $J$ = 9.0, 1.0 Hz, 1H), 3.10 (d, $J$ = 16.5 Hz, 1H), 2.91 (d, $J$ = 16.5 Hz, 1H). $^{13}$C NMR (126 MHz, $CDCl_3$) δ 160.4, 153.7, 140.6, 136.0, 132.9, 131.2, 131.1, 129.6, 129.4, 129.3, 128.9, 128.7, 127.6, 126.4, 124.3, 121.5, 119.0, 110.5, 105.2, 81.8, 49.6, 37.4. IR (neat film, cm$^{-1}$) 3061, 3022, 2947, 2880, 2207, 1596, 1559, 1479, 1460, 1445, 1236, 1217, 1018, 972. HRMS (DART) calculated 353.16539 m/z (found 353.16578 m/z for $C_{24}H_{22}N_2O + NH_3$).

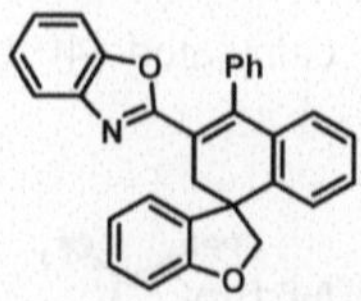


2-(4'-phenyl-2*H*,2'*H*-spiro[benzofuran-3,1'-naphthalen]-3'-yl)benzo[*d*]oxazole (3.114 m)

Prepared according to GP6 using 1-iodo-2-((2-phenylallyl)oxy)benzene (67 mg, 0.2 mmol) and 2-(phenylethynyl)benzo[*d*]oxazole (48 mg, 0.22 mmol). The spirodihydrobenzofuran was purified by flash column chromatography using hexanes:DCM (4:1 v:v) and was obtained as a white solid (71 mg, 0.166 mmol, 83%, > 20:1 rr, M.P.: 84–86 °C).

$^{1}$H NMR (500 MHz, $CDCl_3$) δ 7.65–7.59 (m, 1H), 7.50–7.43 (m, 3H), 7.35–7.31 (m, 1H), 7.31–7.27 (m, 2H), 7.27–7.22 (m, 3H), 7.20 (td, $J = 7.5$, 1.5 Hz, 1H), 7.16 (td, $J = 7.5$, 1.5 Hz, 1H), 7.14–7.10 (m, 1H), 7.09–7.06 (m, 1H), 7.00 (td, $J = 7.5$, 1.0 Hz, 1H), 6.96–6.90 (m, 2H), 4.83 (d, $J = 9.0$ Hz, 1H), 4.46 (d, $J = 9.0$ Hz, 1H), 3.65 (d, $J = 16.5$ Hz, 1H), 3.44 (d, $J = 16.5$ Hz, 1H). $^{13}$C NMR (126 MHz, $CDCl_3$) δ 163.7, 160.5, 150.3, 144.5, 141.2, 140.8, 138.7, 135.3, 132.3, 129.7, 129.5 (broad), 129.1, 128.7, 128.2, 127.6, 127.3, 126.0, 125.1, 124.4, 124.3, 121.2, 120.4, 119.8, 110.2, 110.2, 82.1, 49.9, 37.8. IR (neat film, $cm^{-1}$) 3061, 3024, 2984, 2880, 1618, 1595, 1508, 1480, 1368, 1248, 1217, 1180. HRMS (DART) calculated 428.16505 m/z (found 428.16381 m/z for $C_{30}H_{22}NO_2$).

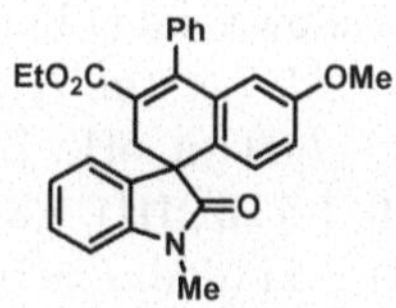


Ethyl 6'-methoxy-1-methyl-2-oxo-4'-phenyl-2'*H*-spiro[indoline-3,1'-naphthalene]-3'-carboxylate (3.115 h)

Prepared according to GP7 using *N*-(2-bromophenyl)-2-(4-methoxyphenyl)-*N*-methylacrylamide (69 mg, 0.2 mmol) and ethyl 3-phenylpropiolate (70 mg, 0.4 mmol). The spirooxindole was purified by flash column chromatography using hexanes: EtOAc (2:1 v:v) and was obtained as a clear colourless oil (77 mg, 0.175 mmol, 88%, >20:1 rr).

$^{1}$H NMR (500 MHz, $CDCl_3$) δ 7.48–7.35 (m, 4H), 7.32–7.27 (m, 3H), 7.01 (td, $J = 7.5$, 1.0 Hz, 1H), 6.93 (d, $J = 7.5$ Hz, 1H), 6.75 (d, $J = 8.5$ Hz, 1H), 6.68 (dd, $J = 8.5$, 2.5 Hz, 1H), 6.46 (d, $J = 2.5$ Hz, 1H), 3.88 (q, $J = 7.0$ Hz, 2H), 3.59 (s, 3H), 3.36 (d, $J = 16.5$ Hz, 1H), 3.34 (s, 3H), 2.93 (d, $J = 16.5$ Hz, 1H), 0.86 (t, $J = 7.0$ Hz, 3H). $^{13}$C NMR (126 MHz, $CDCl_3$) δ 179.0, 168.0, 159.0, 145.0, 142.0, 138.8, 136.6, 133.5, 128.8 (broad), 128.4, 128.4, 128.1, 127.5, 126.6, 124.6, 123.5, 123.0, 115.7, 114.3, 108.5, 60.3, 55.2, 51.3, 35.3, 26.6, 13.5. IR (neat film, $cm^{-1}$) 3013, 2980, 2959, 1709, 1698, 1609, 1568, 1489, 1471, 1371, 1350, 1307, 1250,

1227, 1171, 1092. HRMS (DART) calculated 440.18618 m/z (found 440.18710 m/z for $C_{28}H_{26}$-$NO_4$).

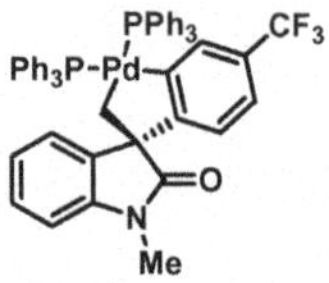

Palladacycle (3.118)

**Note:** All DCM and $CHCl_3$/$CDCl_3$ were passed through a plug of basic alumina prior to use.

To a flame dried 2 dram vial cooled under argon, *N*-(2-bromophenyl)-*N*-methyl-2-(4-(trifluoromethyl)phenyl)acrylamide (1a) (77 mg, 0.2 mmol, 1 equiv), $Pd(PPh_3)_4$ (231 mg, 0.2 mmol, 1 equiv), and $Cs_2CO_3$ (98 mg, 0.3 mmol, 1.5 equiv) were added and allowed to purge for 10 min. PhMe (2 mL, 0.1 M) was added and a Teflon line screw cap was fitted on the two dram vial, sealed with Teflon tape and place in a preheated oil bath at 80 °C for 72 h. The reaction mixture was cooled to room temperature. Once cooled, the reaction was passed through a plug of Celite into a flask using minimal $CHCl_3$. Once plug of Celite becomes colourless, an excess of $CHCl_3$ is passed through the plug into a new flask (3.111j has very low solubility). The solution was concentrated *in vacuo* and the palladacycle was recrystallized with $Et_2O$ and $CHCl_3$ to obtain a pale yellow solid (135 mg, 0.144 mmol, 72%).

$^1$H NMR (600 MHz, $CDCl_3$) δ 7.57–7.49 (m, 7H), 7.36–7.30 (m, 1H), 7.29–7.22 (m, 10H), 7.22–7.17 (m, 4H), 7.15–7.09 (m, 6H), 7.09–7.03 (m, 7H), 7.00 (td, $J$ = 7.5, 1.0 Hz, 1H), 6.85 (dd, $J$ = 8.0, 1.5 Hz, 1H), 6.81 (d, $J$ = 7.5 Hz, 1H), 6.46 (dd, $J$ = 8.0, 2.5 Hz, 1H), 3.25 (m, 3H), 2.11 (ddd, $J$ = 10.0, 9.0, 4.5 Hz, 1H), 2.05 (ddd, $J$ = 10.0, 7.5, 5.0 Hz, 1H).

*Due to the insolubility and complexity of the spectra, some of the $^{13}$C NMR peaks were not distinguishable.* $^{13}$C NMR (126 MHz, $CDCl_3$) δ 181.0 (d, $J$ = 5.0 Hz), 162.3, 143.0, 138.0 (d, $J$ = 5.0 Hz), 136.5, 134.5 (d, $J$ = 13.0 Hz), 134.2 (d, $J$ = 12.5 Hz), 133.7 (d, $J$ = 31.5 Hz), 133.0 (d, $J$ = 34.2 Hz), 129.5, 129.3, 128.0 (d, $J$ = 9.5 Hz), 127.8 (d, $J$ = 9.5 Hz), 126.8, 124.3, 121.9, 119.9, 107.3, 68.75 (dd, $J$ = 7.0, 5.0 Hz), 47.1 (dd, $J$ = 90.5, 7.5 Hz), 26.4. $^{31}$P NMR (243 MHz, $CDCl_3$) δ 25.93 (d, $J$ = 24.0 Hz), 24.46 (d, $J$ = 24.0 Hz). $^{19}$F NMR (470 MHz, $CDCl_3$) δ -62.0. IR (neat film, cm$^{-1}$) 3053, 3001, 2918, 1713, 1698, 1609, 1433, 1369, 1316, 1221, 1154. HRMS (DART) calculated 934.1820 m/z (found 934.1820 m/z for $C_{53}H_{42}F_3NOP_2Pd$).

# References


1. (a) Yoon H, Lossouarn A, Landau F, Lautens M (2016) Org Lett 18: 6324. (b) Yoon H, Rölz Landau F, Lautens M (2017) Angew Chem Int Ed 56:10920
2. (a) Shilov AE, Shul'pin GB (1997) Chem Rev 97:2879. (b) Labringer JA, Bercaw JE (2002) Nature 417:507
3. For selected reviews and references therein, see: (a) Yu J-Q, Shi Z (2010) C–H activation; Springer, Berlin Heidelberg. (b) Dixneuf PH, Doucet H (2016) C–H bond activation and catalytic functionalization I; Springer, Berlin Heidelberg
4. Lafrance M, Fagnou K (2006) J Am Chem Soc 128:16496
5. (a) For an overview, see: Lapointe D, Fagnou K (2010) Chem Lett 39:1118. (b) Gorelsky SI, Lapointe D, Fagnou K (2008) J Am Chem Soc 130:0848. (c) Gorelsky SI, Lapointe D, Fagnou K (2012) J Org Chem 77:658
6. Waterman R (2013) Organometallics 32:7249
7. Perutz RN, Sabo-Etienne S (2007) Angew Chem Int Ed 46:2578
8. For a comprehensive review and references therein, see: (a) Chen X, Engle KM, Wang D-H, Yu J-Q (2009) Angew Chem Int Ed 48:5094. (b) Lyons TW, Sanford MS (2010) Chem Rev 110:1147. (c) Gensch T, Hopkinson MN, Glorius F, Wencel-Delord J (2016) Chem Soc Rev 45:2900. (d) Sambiago C, Schönbauer D, Blieck R, Dao-Huy T, Pototschnig G, Schaaf P, Wiesinger T, Zia MF, Wencel-Delord J, Besset T, Maes BUW, Schnürch M (2018) Chem Soc Rev 47:6603. https://doi.org/10.1039/c8cs00201k
9. For the initial catalytic reports, see: (a) Cho J-Y, Iverson CN, Smith III MR (2000) J Am Chem Soc 122:12868. (b)Cho J-Y, Tse MK, Holmes D, Maleczka Jr RE, Smith III MR (2002) Science 295:305. (c) For a review on Ir-catalyzed borylation, see: Hartwig JF (2011) Chem Soc Rev 40:1992
10. (a) For selected examples of Rh-catalyzed silylation, see: Cheng C, Hartwig JF (2014) Science 343:853. (b) For Pt-catalyzed silylation, see: Tsukada N, Hartwig JF (2005) J Am Chem Soc 127:5022
11. For selected examples of oxidative Pd-catalyzed arylation, see: (a) Hickman AJ, Sanford MS (2011) ACS Catal 1:170. (b) Sun C-L, Li B-J, Shi Z-J (2010) Chem Commun 46:677
12. (a) Ball LT, Lloyd-Jones GC, Russell CA (2012) Science 337:1644. (b) Ball LT, Lloyd-Jones GC, Russel CA (2014) J Am Chem Soc 136:254
13. For selected recent examples of the Fujiwara-Moritani reaction, see: (a) Zhang Y-H, Shi B-F, Yu J-Q (2009) J Am Chem Soc 131:5072. (b) Wang P, Verma P, Xia G, Shi J, Qiao JX, Tao S, Cheng PTW, Poss MA, Farmer ME, Yeung K-S, Yu J-Q (2017) Nature 551:489. (c) Chen H, Wedi P, Meyer T, Tavakoli G, van Gemmeren M (2018) Angew Chem Int Ed 57:2497
14. Catellani M, Frignani F, Rangoni A (1997) Angew Chem Int Ed 36:119
15. For recent reviews and references therein, see: (a) Ye J, Lautens M (2015) Nature Chem 7:863. (b) Della Ca' N, Fontana M, Motti E, Catellani M (2016) Acc Chem Res 49:1389
16. Lautens M, Piguel S (2000) Angew Chem Int Ed 39:1045
17. Wang J, Li R, Dong Z, Liu P, Dong G (2018) Nature Chem 10:866
18. Ames DE, Bull D (1982) Tetrahedron 38:387
19. Albicker MR, Cramer N (2009) Angew Chem Int Ed 48:9139
20. Nakanishi M, Katayev D, Besnard C, Kündig EP (2011) Angew Chem Int Ed 50:7438
21. For recent reviews and references therein, see: (a) Baudoin O (2017) Acc Chem Res 50:1114. (b) Newton CG, Wang S-G, Oliveira CC, Cramer N (2017) Chem Rev 117:8908. (c) He J, Wasa M, Chan KSL, Shao Q, Yu J-Q (2017) Chem Rev 11: 8754
22. (a) Grigg R, Fretwell P, Meerholtz C, Sridharan V (1994) Tetrahedron 50:359. (b) Brown D, Grigg R, Sridharan V, Tambyrajah V (1995) Tetrahedron Lett 36:8137
23. Campo MA, Larock RC (2002) J Am Chem Soc 124:14326
24. Huang Q, Fazio A, Dai G, Campo MA, Larock RC (2004) J Am Chem Soc 126:7460
25. Lu Z, Hu C, Guo J, Li J, Cui Y, Jia Y (2010) Org Lett 12:480
26. Bunescu A, Piou T, Wang Q, Zhu J (2015) Org Lett 17:334

27. For a comprehensive review on Pd-catalyzed C–H functionalization in a domino process, see: Mehta VP, García-López J-A (2017) ChemCatChem 9:1149
28. Sickert M, Weinstabl H, Peters B, Hou X, Lautens M (2014) Angew Chem Int Ed 53:5147
29. Ye J, Shi Z, Sperger T, Yasukawa Y, Kingston C, Schoenebeck F, Lautens M (2017) Nature Chem 9:361
30. Mortier J (2016) Arene Chemistry reaction mechanisms and methods for aromatic compounds. Wiley, Inc., New Jersey
31. Roberts JD, Simmons HE, Carlsmith LA, Vaughan CW (1953) J Am Chem Soc 75:3290
32. Himeshima Y, Sonoda T, Kobayashi H (1983) Chem Lett 12:1211
33. For a review and references therein on Pd-catalyzed reactions involving arynes, see: Worlikar SA, Larock RC (2011) Curr Org Chem 15:3214
34. Peña D, Escudero S, Pérez D, Guitián E, Castedo L (1998) Angew Chem Int Ed 37:2659
35. Retbøll M, Edwards AJ, Rae AD, Willis AC, Bennett MA, Wenger E (2002) J Am Chem Soc 124:8348
36. Parthasarathy K, Han H, Prakash C, Cheng C-H (2012) Chem Commun 48:6580
37. Peña D, Cobas A, Pérez D, Guitián E (2002) Synthesis 10:1454
38. Pérez-Gómez M, García-López J-A (2016) Angew Chem Int Ed 55:14389
39. Medina JM, Mackey JL, Garg NK, Houk KN (2014) J Am Chem Soc 136:15798
40. Portscheller JL, Malinakova HC (2002) Org Lett 4:3679
41. Franzoni I, Yoon H, García-López J-A, Poblador-Bahamonde AI, Lautens M (2017) Chem Sci 9:1496
42. Davies DL, Macgregor SA, McMullin CL (2017) Chem Rev 117:8649
43. (a) Pérez-Gómez M, Hernández-Ponte S, Bautista D, García-López J-A (2017) Chem Commun 53:2842. (b) Shao C, Wu Z, Ji X, Zhou B, Zhang Y (2017) Chem Commun 53:10429
44. Rodríguez JF, Marchese AD, Lautens M (2018) Org Lett 20:4367

The manufacturer's authorised representative in the EU is Springer Nature Customer Service Centre GmbH, Europaplatz 3, 69115 Heidelberg, Germany. If you have any concerns regarding our products, please contact ProductSafety@springernature.com

Printed and bound by CPI Group (UK) Ltd, Croydon, CR0 4YY
03/10/2026
02227545-0004